CULTURAL FACTORS IN SYSTEMS DESIGN

Decision Making and Action

INDUSTRIAL AND SYSTEMS ENGINEERING SERIES

Series Editor
Gavriel Salvendy

PUBLISHED TITLES:

Cultural Factors in Systems Design: Decision Making and Action
Robert W. Proctor, Shimon Y. Nof, and Yuehwern Yih

Handbook of Healthcare Delivery Systems
Yuehwern Yih

CULTURAL FACTORS IN SYSTEMS DESIGN

Decision Making and Action

Edited by

Robert W. Proctor   Shimon Y. Nof   Yuehwern Yih

CRC Press is an imprint of the
Taylor & Francis Group, an **informa** business

CRC Press
Taylor & Francis Group
6000 Broken Sound Parkway NW, Suite 300
Boca Raton, FL 33487-2742

International Standard Book Number: 978-1-4398-4646-9 (Hardback)

Library of Congress Cataloging-in-Publication Data

Cultral factors in systems design : decision making and action / [edited by] Robert W. Proctor, Shimon Y. Nof, Yuehwern Yih.
p. cm. -- (Industrial and systems engineering series)
Includes bibliographical references and index.
ISBN 978-1-4398-4646-9 (hardcover : acid-free paper)
1. System design--Decision making. 2. Decision making--Social aspects. I. Proctor, Robert W. II. Nof, Shimon Y., 1946- III. Yih, Y. (Yuehwern)

QA76.9.S88C85 2012
003--dc23 2011038266

Visit the Taylor & Francis Web site at
http://www.taylorandfrancis.com

and the CRC Press Web site at
http://www.crcpress.com

Contents

SECTION III International Partnership and Collaboration

SECTION IV Decision Making in Healthcare

SECTION V Specific Applications

SECTION VI Conclusion

Foreword

I established the International Symposium on Frontiers in Industrial Engineering in my name upon retirement from the School of Industrial Engineering at Purdue University, where I served as a faculty member for many years. The symposium is to be held every two years to highlight important topics in industrial engineering. Invited speakers are to represent an international group of academics and professionals from a variety of disciplines who work in the chosen topic area. The objective of the symposium is to present scientific, technical, and professional expertise and insight on emerging areas of industrial engineering.

The first symposium was organized by a three-person committee, consisting of professors Robert Proctor (chair), Shimon Nof, and Yuehwern Yih, the editors of this volume. This book, which contains chapters written by most of the invited speakers, provides cutting-edge knowledge of research and practice on the critical topic of cultural factors in decision making and action. It also provides researchers and practitioners with current trends and future directions of the field. This book, being the first in the series, sets a high standard for the volumes that are to follow. My sincere thanks and appreciation go to the editors for their outstanding work in organizing the symposium and this edited volume.

Gavriel Salvendy

Preface

In April 2010, the first Gavriel Salvendy International Symposium on Frontiers in Industrial Engineering was held at Purdue University, focusing on the theme of "Cultural Factors in Decision Making and Action." The purpose of the symposium was to bring together an interdisciplinary group of experts in a variety of areas relating to cultural factors in decision making, with the goal of fostering innovations in the understanding of how cultural differences influence decision making and action. This book is an outgrowth of that symposium, with the chapters based on invited presentations for the symposium.

The authors of the chapters in this book are representative of the diversity of interests and viewpoints that characterize the current state of decision making and cultural research. The authors have international backgrounds and experiences (coming from Canada, China, Great Britain, India, Italy, and Singapore, as well as the United States) and have worked both inside and outside of academia. They represent a variety of different disciplines (including aviation, computer science, health sciences, industrial engineering, management, psychology, and social and decision sciences), and they specialize in areas ranging from basic decision processes of individuals to decisions made in teams and large organizations to cultural influences on behavior. The chapter authors describe relevant research and experiences from the different perspectives relating to their main areas of expertise. This project is intended to promote increased understanding of how cultural differences influence decision making and action, as well as to facilitate multidisciplinary research on this important topic.

The first section of the book contains an introductory chapter based on the keynote address, which was given by William B. Rouse, a professor in the School of Industrial and Systems Engineering at the Georgia Institute of Technology, with joint appointment in the College of Computing, who serves as executive director of the university-wide Tennenbaum Institute for Enterprise Transformation. Rouse's chapter, "Cultures, Situations, and Behaviors: Theories and Models That Explain Why Best Laid-Plans Go Awry," examines conditions that can yield unanticipated behaviors from complex, adaptive sociotechnical systems. Rouse combines academic knowledge with professional experience in providing answers to two questions: (1) Why are targeted populations often not able to take full advantage of ventures intended to lead to economic development? (2) How do different cultural groups, with associated sources of conflict, emerge?

Section II contains four chapters representing "Contemporary Perspectives on Decision Making and Culture." In Chapter 2, "Explaining Cultural Differences in Decision Making Using Decision Field Theory," Matthew and Busemeyer note that there are marked differences between the ways in which East Asians and Westerners make decisions, and they raise the question of whether any formal model can be applied to all decision makers. To answer this question, they propose and evaluate a dynamic model of decision making, called multiattribute decision field theory. In Chapter 3, "Dynamic Decision Making and Cultural Affiliation," Gonzalez and

Martin also treat decision making as a dynamic process. They adopt an experimental approach to investigating cultural influences on decision making in interpersonal interactions, employing both context-specific computer games and more general decision making tasks.

Whereas Chapters 2 and 3 apply theory and methods of individual and interpersonal decision making to cultural issues, Chapters 4 and 5 are based primarily in cultural perspectives. In Chapter 4, "Why Do People Think Culturally When Making Decisions? Theory and Evidence," Tong and Chiu argue that although there are cultural differences in social judgment and decision making, the psychological processes engaged when arriving at decisions seem to be the same. Their view is that cultural influences on judgments and decisions are probabilistic. In Chapter 5, "Cross Cultural Decision Making: Impact of Values and Beliefs on Decision Choices," Balasubramanian also examines the complexity of cultural differences, emphasizing that consideration of beliefs and values seems to be central to those differences. He takes into consideration factors such as views of wisdom and its relation to age, gender differences, sharing versus ownership, and ends versus the means to achieve the ends.

The four chapters in Section III concern issues in "International Partnership and Collaboration." In Chapter 6, "Cultural Factors, Technology and Operations in Developing Countries: Two Case Studies," Upton reports case studies of successful initiatives conducted in India, examining the decision-making principles underlying their successes. The case studies illustrate the importance of developing decision-making principles based on understanding of the local culture. Chapter 7 by de Bedout, "Building Global Teams for Effective Industrial Research and Development," emphasizes that it is also important to take into account cultural differences when implementing global teams. The chapter provides as a specific case the globalization of research and development operations at General Electric Corporation.

In Chapter 8, "Cultural Factors: Their Impact in Collaboration Support Systems and on Decision Quality," Nof pursues further the issue of team decision making and collaboration, focusing mainly on the role of organizational and work group cultures in determining the effectiveness of collaborative e-work. In Chapter 9, "Effects of Group Orientation and Communication Style on Making Decisions and Interacting with Robots," Rau and Li indicate that the interactions in team decisions can involve not only multiple persons but also robots. They emphasize the importance of understanding cultural differences in interactions with social robots, since use of robots in various settings is becoming increasingly widespread.

Section IV contains four chapters that focus on "Decision Making in Healthcare." In Chapter 10, "Deliberation and Medical Decision Making," Flynn, Khan, Klassen, and Schneiderhan discuss the role of deliberate thought processing when making medical decisions. They summarize research on decision making in the social science and medical literatures, showing that the best insight is gained through combination of insights from each field. In Chapter 11, "Design of Performance Evaluation and Management Systems for Territorial Healthcare Networks of Services," Villa and Bellomo discuss implementation of real-world systems for making decisions about healthcare. They emphasize the importance of having models of the healthcare system and methods for system performance analysis to enable the best weighting of reduction in healthcare costs and adequacy of healthcare services.

In Chapter 12, "Decision Making in Healthcare System Design: When Human Factors Engineering Meets Healthcare," Carayon and Xie deal with the problem of integrating the culture of human factors engineers with that of healthcare professionals. They describe the different values of the human factors culture and the healthcare culture, providing recommendations for how the two can be integrated. Chapter 13, "Cultural Factors in the Adoption and Implementation of Health Information Technology," by Zafar and Lehto, discusses the potential benefits of adopting new health information technologies and possible barriers to doing so. They explore factors that influence who will likely adopt new healthcare technologies and consider ways in which new technology may clash with the cultural norms of providers and other stakeholders in the healthcare system.

Section V, "Specific Applications," includes applications in the areas of information privacy and aviation. In Chapter 14, "Cultural Factors and Information Systems: An Application to Privacy Decisions in Online Environments," Farahmand applies contemporary decision-making theory to privacy decisions in online environments. He notes that standard economic models of decision making do not incorporate cultural differences, even though cultures differ in their value estimation judgments and in the extent to which social and contextual information is taken into account. In Chapter 15, "Factors Influencing the Decisions and Actions of Pilots and Air Traffic Controllers in Three Plausible NextGen Environments," Vu, Strybel, Battiste, and Johnson describe research they have been conducting on factors that play a role in the decisions made by pilots and air traffic controllers. Vu et al. discuss how changes being made in the air traffic management system will alter the tasks and interactions of pilots and controllers, incorporating many advanced technologies. They consider how alternative concepts of operations that change the responsibilities assigned to pilots and controllers affect their decisions and actions.

In the concluding chapter of the book, Proctor, Vu, and Salvendy integrate issues covered in the individual chapters, organizing them around questions that provided the basis for a roundtable session held at the end of the symposium. This chapter highlights some of the areas of agreement among authors of the previous chapters and notes issues in need of further investigation.

Cultural factors, in both the narrow sense of different national, racial, and ethnic groups, and in the broader sense of different groups of any type, play major roles in individual and group decisions. With increasing globalization of organizations and interactions among people from various cultures, better understanding of how cultural factors influence decision making and action is a necessity. It is our hope that the range of thought on these matters presented in this volume will lead the way in this process.

We would like to acknowledge the following individuals for assistance in conducting the symposium and in preparing this book. First, we thank Gavriel Salvendy for endowing the symposium series and assisting in organization of the first symposium. Without his support and exhortations, the entire project would not have been possible. We also thank Joseph Pekny, interim Head of the School of Industrial

Engineering at Purdue, for providing the opening remarks for the symposium and both financial support and personal encouragement to the organizers. We appreciate the efforts that Daniel Folta, director of development for the School of Industrial Engineering, devoted to promoting the symposium. We express gratitude to John Edwardson, chairman and CEO of CDW Corporation, for giving the initial address at the symposium. Also, the following individuals who are not authors of chapters in this volume either chaired a session or participated in the roundtable session: Steven Landry, Nelson Uhan, Sang Eun Woo, and Ji Soo Yi from Purdue University and Louis Tay from the University of Illinois. We are grateful to Liang (Leon) Zeng and Chung-Yin (Joey) So, graduate students in industrial engineering, for transcribing the session. Finally, our warm thanks go to Erica Wilson and Geni Greiner of the Conference Division at Purdue University for coordinating the symposium and seeing that everything went smoothly, and to Cindy Carelli, senior acquisitions editor for CRC Press, for her support of this book.

Robert W. Proctor
Shimon Y. Nof
Yuehwern Yih

Editors

Robert W. Proctor is Distinguished Professor of Psychological Sciences at Purdue University. Dr. Proctor has a courtesy appointment in the School of Industrial Engineering and is cocoordinator of the interdisciplinary human factors program. Dr. Proctor teaches courses in human factors in engineering, human information processing, attention, and perception and action. He is faculty advisor of the Purdue student chapter of the Human Factors and Ergonomics Society. Dr. Proctor's research focuses on basic and applied aspects of human performance. He has published over 200 articles on human performance and is author of numerous books and book chapters. His books include *Human Factors in Simple and Complex Systems* (1st and 2nd editions, coauthored with Trisha Van Zandt), *Skill Acquisition and Human Performance* (coauthored with Addie Dutta), and *Handbook of Human Factors in Web Design* (1st and 2nd editions, coedited with Kim-Phuong L. Vu). Dr. Proctor is currently editor of the *American Journal of Psychology*. He is a fellow of the American Psychological Association and the Association for Psychological Science and an honorary fellow of the Human Factors and Ergonomics Society.

Shimon Y. Nof is professor of industrial engineering at Purdue University, and held visiting positions at MIT and at universities in Chile, the European Union, Hong Kong, Israel, Japan, and Mexico. He is the director of the NSF-industry supported PRISM Center (production, robotics, and integration software for manufacturing and management), recent chair of the IFAC Coordinating Committee "Manufacturing & Logistics Systems," recent president of IFPR (International Federation of Production Research), fellow of the IIE (Institute of Industrial Engineers), and inaugural member of Purdue's Book of Great Teachers. He is the author, coauthor, and editor of 10 books, including the *Handbook of Industrial Robotics* (1st and 2nd editions), the *International Encyclopedia of Robotics* (both winners of the "Most Outstanding Book in Science and Engineering"), *Information and Collaboration Models of Integration*, *Industrial Assembly*, and *Springer Handbook of Automation*.

Dr. Yuehwern Yih is a professor of the School of Industrial Engineering at Purdue University and director of smart systems and the operations laboratory. She is a faculty scholar of the Regenstrief Center for Healthcare Engineering. Dr. Yih's expertise resides in system and process design, monitor, and control to improve its quality and efficiency. Her research has been focused on dynamic process control and decision making for operations in complex systems (or systems in systems), such as healthcare delivery systems, manufacturing systems, supply chains, and advanced life support system for mission to Mars. Dr. Yih has published over 100 journal papers, conference proceedings, and book chapters on system design and operation control. Her recent book, *Handbook of Healthcare Delivery Systems*, was published in December 2010. She is an IIE fellow.

Contributors

Parasuram Balasubramanian, PhD
Chief Executive Officer
Theme Work Anaytics Pvt. Ltd.
Bangalore, India

Vernol Battiste, MS
Flight Deck Display Research Laboratory
San Jose State University
San Jose, California

and

NASA Ames Research Center
Human Systems Integration Division (TH)
Moffett Field, California

Dario Bellomo, PhD
Azienda Sanitaria Locale
Asti, Italy

Jerome R. Busemeyer, PhD
Psychological and Brain Sciences
Indiana University
Bloomington, Indiana

Pascale Carayon, PhD
Department of Industrial and Systems Engineering
University of Wisconsin
Madison, Wisconsin

Chi-Yue Chiu, PhD
Nanyang Business School
Nanyang Technological University
Singapore

Juan M. de Bedout, PhD
Power Conversion Systems Organization
GE Global Research
Niskayuna, New York

Fariborz Farahmand, PhD
CERIAS
Purdue University
West Lafayette, Indiana

Kathryn E. Flynn, PhD
Department of Psychiatry and Behavioral Sciences
Duke University School of Medicine
Durham, North Carolina

Cleotilde Gonzalez, PhD
Department of Social and Decision Sciences
Carnegie Mellon University
Pittsburgh, Pennsylvania

Walter W. Johnson, PhD
NASA Ames Research Center
Human Systems Integration Division (TH)
Moffett Field, California

Shamus Khan, PhD
Department of Sociology
Columbia University
New York, New York

Amy Klassen, MA
Department of Sociology
University of Toronto
Toronto, Ontario, Canada

Mark R. Lehto, PhD
School of Industrial Engineering
Purdue University
West Lafayette, Indiana

Ye Li, PhD
Department of Industrial Engineering
Tsinghua University
Beijing, China

Jolie M. Martin, PhD
Department of Social and Decision Sciences
Carnegie Mellon University
Pittsburgh, Pennsylvania

Mervin R. Matthew, PhD
Psychology Department
DePauw University
Greencastle, Indiana

Shimon Y. Nof, PhD
School of Industrial Engineering
Purdue University
West Lafayette, Indiana

Robert W. Proctor, PhD
Department of Psychological Sciences
Purdue University
West Lafayette, Indiana

Pei-Luen Patrick Rau, PhD
Department of Industrial Engineering
Tsinghua University
Beijing, China

William B. Rouse, PhD
H. Milton Stewart School of Industrial and Systems Engineering
Georgia Institute of Technology
Atlanta, Georgia

Gavriel Salvendy, PhD
School of Industrial Engineering
Purdue University
West Lafayette, Indiana

and

Department of Industrial Engineering
Tsinghua University
Beijing, China

Erik Schneiderhan, PhD
Department of Sociology
University of Toronto
Toronto, Ontario, Canada

Thomas Z. Strybel, PhD
Department of Psychology
California State University, Long Beach
Long Beach, California

Yuk-Yue Tong, PhD
School of Social Sciences
Singapore Management University
Singapore

David M. Upton, PhD
Saïd Business School
University of Oxford
Oxford, United Kingdom

Agostino Villa, PhD
Dept. of Manufacturing Systems & Business Economics
Politecnico di Torino
Torino, Italy

Kim-Phuong L. Vu, PhD
Department of Psychology
California State University, Long Beach
Long Beach, California

Anping Xie, MS
Department of Industrial and Systems Engineering
University of Wisconsin
Madison, Wisconsin

Yuehwern Yih, PhD
School of Industrial Engineering
Purdue University
West Lafayette, Indiana

Atif Zafar, MD
Health Information and Translational Sciences
Indiana University School of Medicine
Indianapolis, Indiana

Section I

Introduction

Understanding Cultures, Situations, and Behaviors

1 Cultures, Situations, and Behaviors

Theories and Models That Explain Why Best-Laid Plans Go Awry

William B. Rouse

CONTENTS

INTRODUCTION

Despite being well intentioned and well planned, attempts to intervene in other cultures can yield unanticipated results. More specifically, initiatives whose objectives are to help targeted populations can result in behaviors other than those expected, and consequences different from those sought. This chapter proposes explanations for this phenomenon and suggests means for better anticipating results and, therefore, enabling the desired results.

I hasten to emphasize that my goal is to elucidate the conditions under which unexpected behaviors can emerge from complex adaptive sociotechnical systems rather than to predict whether such behaviors will or will not emerge (e.g., Rouse 2000, 2008). Thus, the outcomes of interest are qualitative estimates of the risks of unexpected behaviors given the presence or absence of various environmental characteristics.

The ideas and proposals in this chapter have two sources. First, a focused set of key publications are cited. These publications provide in-depth evidence for the concepts presented. In fact, several of these publications provide, in themselves, rich reviews of the literature from the behavioral and social sciences, as well as management. As might be imagined, the overall literature from which these are drawn is immense.

The second source is experiential; in point of fact, drawing upon my personal experiences in Africa (South Africa and Zimbabwe), Asia (India and Vietnam), Eastern Europe (Hungary and Ukraine), and Latin America (Bolivia and Mexico). My experiences in these countries were centered on economic development, which can be considered a key element of "nation building." Most of these experiences involved providing advice, training, and seed capital to encourage and support new ventures that were envisioned as key means to economic development. In many of these situations, the targeted population was unable to take full advantage of these offerings. One question of interest concerns why such an outcome would result.

Another question of interest concerns the ways in which different cultural groups emerge, including possible sources of conflict and competition among these groups. I employ two classic theories to outline a dominant source of group formation as well as conflict and competition. The resulting model provides fresh insights into how conflict can be defused.

CULTURES, BEHAVIORS, AND SITUATIONS

The first question is why were people unable to take full advantage of the economic development assistance offered? Figure 1.1 provides some high-level insights to answering this question. The consequences we sought were venture formation and economic growth. The behaviors we expected were learning about best practices

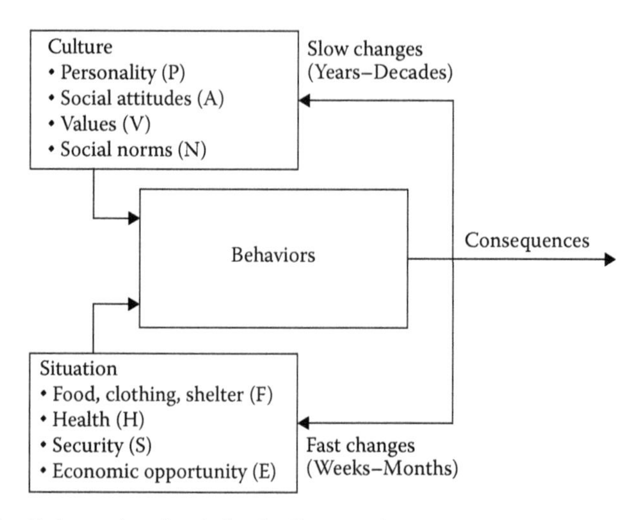

FIGURE 1.1 Culture, situation, behavior framework.

in venture formation and management, as well as proactive pursuit of the resources available to enable these pursuits. Instead, we encountered the following types of responses:

- "The training you propose has great potential but people are much more concerned with whether they will be able to eat today." (Bolivia)
- "The training you are delivering is very valuable, but the students value even more the lodging and meals they receive while being trained." (South Africa)
- "The venture capital you offer is indeed a great opportunity, but the people you are addressing prefer a safe government job." (Africa, Eastern Europe)
- "The idea of a company achieving profits is intriguing, but why would one want to achieve these ends?" (Eastern Europe)
- "Starting a high-tech business would be intriguing and what my education has prepared me for, but family and childbearing are my top priorities." (Mexico)

From the framework in Figure 1.1, it can be seen that two broad classes of factors influenced the behaviors observed in these situations. One class concerns the situation of the targeted population, denoted by S = S (F, H, S, E). People who are very hungry, and uncertain about their next meal, are more likely to focus on food than learning about auto repair. People who are insecure and uncertain about the economy and political situation are more likely to want a safe job than the thrills of entrepreneurship.

The other class of factors concerns cultural factors, denoted by C = C (P, A, V, N). People whose social values place priority on family and children are more likely to focus on this rather than venture formation, especially women. People whose social norms do not include making money (profits) from other members of society are less likely to be compelled by opportunities to gain the "first mover's advantage."

In general, as indicated in Figure 1.1, situational factors can be characterized in terms of an abbreviated Maslow's (1954) hierarchy, whereas cultural factors can be characterized using Stankov's four dimensions (Stankov and Lee 2008), as shown in the figure. There are, of course, other characterizations:

- Hofstede (1980) characterized culture using four dimensions: power distance, individualism, masculinity, and uncertainty avoidance. A fifth dimension, long-term orientation, was later added.
- Klein (2005) employed power distance, dialectical reasoning, counterfactual thinking, risk assessment and uncertainty management, and activity orientation, with particular emphasis on military command and control.
- Rouse (1993, 2007a) developed a needs-beliefs-perceptions model and applied this model in the settings discussed previously, as well as to problems of environmental impacts and organizational change.

The difficulties with these alternatives include being too high level (e.g., Hofstede describes a society rather than individuals) and being too detailed (e.g., Rouse describes the belief structure of particular individuals for specific problems). Klein's dimensions provide a viable alternative to Stankov's, but the discussion in this chapter proceeds using Stankov's characterization.

Another issue of importance concerns how to relate behaviors to cultures and situations, denoted by B = B (C, S). Before proposing an approach to this question, it is important to mention the two feedback loops in Figure 1.1. In the near term, we cannot change culture, but we can change situations. The allocations of resources and the communications (i.e., "messages") associated with these allocations can have fairly rapid impacts. These interventions can then accumulate and enable slower changes. The balance of near-term and longer-term changes is a key strategic issue in nation building.

Approach to Modeling

We would like to be able to model the phenomena in Figure 1.1 for the purpose of better anticipating the behaviors likely to emerge as a function of culture and situation, as well as a function of any planned interventions. The modeling process should begin at the right of Figure 1.1. The first concern is the consequences to be sought. For political and economic development, consequences might be stated in terms of political bodies and processes formed, or perhaps in terms of economic goals such as jobs created and companies formed.

The next step involves formulating plans or scenarios for achieving these consequences. These scenarios can be expressed in terms of the behaviors B (B_1, B_2, B_3, ... B_N) needed from the targeted population in order to achieve the consequences sought, as well as the resources (R) and messages (M) intended to prompt these behaviors. Equation 1.1 indicates a notional relationship between behaviors and cultures, situations, resources, and messages:

$$B\,(B_1,\ B_2,\ B_3,\ \ldots\ B_N) = \text{Function}\ [C\,(P,\ A,\ V,\ N),\ S\,(F,\ H,\ S,\ E),\ R,\ M] \quad (1.1)$$

To instantiate this equation, one must choose the situational and cultural factors of interest, as well as the measures associated with these factors. In choosing factors, one must differentiate between cultures in general versus characteristics of particular individuals, cultures, and situations. Hofstede (1980), Klein (2005), and Stankov and Lee (2008) provide insights into different cultures. However, given that the variability of individual differences is typically much larger than the variability of cultural differences (see Chapter 4), one should choose the factors associated with a particular set of individuals. Once the factors are chosen, Stankov and Lee, as well as others, provide some guidance with regard to measures (see Tables 1 and 2 in Stankov and Lee 2008).

With variables and metrics chosen, the next step involves defining relationships among variables. Defining relationships among the situational and cultural factors and the behaviors sought can be determined from relationships derived from research studies such as those cited above. In many cases, this knowledge is limited to directional relationships, for example, X increases with increasing Y, but decreases with increasing Z.

Such directional relationships can be elaborated using parametric relationships. Linear, exponential, and sigmoidal relationships are often sufficient to develop an initial model. Choices among these types of relationships can be based on context-specific knowledge of limiting phenomena such as diminishing returns or asymptotic behaviors. Sensitivity analysis can be used to understand which parameterized relationships have the greatest impact and, therefore, warrant empirical assessment.

With these relationships specified, one can then link all these elements into an overall object-oriented model and compute answers to several types of questions:

- What is the probability that the set of behaviors needed will, in fact, be exhibited by the target population?
- What factors are likely to be the greatest impedances to exhibiting the desired behaviors?
- How should resources be allocated to enhance the probability of success and/or mitigate inhibiting factors?
- What messages are likely to best align with the cultural factors that may inhibit the desired behaviors?

Note that the model developed in this way need not be limited to simple analytic formulations. Another approach would involve organizational simulation (Rouse and Bodner 2009; Rouse and Boff 2005). Such simulations might include elements of discrete event, systems dynamics, or agent based representations, or possibly combinations of multiple representations.

Benefits of Approach

A primary benefit of the types of analysis envisioned here would be to provide insights into factors that are likely to compromise the success of the initiative of interest due to the targeted population not responding as expected and needed. The projected probability of their likely response will be much less important than insights into the basis of their responses and evaluations of mitigations (e.g., resources and messages) that are most likely to enhance the possibilities of success. The "bottom line" is improved planning and greater success in plan execution.

Another primary benefit of this approach is increased understanding of the sensitivity of the behaviors of the targeted population to cultural and situational factors, as well as allocations of resources and communication of messages. This will enable identification of high-impact relationships for which better data are needed, possibly leading to new experimental and field studies. This would enable deeper elaboration of the nature of Equation 1.1, perhaps in terms of a "catalog" of relationships among variables and the conditions under which relationships have been assessed. I can imagine a compendium much like Boff and Lincoln's (1988) *Engineering Data Compendium*.

GROUP FORMATION AND CONFLICT

When we attempt to assist the types of groups discussed in the introduction to this chapter, we often find that these groups have a cultural history that has resulted in their being in conflict with other groups. Consequently, a portion of the resources must be devoted to security and remediation of the perceived sources of conflict. In some cases, destructive conflict can be transformed into healthy competition.

Reflecting on these phenomena brought to mind a classic book by Richard Neibuhr, *The Social Sources of Denominationalism*, published in 1929. I uncovered this fascinating work in my research for *Catalysts for Change* (Rouse 1993), a book that considers the role of belief systems in processes of change and innovation.

Neibuhr's theory of denominationalism can be summarized, albeit somewhat simplistically, in the pithy phrase, "Castes make outcastes, and outcastes make castes." This statement captures, for me, the dynamics of an important source of group formation and conflict. As social networks (Burt 2000) become stronger, there are inevitably people who feel—or are made to feel—that they do not belong and thus they depart. They become outcastes.

This process tends to be quite subtle, although throughout the centuries people's abilities to suppress outcastes has certainly had extremes (e.g., Foucault 1995). Nevertheless, it is seldom that people lose their club membership cards and cannot get through the door. Instead, the increased cohesiveness and economic cooperation of the "in group," and their consequent prosperity (Granovetter 2005), results in people being excluded because they do not believe, cannot compete, or feel uncomfortable with the trappings of the strong dominant social network.

Thus, there are some people who are caste out, and many more people who never have belonged and who consequently become increasingly distant from the mainstream. The members of this collection of outcastes eventually affiliate with each other and seeds of cooperation are sown. Neibuhr suggests that the pursuit of social justice is a driving force of this cooperation—the outcastes feel they have been wronged by the castes. This may explain, for example, why the Pilgrims set sail for North America in the early seventeenth century.

The affiliation of outcastes leads, over time, to the emergence of a new caste that, via economic cooperation, brings prosperity with the strengthening of the new social network. The values of the new caste eventually morph to salvation rather than social justice, as they seek assurance that their increasing wealth is justified and blessed. The success of the new caste inevitably leads to some people feeling excluded and becoming outcastes. Thus, the casting out of those who do not fit in continues providing fertile ground for the emergence of new castes.

The Pilgrims left England to seek religious freedom. Once settled in Massachusetts, they created an increasingly strong social system and eventually achieved economic success. They came to be intolerant of beliefs that did not fit into their system. Consequently, their Massachusetts caste began to create outcastes. These outcastes eventually migrated to what would become Rhode Island, seeking religious freedom, just as the Pilgrims sought this in leaving England (Philbrick 2007).

Thus, Neibuhr's theory suggests a growing set of "denominations," as is evident for Christians, Jews, Muslims, Buddhists, Republicans, Democrats, and so on. He observes that the more mature castes seek sustained economic success; the less mature seek social justice. There are ample opportunities for conflicts among castes. This brings us to another useful theory.

Edward Jones was a leading thinker in the creation of what is known as attribution theory (Jones and Harris 1967). Attribution theory is concerned with how people attribute cause to observed behaviors and events. Perhaps the best-known element of this theory is the fundamental attribution error. Put simply, people attribute others' success to luck and their own success to hard work. Similarly, they attribute others' failure to a lack of hard work and their own failure to bad luck.

This phenomenon of attribution sets the stage for castes to misperceive each other. In particular, members of a mature caste may perceive members of an immature caste as indolent, not working hard enough to succeed. At the same time, members of an immature caste may perceive members of a mature caste as blessed by birthright rather than having worked for all they have achieved. These differing perceptions can provide a strong motivation for conflict.

The next section formalizes the notions introduced in these introductory comments (Rouse 2007b). This formalization enables us to reach fairly crisp conclusions that explain a great deal regarding the implications of these two theories.

Castes and Outcastes

The number of members in the ith caste, NC_i, is given by

$$NC_i(t+1) = (1 + BC_i - DC_i)\, NC_i(t) + NA_i(t) - ND_i(t) \tag{1.2}$$

where BC_i and DC_i are the birth and death rates, respectively, of the caste, and NA_i and ND_i are the number of arrivals to and departures from the caste due in the former case to recruiting and in the latter case to voluntary or forced disaffection, as well as a variety of other causes such as imprisonment, mental health, and so on. Similarly, the number of members in the jth set of outcastes, NO_j, is given by

$$NO_j(t+1) = (1 + BO_j - DO_j)\, NO_j(t) + NA_i(t) - ND_i(t) \tag{1.3}$$

where BO_j and DO_j are the birth and death rates, respectively, of this set of outcastes, and NA_j and ND_j are the number of arrivals to and departures from the caste.

The process of castes spawning outcastes is depicted in Figure 1.2. This figure indicates that not all outcastes are the same. Outcaste Africans and Asians are unlikely, if only due to geography, to belong to the same outcaste population. Thus, at a particular point in time, there are N castes and M sets of outcastes.

Assuming that departures constitute a fraction of the size of a caste, denoted by PD_i for the ith caste, Equation 1.2 becomes

$$NC_i(t+1) = (1 + BC_i - DC_i - PD_i)\, NC_i(t) + NA_i(t) \tag{1.4}$$

Departures from sets of outcastes to form a new caste require a critical mass of similar outcastes, a tendency for outcastes to affiliate with each other and, consequently, the emergence of a new caste. Thus, the emergence of a new caste from the jth set of outcastes relates to the number of outcastes NO_j and the probability of members of this set of outcastes affiliating, PA_j. It could reasonably be argued that the probability of a new caste, denoted by k, being formed, PN_k, might be approximated by

$$PN_k = (1 - e^{-\lambda PA_j NO_j}) \tag{1.5}$$

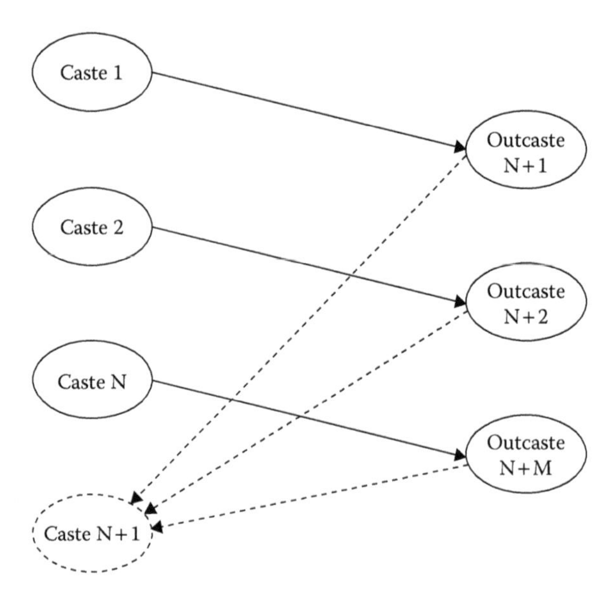

FIGURE 1.2 Castes make outcastes and outcastes make castes.

where λ provides a means for calibrating Equation 1.5 to yield reasonable results. Thus, as the probability of affiliation and the number of outcastes increase, the probability of a new caste increases.

As shown in Figure 1.2, one or more of the sets of outcastes can form a new caste. The dotted lines indicate that the N + 1st caste can emerge from any set of outcastes. While it is certainly possible that two or more sets of outcastes can form a single new caste, this would seem unlikely.

Equations 1.2 through 1.5 and Figure 1.2 define the dynamics whereby castes grow, outcastes depart, and new castes are formed. Castes are constantly shedding members, and these outcaste members eventually form and/or affiliate with newly emerging castes.

Conflict and Competition

Members of a given caste cooperate and subsequently prosper. Mature castes compete with one another, as we see with developed economies in the Americas, Asia, and Europe. Their ensuing battles are often economic, but not necessarily. In contrast, immature castes often conflict, both with each other and with mature castes. This can be understood using Neibuhr's theory of denominationalism.

As indicated in the discussion of group formation and conflict, the typical focus of immature castes is social justice. They feel wronged by the mature castes from which they were outcaste and seek redress of their grievances. The cooperation of members of the immature castes in pursuit of this cause eventually leads to economic success. As their economic success grows, they eventually shift focus from social justice to salvation, seeking assurance that their economic success is justified and blessed. Thus, they shift from seeking change to seeking preservation of the status quo.

Jones' attribution theory provides a basis for understanding the conflicts that emerge between mature and immature castes. The fundamental attribution error causes mature castes to view the lack of success of members of immature castes to be due to their lack of effort. Similarly, they view their own success as due to hard work, not just luck in belonging to a mature caste. In contrast, members of the immature caste view their lack of success as due to the oppression by the mature caste, the success of whose members is due to accidents of birth.

Of particular importance is that, while the mature castes attempt to maintain the status quo, immature castes try to foster change. They do not want things to stay the way they are. They want social justice in terms of redistribution of power and resources. Thus, the clash between mature and immature castes seems inevitable.

One might expect castes to move inexorably to maturity, M_C, perhaps following

$$M_C(t+1) = \alpha M_C(t) + (1-\alpha)\, G_E(t) \tag{1.6}$$

As α increases, the maturity (or immaturity) of a caste has high inertia. As α decreases, caste maturity tracks global economic growth, G_E. An elaboration of Equation 1.6 would probably also include local (nonglobal) drivers of economic growth. Unfortunately, monotonically growing maturity is not guaranteed (e.g., see Lewis 2003).

Model Behavior

Equations 1.2 through 1.6 might be computationally modeled as a bifurcating dynamic system (Hideg 2004) that spawns new systems, or as an agent-based system (Terna 1998) that spawns new agents, each of which behaves according to the dynamic processes with specified by these equations. With either approach, we would have to be able to represent the expanding state space as new castes are formed, as well as a potentially contracting state space as sets of outcastes disappear.

However, rather then focus on computationally modeling Equations 1.2 through 1.6, we can easily imagine the course of NC and NO under reasonable assumptions. If $BC - DC - PD < BO - DO$, which is typically the case, NO will grow more rapidly than NC. Thus, the outcaste population will, over time, likely surpass the caste population. It will take longer if PD is small, but this will nevertheless happen eventually.

The growth of NO will lead to formation of new castes, sooner for large PA and later for small PA. These castes will certainly be immature. Consequently, conflicts will frequently emerge. If α is low, these castes will quickly reach maturity and conflict will be replaced by competition. If the reverse holds (i.e., high α), then conflict will be long lasting.

This suggests that strategies to minimize conflicts should minimize BO, maximize DO, and/or decrease α. Interestingly, there is strong evidence that economic growth decreases BO. Thus, decreasing α can provide two means of decreasing conflict. In contrast, maximizing DO, naturally or otherwise, often seems to be the consequence of not paying attention to the other two strategies.

Implications

Beyond offering a possible explanation for why the world is inherently rife with groups and conflicts, the line of reasoning outlined here can provide insights into how we should think about systems of social systems. This formulation provides an archetype where the component systems have inherently conflicting objectives. The overall system includes all of humanity and the component systems include the castes and outcastes.

In a corporate setting, castes might be functional areas such as marketing and engineering. Another illustration is an established technology corporation as a caste, and small technology startups as the result of outcastes departing. The key is to understand the underlying basis for conflicts between mature and immature castes, appreciate that such conflicts are natural, and devise strategies to move castes beyond these conflicts towards healthy competition.

Thus, systems of systems that involve social systems often have inherently conflicting objectives across component systems. These conflicts cannot be designed out of the overall system. Instead, incentives and inhibitions need to be designed to foster creative competition rather than destructive conflict. The model elaborated here would suggest that providing all castes the benefits of economic growth might be an excellent core strategy.

CONCLUSIONS

This chapter has outlined an approach to understanding why people may not behave as expected and, subsequently, undermine the successful pursuit of desired consequences. This approach embodies a blend of theory and heuristics to provide a pragmatic framework for analysis. The overall goal is to gain insights into likely threats to success and evaluate ways to mitigate these threats.

I also considered situations where groups conflict and resulting behaviors are not only unexpected; they are very much undesirable. Members of cultures interact to create castes and outcastes. The possibility that desired behaviors will be exhibited can be undermined by animosities between rival groups and tribes, for example. This phenomenon, by the way, is not limited to developing countries.

An excellent example is the emergence of the *mujahideen* who conflicted with the Soviets in Afghanistan and then morphed into the Taliban, whom the United States is now confronting. It would be a stretch to suggest that the models presented here could have predicted this specific series of outcomes. It could, however, have been useful to use the castes-to-outcastes-to castes dynamic to explore the possible emergent groups. We might not know what would emerge, but we could safely predict that a new group would emerge, quite likely with a different orientation.

The best-laid plans of mice and men go oft awry (Rouse 1994). Careful consideration of the abilities, limitations, and inclinations of those expected to execute these plans can provide insights into why plan execution may falter, as well as how to support execution to increase the chances of plans succeeding (Rouse 2007a). Understanding cultures, situations, and behaviors is central to such insight and support.

ACKNOWLEDGMENTS

The author gratefully acknowledges the comments and suggestions of Dr. Kenneth R. Boff on an earlier draft of this chapter.

REFERENCES

Boff, K. R., and J. E. Lincoln. 1988. *Engineering Data Compendium: Human Perception and Performance*. Dayton, OH: Harry G. Armstrong Aerospace Medical Research Laboratory.

Burt, R. S. 2000. The network structure of social capital. In *Research in Organizational Behavior*, ed. R. I. Sutton and B. M. Staw, Vol. 22, 345–423. Greenwich, CT: JAI Press.

Foucault, M. 1995. *Discipline & Punish: The Birth of the Prison*. New York: Vintage.

Granovetter, M. 2005. The impact of social structure on economic outcomes. *J Econ Perspect* 19:33–50.

Hideg, E. 2004. Foresight as a special characteristic of complex social systems. *Interdiscip Description Complex Syst* 2:79–87.

Hofstede, G. 1980. Motivation, leadership, and organization: Do American theories apply abroad. *Organ Dyn* 9:42–63.

Jones, E. E., and V. A. Harris. 1967. The attribution of attitudes. *J Exp Soc Psychol* 3:1–24.

Klein, H. A. 2005. Cultural differences in cognition: Barriers in multinational collaborations. In *How Professionals Make Decisions*, ed. H. Montgomery, R. Lipshitz, and B. Brehmer. Mahwah, NJ: Lawrence Erlbaum Associates.

Lewis, B. 2003. *What Went Wrong? The Clash between Islam and Modernity in the Middle East*. New York: Harper.

Maslow, A. H. 1954. *Motivation and Personality*. New York: Harper.

Neibuhr, H. R. 1929. *The Social Sources of Denominationalism*. New York: Henry Holt.

Philbrick, N. 2007. *Mayflower: A Story of Courage, Community, and War*. New York: Penguin.

Rouse, W. B. 1993. *Catalysts for Change: Concepts and Principles for Enabling Innovation*. New York: Wiley.

Rouse, W. B. 1994. *Best Laid Plans*. Englewood Cliffs, NJ: Prentice Hall.

Rouse, W. B. 2000. Managing complexity: Disease control as a complex adaptive system. *Inf Knowl Syst Manag* 2:143–65.

Rouse, W. B. 2007a. *People and Organizations: Explorations of Human-Centered Design*. New York: Wiley.

Rouse, W. B. 2007b. Two theories that explain a lot: The dynamics of cooperation, conflict and competition. *Insight* 10(2):44–6.

Rouse, W. B. 2008. Healthcare as a complex adaptive system. *Bridge* 38(1):17–25.

Rouse, W. B., and D. A. Bodner. 2009. Organizational simulation. In *Handbook of Systems Engineering and Management*, ed. A. P. Sage and W. B. Rouse, 2nd ed., chap. 19. New York: Wiley.

Rouse, W. B., and K. R. Boff, eds. 2005. *Organizational Simulation: From Modeling and Simulation to Games and Entertainment*. New York: Wiley.

Stankov, L., and J. Lee. 2008. Culture: Ways of thinking and believing. In *Handbook of Personality Theory and Testing*, ed. G. J. Boyle, G. Matthews, and D. Saklofske, 560–75. Thousand Oaks, CA: Sage Publications.

Terna, P. 1998. Simulation tools for social scientists: Building agent based models with SWARM. *J Artif Societies Soc Simul* 1(2) http://jasss.soc.surrey.ac.uk/1/2/4.html.

Section II

Contemporary Perspectives on Decision Making and Culture

2 Explaining Cultural Differences in Decision Making Using Decision Field Theory

Mervin R. Matthew and Jerome R. Busemeyer

CONTENTS

DIFFERENCES BETWEEN EAST ASIANS AND WESTERNERS

Kim, a newly minted college graduate, managed to go her entire life without ever owning a car. Recent events, namely accepting a job that would require her to work outside of the city in which she lives, have put an end to this state of affairs, and the time has come for her to seek out her personal chariot for the daily grind. Suppose that Kim has a wide range of vehicles available to her. How would she go about choosing between them?

It might be easy to imagine Kim as a college graduate from Boston. She could just as easily be, however, a college graduate from Beijing, and that may make all the difference in the world. Although decision research tends to treat all decision makers as though they use the same processes, a growing body of research shows that there are cognitive differences between how East Asians and Westerners see the world, including how they see the options in a choice set.

Perceptual research shows that East Asians tend to be better than Westerners at recognizing changes in backgrounds, whereas Westerners tend to have the advantage in recognizing differences in focal objects. For example, Kitayama et al. (2003)

showed Japanese and Americans a square frame with a line drawn inside it and then gave them another frame of a different size. They were then asked to draw a line identical to the one they had seen before either in absolute length (the exact same length) or in relative length (proportional to the new frame). Americans performed better on the absolute length task, whereas Japanese were superior on the relative length task.

Boduroglu, Shah, and Nisbett (2009) showed that East Asians were better than Americans at detecting color changes when a layout of a set of colored blocks was expanded to cover a wider region and worse when it was shrunk. East Asians were also slower than Americans at detecting changes in the center of the screen. The explanation for this difference in performance was that East Asians focus their attention more broadly than Westerners, paying attention to the scene as a whole instead of on the focal object.

Masuda and Nisbett (2001) reached a similar conclusion. Japanese and Americans looking at underwater scenes were asked to talk about what they saw. They were also shown objects in their original settings and in new settings during a subsequent test. In describing the underwater scenes, Japanese made more statements about contextual information and relationships than Americans did, and they recognized previously seen objects more accurately when they saw them in their original settings instead of new ones. The change from original to new settings had no such effect on Americans, indicating that only the Japanese found the background integral to their encoding of the objects in the first place.

The differences in perception might be the result of a philosophical difference between how Westerners and East Asians attempt to understand their world. Nisbett et al. (2001) attributed the differences to the West's origins in ancient Greek thought and the East's origins in Confucian thought. Whereas the former focused on learning about the world by finding abstract principles from which to predict future events, the latter emphasized observation of the world and basing estimates of future events directly on past events. The West sees a one/many dichotomy, where each element is seen as more or less autonomous, while the East sees a part/whole dichotomy, with special focus on relationships between the elements rather than the elements themselves. As a result, only Westerners tend to think of an object separate from its context and attributes of that object separate from the object itself. This last point can have a profound effect on how East Asians make choices.

Yates and Lee (1996) pointed out that most decision research in the West focuses on analytical reasoning, the sort of thinking favored by Greek-style thought. With analytical decision making, options are broken down into independent attributes so that the unique contributions of each can be found, and decisions are made by integrating the evaluations of each attribute to arrive at an overall evaluation of the options. However, as Yates and Lee argued, a rule-based decision style is much more in line with Eastern views of the world. With rule-based decisions, all that a decision maker needs to know is that a certain course of action was best under similar circumstances in the past. Use of this decision style in the East contributes heavily to a brand loyalty unmatched in the Western world (Palumbo and Herbig 2000).

Two final important differences exist between how Westerners and East Asians see the world (and, by extension, their decisions). Yates (2010) had Chinese and

American subjects answer closed-ended questions before giving estimates of how likely their answer was to be the correct one. Chinese subjects tended to be more overconfident in their probability judgments than Americans, having a greater gap between their stated probabilities of being correct and their actual probabilities of being correct, and that could lead to more risk-taking behavior. Such risk-taking behavior is typically balanced in Western cultures by risk aversion (Kahneman and Tversky 1979), but Hsee and Weber (1999) found that the Chinese, perhaps as a result of the stronger social support expected in their culture, tend to be risk neutral instead. Not only are East Asians less likely to use analytical thinking, then, but they are also likely to come to far different conclusions than Westerners when they do use it.

It should be clear from looking at the literature that marked differences exist between East Asians and Westerners, even as decision research tends to focus on decisions as though they are made the same way across cultures. Given these differences, is it possible for any formal model to accurately apply to all decision makers? The next segment of this chapter discusses the standard expected utility (EU) theory (Bernoulli 1738/1954), a descendant of the expected value hypothesis (EV). How can EU address cultural differences, and is that explanation adequate?

EV, EU, AND PROSPECT THEORY

EV starts with the assumption that a person is always looking to pick the very best possible option. The most direct way to do that would be to choose the option with the highest global value. For example, assume that there are two options to choose from, expressed in the choice set {A,B}. Assume further that each of the options has two outcomes which can result from choosing it, designated 1 and 2. The global value for Option A, then, could be expressed as $V_A = .5 \times (V_{A,1} + V_{A,2})$, where V_A = the global value of Option A, $V_{A,1}$ = the value of Outcome 1 resulting from choosing Option A, and $V_{A,2}$ = the value of Outcome 2 resulting from choosing Option A. Similarly, the value of Option B could be expressed as $V_B = .5 \times (V_{B,1} + V_{B,2})$ where V_B = the global value of Option B, $V_{B,1}$ = the value of Outcome 1 resulting from choosing Option B, and $V_{B,2}$ = the value of Outcome 2 resulting from choosing Option B. If $V_A > V_B$, Option A would always be chosen. Conversely, Option B would always be chosen if $V_B > V_A$. This method of deciding can be applied to any number of options with any number of outcomes, for example $V_A = (V_{A,1} + V_{A,2} + V_{A,3} + ...)/N$.

This simplified formula for the global values of Options A and B assumes that both outcomes are equally likely. In reality, however, some outcomes might be more likely than others. There is, therefore, a need to weight the outcomes so that their impact on the decision is proportionate to their probability of resulting from whichever option is selected. Adapting the formula for the global value of Option A gives us $V_A = P_1 \times V_{A,1} + P_2 \times V_{A,2}$, where P_1 is the probability given to Outcome 1 and P_2 is the probability given to Outcome 2. Because each of the outcomes account for some percentage of the decision, the probabilities have to sum to 1.0 on a decimal scale. Again, Option A would always be chosen over Option B if $V_A > V_B$, and the converse would also be true. As with the formula for global values, the formula for EV can be applied to any number of options with any number of outcomes.

The magnitude of the increment from the current wealth is irrelevant; an increment of $100 to a person who is broke has the same value as this increment for a person who is a millionaire. These claims conflict with empirical findings such as the St. Petersburg paradox (Bernoulli 1738/1954). In this task, people are asked to name a price they would be willing to pay to play a game where the amount they win comes from how many times a fair coin is flipped before landing on heads. If it lands on heads on the first flip, they win $1; if it lands on tails on the first and heads on the second, they win $2. The gamble proceeds in this fashion, so that it gives them a 1/2 chance of winning $1, a 1/4 chance of winning $2, a 1/8 chance of winning $4, and so on. Although there is only a very small chance that the coin would be flipped a near infinite amount of times before landing on heads, the rare instance when that happens would provide a near infinite award. That gives the gamble an infinite EV, meaning that people should be willing to pay any price to buy into it, but the reality is that people limit the price they would pay to play it regardless.

To get around the problems inherent in assuming EV, Bernoulli (1738/1954) proposed EU instead. Unlike the values of the options on each outcome, the utilities are subjective, allowing for a wide discrepancy between how people see a pair of options. In addressing the St. Petersburg paradox, EU assumes that although the gamble has an infinite value, its utility is far more limited. An infinite amount of money, despite having a higher value than any finite amount, nonetheless would be of no more use to a person than a large finite amount. Paying a large sum to play a gamble with that infinite amount in mind, then, seems unreasonable.

Another application of EU concerns equal differences in value that nonetheless have very different effects on choice behavior. While an option that takes a person from having no money to having 1 million dollars has a very high utility, the same might not be said about an option that takes a person from having 1 billion dollars to 1.001 billion. The extra million in the latter case barely makes a dent because the person might already have way more money than he could ever hope to spend, but it makes all the difference in the world to someone who started with nothing at all. The million dollar sum, then, has a different utility in each of the two conditions.

A few major changes in utility theories have occurred since Bernoulli introduced EU. Savage (1954) applied EU to situations in which the objective probabilities were unknown, and Kahneman and Tversky (1979) introduced prospect theory, the main contribution of which was a shift from using objective probabilities to subjective weights given to those probabilities. That allows an even wider variance in how people see a pair of options, and it also releases the probability judgments from the requirement to sum to 1. For instance, a person might be more sensitive to losses than to gains than to losses, so a .5 chance of losing $100 might actually feel like a .75 chance. That does not, however, translate into the person feeling like a .5 chance of winning $100 is the same as a .25 chance. Kahneman and Tversky also pointed out that this pattern, where the prospect of a loss looms larger than the prospect of an equivalent gain, is typical, at least, in the West. Later, Tversky and Kahneman (1992) extended prospect theory, which was originally limited to binary choice sets, to decisions involving multiple options. This is the extension of EU in use today.

The extended versions of EU afford two means of accounting for differences between cultures. On one hand, it can be the case that the subjective decision weights vary between cultures, something that can explain why Chinese subjects tended to be more extreme with their probability judgments than Americans (Yates 2010). East Asians' tendency to be more risk-neutral than Westerners (Hsee and Weber 1999) can be explained via the subjective utilities, where losses have to be far greater in order for them to bother the decision maker. Because it focuses only on probabilistic events, however, EU remains silent on how differences in brand loyalty can affect decisions, and it also says nothing about how decisions can be made through the holistic processing typical of East Asians. Another method, similar to EV and EU, seems more applicable to preferential choice problems such as car purchases.

WEIGHTED ADDITIVE RULES

Keeney and Raiffa (1976) characterized problems of preferential choice as involving tradeoffs between attributes of the options being compared. For example, Roe, Busemeyer, and Townsend (2001) broke cars down along the attribute dimensions of quality and economy. Using this framework, each option would have a particular value associated with its attributes, for example, Car A might have a value of 8 on quality and a value of 2 on economy. That could be expressed generically with the equation $V_A = V_{A,1} + V_{A,2}$, where V_A = the global value of Option A, $V_{A,1}$ = the value of Option A on Attribute 1, and $V_{A,2}$ = the value of Option A on Attribute 2. Just as with EV and EU, there can be any number of options and any number of attributes.

Continuing the parallel with EV and EU, the values of the attributes on each of the options can be weighted according to their subjective importance to the decision maker. That gives us the formula $V_A = W_1 \times V_{A,1} + W_2 \times V_{A,2}$, where W_1 is the weight given to Attribute 1 and W_2 is the weight given to Attribute 2. Although these weights serve a function similar to those of prospect theory, they share a restraint with the probabilities used in EV. Specifically, because the weights all represent how much each attribute contributes to the total value of the options, the sum of the weights has to be equal to 100%, or 1.0 in decimal form. If there are only two options and the weight placed on the first is 0.7, the remaining option is bound to have a weight of 0.3.

Finally, just as the values might have different utilities for different people in probabilistic choice, they might also have different utilities for different people in choice problems such as Kim's, described at the beginning of the chapter. The difference between a car that can be obtained for free (say, as a prize for winning on a game show) and one that costs \$10,000 might loom far larger than the difference between a car that costs \$200,000 and another that costs \$210,000. With that in mind, weighted additive models tend to weight the utilities of the attributes' values instead of the values themselves: $U_A = W_1 \times U_{A,1} + W_2 \times U_{A,2}$, where U_A is the utility of Option A, W_1 is the weight given to Attribute 1, $U_{A,1}$ is the utility of the value of Option A on Attribute 1, and so on. The proper name for these models, then, is weighted additive utility models (WADD) (Keeney and Raiffa 1976).

EV and EU afforded two means of explaining cultural differences, and the same applies to WADD models. People from one culture might have a higher utility for economy, for instance, or they could place more importance on a car's economy than on its quality. Either of these would give the car with the greatest economy the best chance to be chosen. However, unless all of the weight is placed on economy, a car that has lower economy might be able to compensate by having very high quality. People might be willing to pay more for a car that has all of the most state-of-the-art features, cooks food for them, and changes the baby's diapers. Even though options are broken down by attribute, selection of the option with the highest utility on the most preferred attribute is not a given.

WADD models tend to be somewhat bottom-up in their processing. That is, they begin with the idea that there is no opinion of the options and then have the opinion of the options form as a result of integrating the weighted utilities. This does not match up well with the predecision bias that brand loyalty introduces, casting doubt on the usefulness of these models for modeling cultural differences. Also, since the overall utilities of the options come from integrating individual attributes, WADD models might be even more limited when dealing with a population that understands its options holistically. Although better than just using EU, then, WADD models still leave a lot to be desired when it comes to modeling the differences between East Asians and Westerners. Once again, their use of only two parameters limits their ability to capture the tendencies of varying groups.

DECISION FIELD THEORY AND MULTIALTERNATIVE DECISION FIELD THEORY

EU and WADD do not claim to be process models. Instead, they aim to offer a snapshot of the moment of decision. However, the means to the end is, in the study of decision making, as important as the end itself. Decision field theory (DFT) (Busemeyer and Townsend 1993) is a process model that offers a better opportunity to model cultural differences. Four new parameters are added to model the process: the initial bias (z), mean stochastic drift rate (δ), drift variance (σ^2), and threshold bound (θ). These parameters can account for differences between cultures in terms of decision processes.

DFT envisions decision making as a stochastic process in which affective valences for an option are accumulated over time through sequential sampling. Imagine that there are two options from which to choose, each of which has two attributes on which it can be judged. The options start off with some initial chance z of being selected. At each time step, the decision maker has some probability (P) of paying attention to an attribute, and he compares both of his options with respect to their utility (U) on whichever attribute he pays attention to. Whichever option is superior with respect to that attribute rises in preference according to the mean stochastic drift rate (δ), and that rise is tempered by some interference from other irrelevant factors, represented by the drift variance (σ^2). Because the model uses preferences, an advantage for one option has to be a disadvantage for the other, and the preference states thus sum to zero at each time step. This process repeats itself until one of the options accumulates enough preference to reach the threshold bound (θ), at which point that option is chosen. Generally, the lower the threshold bound, the sooner the decision is made, and vice-versa.

Like many decision theories, DFT was originally designed to deal with binary choice under uncertainty. Townsend and Busemeyer (1995) extended it to deal with the sort of problem faced by Kim but still restricted it to binary choice, as did Diederich (1997), who extended it to situations of binary choice with more than two attributes. The full power of DFT, however, was revealed by Roe, Busemeyer, and Townsend (2001), when it was extended to situations with not only multiple attributes but also multiple options. This more recent version of DFT, multialternative decision field theory (MDFT) accounts for decisions made amongst any number of options with any number of attributes. It also holds the greatest promise for modeling cultural differences.

Just as in the original DFT, the valences have to sum to zero at each time step. However, because there are multiple options, there exists the possibility that all but one of the other options receives a boost when compared along a particular attribute. To allow for a rank ordering among alternatives that all receive a boost from their utilities on a particular attribute, MDFT incorporates the mechanism of lateral inhibition (McClelland and Rumelhart 1981). In its original conceptualization, lateral inhibition applied to neurons which, when stimulated, acted against stimulation of neighboring neurons. MDFT applies this to options such that favored options act against favoring similar options. Just as lateral inhibition in neurons is strongest between neurons closest to one another, lateral inhibition in MDFT is strongest between options that have the greatest similarity. This mechanism not only prevents a person from selecting multiple options but also allows MDFT to explain the effects which will be discussed in the next section.

MDFT offers various avenues through which to address cultural differences. In addition to the weights and utilities used to explain East Asians' risk neutrality and overconfidence, MDFT can also explain the effect of brand loyalty by appealing to z. Differences in the amount of time needed to make decisions can be explained by θ; Americans might have a higher threshold than East Asians. They can also be explained by δ or by σ^2; a higher drift rate or less interference from other factors would allow the decision maker to reach θ sooner. In particular, a lower amount of interference would be expected if attention did not shift between attributes, something that fits with the claims of Nisbett et al. (2001). However, Nisbett et al. (2001) left one rather large detail to be accounted for.

Like WADD models, MDFT assumes that decision makers break options down into independent attributes for comparison purposes. That assumption may very well be true with people in the West, but Nisbett et al. (2001) provided ample argument that it may be untrue with East Asians. With that in mind, how can MDFT apply to East Asians at all? Where is there room for attention switching and compensatory strategies in a group that processes information holistically? Would MDFT, already carrying more parameters than the other models, need yet another to account for cultural differences?

It turns out that holistic processing of information is still within the scope of MDFT. The model looks at comparisons between attributes, but the attributes need not be nonconfigural. In the example of buying a car, the attribute of economy can encompass many more specific attributes, for example, how many miles per gallon a car can travel on the highway or how much an owner would need to pay for repairs. To address holistic processing in East Asians, MDFT assumes that in addition to more specific attributes, decision makers can also pay attention to special relationships between attributes. The attention weights for such relationships would be low for Westerners, explaining why more specific attributes have more of an effect, but those same weights would be high for East Asians, reducing the role of more specific attributes as a result.

This same explanation can partially be applied to WADD models. The configural attribute can be treated as another attribute in WADD models, and the argument can be made that a Chinese decision maker would put more weight on that configural attribute than an American would. Be that as it may, WADD models are still deterministic, so that would leave a Chinese decision maker with no choice but to pick the option with the highest utility on the configural attribute. With cultural differences as well as within cultures, the predictions of MDFT resemble more closely empirical results than those of WADD models.

CONTEXT EFFECTS

With regard to preferential choice, an assumption of WADD models and utility models is that the options maintain their order of preference no matter what other options are introduced. This is a consequence of the way the options are ranked in the first place: if Option A is better than Option B on the preferred Attribute 1, introducing a new option will have no effect on Option A's superiority in the A-to-B comparison. However, three effects in the preferential choice literature violate this assumption. These effects—the similarity effect (Tversky 1972), attraction effect (Huber, Payne, and Puto 1982) and compromise effect (Simonson 1989)—defy utility models. They can, however, be explained using MDFT, as will be shown shortly.

East Asians' greater sensitivity to context should logically make them more susceptible to context effects. However, this may not be easy to explain through typical means. Explanations for context effects tend to focus on consideration of independent attributes (e.g. Tversky 1972), something that does not apply as well to East Asians as to Westerners. Context effects can still exist, though, via a different mechanism for East Asians. We will discuss the three context effects, how MDFT explains their emergence with Westerners, and how MDFT can be interpreted to explain their emergence or lack thereof with East Asians.

SIMILARITY EFFECT

The first context effect, similarity, involves introducing a third option, S, which is more similar to one of the original two options than to the other. For instance, if $V_{A,1} = 2$ and $V_{B,1} = 8$, $V_{S,1}$ might be 3 or 4. A comparable distribution would be found on the other attributes as well, such that for any attribute on a continuous scale, the values for Option S would be closer to those for Option A than to those for Option B. In that case, Option S would be seen as similar to Option A but not Option B. To achieve the similarity effect, there can be no clear dominance between Option S and Option A, that is, each must have a higher value than the other on at least one attribute.

According to WADD models, adding a new option should have no impact on preference between the original two. If consumers choose Option A half of the time and Option B the other half, introducing Option S should not affect Options A and B such that one becomes preferred over the other. Tversky (1972) found that WADD models did not hold up in this situation. While Option B might retain its 50%, for instance, consumers might split the difference between Options A and S such that Option A carries 25% of the consumers while Option S does the same. Option B, then, becomes the preferred option in the ternary choice despite being merely an equal in the binary choice.

MDFT approaches the similarity effect by appealing to the attention-switching mechanism. Recall that decision makers, instead of paying attention to all attributes at a given moment in time, focus on only one attribute at any given time step and accumulate preferences over time. Suppose that Options A and S are superior to Option B on Attribute 1 but inferior to Option B on Attribute 2. Suppose further that Option A is slightly superior to Option S on Attribute 1 but slightly inferior on Attribute 2. When a decision maker focuses on Attribute 2, Option B always comes out as the preferred option because its clear advantage on that attribute raises the preference state for that option while lowering the preference state for the other two. This would have more of a negative impact on Option A than on Option S, however, because Option S is superior to Option A on Attribute 2. On the other hand, focusing on Attribute 1 would raise the preference state for both Option A and Option S, and while it would raise the preference state for Option A more, Option S would be helped in the long run by not having had its preference state lowered as much as Option A when Attribute 2 is considered. The overall effects of its preference state having less variance than Option A is that Option S is chosen as often as Option A. Those two options split the benefit of Attribute 1's importance while Option B, with no other option close to it, reaps the full benefit of Attribute 2's importance. New options that are similar to but not dominated by a prior option, then, hurt that prior option but not prior options dissimilar to it.

Appealing to attention switching in this fashion creates a problem for MDFT when it comes to modeling the similarity effect in East Asian cultures. The reason for this is simple: with regard to individual attributes, East Asian cultures are thought to engage in far less attention-switching than Western cultures. If a decision maker considers options holistically instead of breaking them down into independent attributes, what are the grounds for claiming that Option S is more similar to Option A than to Option B? That a car has the same color as another seems like it should be irrelevant if the color of the car has no specific impact on the value of the car as a whole. This last sentence, though, provides exactly the grounds needed for the comparison.

Recall that Masuda and Nisbett (2001) found that East Asians were better than Westerners at identifying information about contextual relationships, while worse at answering questions about a focal object. If attention switching takes place with East Asians, it more likely involves shifting attention between various relationships instead of between individual elements. Surely, a Japanese person shown a green car with a beige interior can identify that car's color as "*midori*," but he would place value on the fact that that the color meshed well with the countryside where the car would be driven instead of evaluating it on its own. The similarity in question for East Asians pertains not to individual attributes of an object but to that product's relationship to the whole.

Imagine a choice between three cars: Car A, which costs little to purchase and maintain but which has no luxury features; Car B, which costs a lot in both initial expense and maintenance but has luxury features galore; and Car S, which costs slightly more than Car A but has a few luxury features. Imagine also that the cars are shown in two different contexts, where Car A is only ever seen driving around in the countryside while Cars B and S are seen cruising down a city street. For a Westerner, who separates the object from its surroundings, this difference in context has little impact on his evaluations of the cars themselves, and he sees Car S as most similar to Car A because of the similarities in luxury features and cost. An East Asian, however, pays attention to the context and sees Car S as most similar to Car B because he can more easily imagine both of those cars together. As a result, Westerners are more likely to favor Car B in this ternary choice because Cars A and S split the benefit of costing less than Car B while also sharing the disadvantage of not matching Car B in terms of luxury features. East Asians would be likelier to favor Car A in this example because people focusing more on driving in the city would split their preference between Cars B and S while Car A would have no rival for those focusing more on driving in the countryside.

The similarity effect both for Westerners and East Asians depends on two options in a choice set being similar to each other without one being dominant over the other. A slight adjustment to this relationship, however, leads into an effect altogether different from the one just discussed.

Attraction Effect

The second context effect, the attraction effect (Huber, Payne, and Puto 1982), involves introducing a different third option, D, which is more similar to one of the original two options than to the other and dominated only by that similar option. For instance, if $V_{A,1} = 2$ and $V_{B,1} = 8$, $V_{D,1}$ might be 1. Just as with the similarity effect, for any attribute on a continuous scale, the values for Option D would be closer to those for Option A than to those for Option B. What separates the attraction effect from the similarity effect is that Option D has to be inferior to the similar option on at least one attribute and no better than equal on all others. That leaves no tradeoff to be made, and Option D has no chance of ever being favored.

The result of introducing Option D shifts the balance between Options A and B, something it would also do in cases involving the similarity effect. Here, however, instead of lowering the chance that Option A would be selected, the dominated Option D actually increases those chances. In other words, if Option A received 50% of the preference before Option D was introduced, it receives more than 50% after said introduction. This violates the regularity principle of WADD models, which states that introducing a new option to a choice set can never raise the chances of another item in that set being selected (in the similarity effect, Option B benefits solely by Option S detracting from Option A).

MDFT explains the attraction effect by means of lateral inhibition. Consider the evolution of the process as a series of pair-wise comparisons. Suppose that Option A is superior to Option D on Attribute 1 and equal to it on Attribute 2. Further, suppose that Option D is superior to Option B on Attribute 1 but inferior to it on Attribute 2.

Finally, suppose that Option B is inferior to Option A on Attribute 1 but superior to it on Attribute 2. In half of its comparisons with Option A, Option B comes out on top (whenever Attribute 2 is the focus). This is also true for Option B's comparisons with Option D. However, while losing to Option B half the time (when Attribute 2 is the focus), Option A never loses to Option D because it is at least as good on every attribute while being better on at least one. The net result is that Option A receives a boost no matter which attribute is focused on while Options B and D receive boosts only when Attributes 2 and 1 are focused on, respectively. Option A's dominance over Option D makes it seem like an even stronger option, making it even more likely to be selected than it was in the binary choice set.

With regard to East Asians, the attraction effect can seemingly be expressed in the same way that the similarity effect was expressed in MDFT. Instead of comparing options with regard to attributes, the options are compared with regard to how they relate to their context, and attention is switched between those comparisons the same way. If, in the three car example, Car A dominates Car D while no dominance relationship exists between Cars A and B or Cars B and D, Car A should come out as the strongest option. The dominance relationship between Cars A and D gives Car A a boost unmatched by Car B.

Two special problems occur with East Asians with regard to the attraction effect, however. Nisbett et al. (2001) discussed yet another contrast that Westerners have with East Asians, namely that the former believes in and seeks a single correct answer while the latter is more open to the possibility that there might be several correct answers to a problem. The concept of dominance, where one option will always be better than another, is less likely to play a role in East Asian decisions. The second problem is that lateral inhibition is also less likely to play a role because East Asian thought does not embrace the concept of contradiction (Nisbett et al. 2001). Put another way, Western thought demands that the preference states for all options sum to zero because only one option can be preferred, but since Eastern thought allows for all options to be preferred simultaneously, that restriction has to be different for East Asians.

Those problems lead to the conclusion that East Asians are less susceptible to the attraction effect than are Westerners, and they may be immune to it altogether. Using the car example, suppose that there are three cars, one of which (Car A) is shown driving through a city and two (Cars B and D) that are shown driving in the countryside. Suppose also that one of the cars shown in the countryside (Car D) is having all sorts of trouble navigating the dirt roads while the other (Car B) has no trouble to speak of. While both of the cars in the countryside would be similar according to the principles discussed with regard to East Asians and the similarity effect, no Westerner would choose Car D over Car B. That should make Car B dominant over Car D and therefore subject to the benefits of the attraction effect—but not for East Asians. Inferior though it might be, Nisbett et al.'s (2001) analysis implies that Car D would still be seen as a viable candidate for reasons the authors' Western brains struggle to comprehend. That being the case, Car B would receive no boost, and if it was preferred no more than 50% of the time in the binary choice with Car A, it would be guaranteed not to rise significantly above 50% in the ternary choice with Car D included.

The East Asian outlook on conflict dooms the attraction effect in those cultures (or, at least, dulls it). Even so, it can be helpful with regard to context effects overall. This is particularly true with regard to the third context effect.

Compromise Effect

The third and final context effect, the compromise effect (Simonson 1989), arises when a third option, Option C, is added to a binary choice set consisting of Options A and B, such that it falls midway between the two in terms of its value on the important attributes. For example, if $V_{A,1} = 2$ and $V_{B,1} = 8$, $V_{C,1}$ would be 5. In order to represent a true compromise between Options A and B, Option C has to fall midway between those options on every attribute. The upshot of this positioning is that Option C can never be too hurt or too helped no matter which attribute is the focus; as a matter of fact, in contrast to the two extremes, Option C is the one most likely to remain absolutely stable no matter what the attention weights given to each attribute.

Of all three context effects, the compromise effect is the hardest for MDFT to explain. Appealing only to attention-switching falls short because, as already stated, Option C neither benefits from nor suffers from the focus on any particular attribute. In fact, if all attributes have equal attention weights and Options A and B each have a 50% chance of being selected in their binary choice set, Option C would end up having a 50% chance of being selected if placed in any binary choice set with Option A or B. If attention-switching is the only factor involved, all three options would be equally likely to be selected, that is, Option C would win approximately 33% of time.

What causes the emergence of the compromise effect, according to MDFT, is the inclusion of random noise at each time step. This random noise enters into the deliberation process from decision makers paying attention to matters other than the attributes under consideration during each time step and is reflected in the model's drift variance (the more noise, the greater the variance). When the random noise gives a boost to Option A while the focus is on an attribute favorable to Option B, that boost is tempered by the effect of considering that attribute. The same applies for Option B whenever random noise gives it a boost while the focus is on an attribute favorable to Option A. However, while it gains nothing from consideration of any particular attribute, Option C, the compromise option, also loses nothing from consideration of any particular attribute. With nothing to counteract the enhancing effect of the noise on Option C's favorability, it accumulates enough positive affect to climb above the two extreme options.

Logically, it seems that compromise effect should be the strongest of the context effects among East Asians. Whereas most of the information that a Westerner might detect about his options gets pushed to the wayside as the focal object is analyzed, that information stays very much in play for East Asians. In fact, in line with Boduroglu, Shah, and Nisbett (2009), that information is more likely to be attended to than the focal object itself. Picture this comparison between three cars: Car A runs excellently in the countryside but is a hassle in the stop-and-go driving of the city; Car B is well suited for city driving but gets beaten down by the unpaved roads in the countryside; and Car C is not very well suited for either condition but is just adequate enough for both. Westerners, if they were focused on this aspect of the cars, would be less likely to be influenced by the fact that Car C is also typically driven by their neighbors' son, but this random consideration would be perfectly in line with East Asian attention to social norms (Weber and Morris 2010).

Buyers who live in the countryside and work in the city should be equally likely to pick any of the three options, insomuch as each of the extremes meets one of their daily needs perfectly while the compromise option balances its inability to fully meet either with its minimally acceptable levels of both. However, if Car A suddenly gets a boost because what the neighbor's son drives comes to mind while considering driving in the city, the fact that city driving is hard on that car reduces that boost. A similar pattern would emerge if Car B suddenly got a random boost while considering driving in the countryside. Neither attribute, however, would work against Car C if that happened to be the car driven by the neighbor's son, and so Car C would have a better chance of being picked in the long run.

Outside of the MDFT explanation, the compromise effect has another reason for being stronger among East Asians than among Westerners. This explanation relates to East Asians' tendency to prefer compromise to winner-takes-all debates. As Nisbett et al. (2001) pointed out, the belief in a single right answer dominates Western thinking, so much so that Westerners will pursue even the slightest indication that one option is better than all others. One extreme option, then, would have to be better than the other according to Western logic, whereas that need not be the case for East Asians. As a matter of fact, because of East Asians' preference for harmonious options instead of options that are radically imbalanced, it seems likely that East Asians would begin their deliberation already favoring the compromise option. MDFT would account for this by giving the compromise option a stronger positive initial bias, something that would not be true for Westerners.

The preference that East Asians have for less extreme options would not affect the similarity effect because none of the options in that choice set reflect a true compromise, and the same holds true for the attraction effect. Taken together, the higher drift variance and stronger initial bias toward compromise options should make the compromise effect especially strong in East Asians, and that relationship applies not only to comparisons between East Asians and Westerners but also to in-group comparisons with respect to the three context effects. As far as MDFT goes, the predictions for the compromise effect are the most interesting and offer the most compelling argument for why process models are needed for examining cross-cultural decision making.

FUTURE DIRECTIONS

Although decision research has come around to recognizing different processes used by decision makers, differences between how cultures use those processes have remained relatively unexplored. Even so, attempts to address how cultures use the decision processes have been more developed than attempts to model the differences, most of which center on adjustment of the same limited set of parameters in use in WADD models. MDFT not only shows that the differences across cultures can be described using only one model, but it also produces empirical predictions about what sort of situations would elicit context effects in East Asians. Both of these are beneficial; however, the latter still needs far more support.

According to MDFT, differences in context should have a bigger impact on East Asians than it should on Westerners and should produce context effects in East Asians that would not affect Westerners at all. Interestingly, the model also predicts that situations which would produce context effects in Westerners would not produce them in East Asians because the East Asians would be less sensitive to the individual attributes of the options in their choice set. In order to fully accept that this model can account for intergroup differences, studies need to be conducted which show just this pattern emerging.

One of the biggest obstacles to conducting those studies will be defining a continuous scale through which to judge the similarity of contexts. Studies on the similarity, attraction, and compromise effects have been able to manipulate the relationships between options by assigning a range of values to their attributes. How to do that reliably with contexts, for example, claiming that a car performs almost as well as another in the countryside but slightly worse in the city, is unclear. One method seems to be to simply assign values to the interactions between context and performance, mirroring how car manufacturers describe cars as having one level of fuel efficiency on the highway and another in the city. However, that has limits, for example, a car's fuel efficiency is seen largely as a property of the car, a circumstance that would not lend itself to the thought process typical of East Asians making decisions.

Despite such problems in designing empirical tests, MDFT seems more promising than WADD models for explaining the differences observed between East Asians and Westerners. Its greater number of parameters allows a richer characterization of the decision process and lends itself to predicting differences that have yet to be chronicled. No matter the culture, decision making is a dynamic process, and only a model equipped to handle those dynamics can hope to explain how and why decisions made by Kim from Boston systematically differ from those made by Kim from Beijing.

REFERENCES

Bernoulli, D. 1738. Specimen theoriae novae de mensura sortis. Commentarii Academiae Scientiarum Imperialis Petropolitanae. Translated by Dr. Louise Sommer. 1954. Exposition of a new theory on the measurement of risk. *Econometrica* 22:23–36.

Boduroglu, A., P. Shah, and R. E. Nisbett. 2009. Cultural differences in allocation of attention in visual information processing. *J Cross Cult Psychol* 40:349–60.

Busemeyer, J. R., and J. T. Townsend. 1993. Decision field theory: A dynamic cognition approach to decision making. *Psychol Rev* 100:432–59.

Diederich, A. 1997. Dynamic stochastic models for decision making under time constraints. *J Math Psychol* 41:260–74.

Hsee, C. K., and E. U. Weber. 1999. Cross-national differences in risk preferences and lay predictions for the differences. *J Behav Decis Mak* 12:165–79.

Huber, J., J. W. Payne, and C. Puto. 1982. Adding asymmetrically dominated alternatives: Violations of regularity and the similarity hypothesis. *J Consum Res* 9:90–8.

Kahneman, D., and A. Tversky. 1979. Prospect theory: An analysis of decision under risk. *Econometrica* 47:263–91.

Keeney, R. L., and H. Raiffa. 1976. *Decisions with Multiple Objectives: Preferences and Value Tradeoffs*. New York: Wiley & Sons.
Kitayama, S., S. Duffy, T. Kawamura, and J. T. Larsen. 2003. Perceiving an object and its context in different cultures: A cultural look at New Look. *Psychol Sci* 14:201–6.
Masuda, T., and R. A. Nisbett. 2001. Attending holistically versus analytically: Comparing the context sensitivity of Japanese and Americans. *J Pers Soc Psychol* 81:922–34.
McClelland, J. L., and D. E. Rumelhart. 1981. An interactive activation model of context effects in letter perception: Part 1. An account of basic findings. *Psychol Rev* 88:375–407.
Nisbett, R. E., K. Peng, I. Choi, and A. Norenzayan. 2001. Culture and systems of thought: Holistic vs. analytic cognition. *Psychol Rev* 108:291–310.
Palumbo, F., and P. Herbig. 2000. The multicultural context of brand loyalty. *Eur J Innovation Manage* 3:116–25.
Roe, R. M., J. R. Busemeyer, and J. T. Townsend. 2001. Multialternative decision field theory: A dynamic connectionist model of decision making. *Psychol Rev* 108:370–92.
Savage, L. J. 1954. *The Foundations of Statistics*. New York: Wiley & Sons.
Simonson, I. 1989. Choice based on reasons: The case of attraction and compromise effects. *J Consum Res* 16:158–74.
Townsend, J. T., and J. R. Busemeyer. 1995. Dynamic representation of decision-making. In *Mind as Motion*, ed. R. F. Port and T. van Gelder, 101–120. Cambridge, MA: MIT Press.
Tversky, A. 1972. Elimination by aspects: A theory of choice. *Psychol Rev* 79:281–99.
Tversky, A., and D. Kahneman. 1992. Advances in prospect theory: Cumulative representation of uncertainty. *J Risk Uncertain* 5:297–323.
Weber, E. U., and M. W. Morris. 2010. Culture and judgment and decision making: The constructivist turn. *Perspect Psychol Sci* 5:410–9.
Yates, J. F. 2010. Culture and probability judgment. *Soc Personal Psychol Compass* 4:174–88.
Yates, J. F., and J. W. Lee. 1996. Chinese decision making. In *Handbook of Chinese Psychology*, ed. M. H. Bond, 338–351. Hong Kong: Oxford University Press.

3 Dynamic Decision Making and Cultural Affiliation

Cleotilde Gonzalez and Jolie M. Martin

CONTENTS

DYNAMIC DECISION MAKING AND CULTURAL AFFILIATION

Culture is a complex concept that defines features of an individual's social experience, including beliefs, knowledge, morals, customs, and other habits acquired by humans as members of society (Boellstorff 2006). The role of culture in decision making is poorly understood, and it is often difficult to measure and control for in experimental settings. In this chapter, we describe the difficulty of conducting laboratory experiments to measure cultural affiliation in interpersonal interactions such as conflict resolution, where informational uncertainty and feedback delays preclude fully "rational" strategic behavior. Our tactic for overcoming this difficulty is a combination of top-down and bottom-up approaches. For the top-down approach, we use dynamic and realistic interactive computer games that represent particular cases of conflict. In this case, we present our work with a computer game called PeaceMaker, which we use to observe participant's actions in a realistic simulation of the Israel–Palestine conflict. For the bottom-up approach, we use extensions of typical game theoretic experiments that incorporate temporal and social complexities.

The second part of the chapter presents an example of the top-down approach to examine group identification as an important cultural input to decision making in PeaceMaker. In particular, we focus on in-group favoritism and out-group perspective-taking, both of which are documented cultural phenomena in the social and political psychology literatures. We provide several basic hypotheses for conditions under which one or the other attitude is likely to prevail and present the results of a particular study conducted across a diversity of cultures. Finally, we discuss how the top-down and bottom-up approaches can be pursued in conjunction to enhance our understanding of cultural impacts on dynamic decision making.

LABORATORY EXPERIMENTS OF CULTURAL AFFILIATION IN CONFLICT RESOLUTION

Cultural affiliation is a key factor not only in the development of individual identity, but also in the way that people interact with one another socially. Conflict resolution is one general area of research in which group membership has been shown to be a crucial determinant of strategies and outcomes (Kelman 2008). However, there are a number of complications to isolating the effects of culture on decision making in a dynamic social setting like conflict resolution. Gonzalez, Vanyukov, and Martin (2005) offered a taxonomy of dynamic decision-making features that are often present in conflict resolution. First, decision makers face the *dynamics* of path dependence, meaning that one's available options—and their associated utilities—are influenced by previous decisions, as well as exogenously changing circumstances. A second trademark of conflict resolution is *complexity* in the contingency of individual outcomes on the simultaneous decisions of others. Third, there is inherent *opaqueness* of information and action outcomes in conflict resolution, requiring individuals to adapt to uncertainty. Lastly, feedback loops in conflict resolution are characterized by *dynamic complexity*, such as nonlinear causal relationships and time delays.

In light of these confounds to experimentally isolating the effects of culture on decision making in conflict resolution, we propose a two-pronged approach to examine the particular role of cultural affiliation (Gonzalez et al. 2010). On the one hand, using the top-down approach, we draw conclusions from a realistic case of conflict and decision making. The realistic case of conflict we use is that represented in the interactive computer game PeaceMaker, which places participants in a simulation of the Israel–Palestine conflict. On the other hand, using the bottom-up approach, we draw conclusions from controlled experiments in very simplified versions of conflict situations. These are represented in our own extensions of 2×2 game theory experiments that incorporate dynamic complexity in a controlled way. We used the iterative "prisoner's dilemma" and "chicken" games. We will elaborate on each of these approaches and our current findings in the next sections.

BOTTOM-UP APPROACH USING 2 × 2 GAMES

Perhaps the most common game used to study cooperative versus competitive behavior is the prisoner's dilemma (e.g., Axelrod 1984; Poundstone 1993; Rapoport and Chammah 1965), so called because it arises from a story where each of two suspects are apprehended for their possible role in the same crime and held in separate cells. The police have insufficient evidence to charge either of them with the full crime, and each prisoner must decide simultaneously whether to remain silent ("cooperate" with the other) or turn the other in ("defect" from their pact of silence). If both cooperate, they will each get off with just a minor charge and serve a short jail sentence. If they both defect, they will each get sentences of medium length for partial responsibility. If one defects and the other cooperates, the former goes free and the latter is charged with a long jail term. The two prisoners' best shared outcome is achieved if each cooperates in silence and takes just a short jail sentence. However, regardless of what the other is doing, each is better off defecting and implicating his partner. Under the assumption that Prisoner A cooperates, Prisoner B is better off defecting because he goes free rather than serving a short sentence; under the assumption that Prisoner A defects, Prisoner B is again better off defecting since he then gets a medium instead of long jail sentence. Thus, in an abstract game with this structure, traditional economic theory predicts via the concept of Nash equilibrium (best response to one another's actions) that both individuals will follow the dominant strategy of defection (Nash 1950). However, in the more complicated case of repeated interaction, even self-interested players who wish to maximize their own payoffs may maintain cooperation (Aumann 1959). One strategy that performs particularly well in a computerized tournament of the repeated game is "tit-for-tat," where one plays the same move as their opponent's most recent move, in effect rewarding cooperation with cooperation and punishing defection with defection (Axelrod 1984).

With a different payoff structure, the game of chicken is used to study the collective ability of two individuals to coordinate their behavior. As in the prisoner's dilemma, each player has two possible actions from which to choose, but the analogy here reflects different incentives. Two people drive towards one another on a collision course, and each can decide whether to stay on the road or swerve, thus being a "chicken." The best outcome for each person individually, is to be the one to stay on the road while the other swerves, but if each tries to achieve this outcome they will both realize the worse fate of collision. An intermediate outcome is achieved if both swerve, which provides a better outcome for both players. The interesting aspect of chicken is that players do best collectively if one stays the course and the other swerves, but they need some mechanism to determine who will take on which role, and each wants to avoid being the chicken, thus increasing their proclivity toward the collision outcome. In a single shot game, either outcome where both players take opposite actions is a Nash equilibrium, but in actuality some decision mechanism is required to allocate roles. Often, cultural norms help to dictate some means of coordination across repetitions of interaction like this through fairness concerns (e.g., "You give me this one, and I'll give you the next one") or status (e.g., "My greater power in another domain suggests that I should be the one to reap greater rewards in this setting also").

There have been several attempts to incorporate fairness concerns and other apparent deviations from rationality into economic theory (Camerer and Thaler 1995; Rabin 1993). However, a unification of these models will be difficult—if not impossible—given the substantial empirical evidence for subtle cultural influences on behavior in simple games. For example, Hoffman, McCabe, and Smith (1998) hypothesized that the prevalence of reciprocity and prosocial outcomes in lab experiments can be attributed to biological mechanisms for identifying and punishing cheaters in social exchange. Furthermore, Bohnet and Zeckhauser (2004) show that individuals placing trust in one another do not treat this strictly as a risky bet, but rather experience "betrayal aversion" and require greater assurance that their counterpart will act in a trustworthy manner to take the chance than they would require from a purely probabilistic outcome. Cultural factors become increasingly important when anonymity, and thus social distance, between individuals is reduced even in a very minimal way. In one study, silent identification without communication increased solidarity in a prisoner's dilemma (Bohnet and Frey 1999); in another, knowing a counterpart's family name was sufficient enough to increase offers in a "dictator game" (Charness and Gneezy 2000). Extending the implications of these studies to the real world, Habyarimana et al. (2007) explained the low levels of public goods provision within diverse ethnic communities in part by the difficulty of enforcing cooperation through the threat of social sanction. Building upon similar themes, the objective in this research is to understand precisely how culture influences individual decision making about which actions to take in abstract settings, so that we can make predictions about basic patterns of interaction in real environments with similarly structured payoffs.

Current Experiments Under the Bottom-Up Approach

Our studies under the bottom-up approach are inspired by the typical game theoretic abstraction of real-world relationships into a context-free environment, but include many possible extensions to the standard paradigm so that we can explore aspects of culture that are ignored by purely economic research. For this purpose, we have developed a flexible software tool that allows several individuals to interact with one another remotely via the Internet. We can control the instructions that are given to each player, the information that they receive about one another, the modes of communication between them, and the structure of the game (e.g., number of trials, payoffs in each trial depending on player actions, and so forth). One of the major benefits of this approach is that we can successively build greater complexity of cultural variables into the interaction once we understand behavior in a simpler setup.

Factors relevant to cultural identification are the perceived similarity of one's opponent, the framing of the game as cooperative or competitive, the amount of information received about one's own and the opponent's accumulated payoffs across multiple trials (for the sake of social comparison), the extent of communication with the opponent, and the "meta-knowledge" about how much information the opponent has. Through variations of the game setup, we can create mappings of actual case studies that help us to both explain and predict behavior in the real world. We can also then test the sensitivity of participant behavior to different changes in the game and opponent relationship.

In our studies, we characterize pairs by their degree of overall cooperation and accumulated joint payoffs, and see how these are predicted by individual differences in demographic makeup, personality, and other measures. However, we would also like to understand the evolution of cooperation over time, thus assessing the trajectory of the proportion of each type of action taken at each time step across all pairs of participants. Although not addressed in this chapter, our research program also examines the cognitive mechanisms driving behavior by comparing decision making by human players to that of an instance-based learning (IBL) cognitive model which operates by storing instances of actions and their corresponding payoffs (Gonzalez, Lerch, and Lebiere 2003; Gonzalez and Lebiere 2005). We use the basic ACT-R cognitive architecture (Anderson and Lebiere 1998), and the newly developed IBL models and tools (e.g., Gonzalez and Dutt 2011; Lejarraga, Dutt, and Gonzalez, in press). IBL is a straightforward model, where experiences are encoded in memory, and future choices depend on the value of the observed outcome and the probability of retrieval of that outcome from memory (Gonzalez, Lerch, and Lebiere 2003). The probability of retrieval is determined by the level of memory activation of previous instances. Prior to these experiments, we verified that our ACT-R model was able to predict basic patterns of behavior demonstrated in prior research (Rapoport, Guyer, and Gordon 1976).

An example of an initial experiment under the bottom-up approach involves a straightforward design to explore avenues for future variation with the greatest potential for meaningful cultural variation. Participants from the Center for Behavioral Decision Research subject pool at Carnegie Mellon University were recruited to adjacent computer labs during one-hour sessions, but were given no information about one another or the nature of the task they would engage in. The participants for each session were seated in two separate labs so that they could not see each other, and were given instructions via computer that they would be randomly matched with another participant in the other lab to play an interactive game where their payoffs would depend on their own and their opponent's simultaneous actions, with each point earned translating to $.01 at the end of the game (allowing them the possibility of earning more than the $10 base payment if their earnings exceeded this amount).

Once they were logged in and the software matched them with an opponent in the opposite lab, the first game began, consisting of 200 trials of a repeated prisoner's dilemma with the payoffs shown in Table 3.1, followed by 200 trials of chicken

TABLE 3.1
Payoffs for Two Players in the Prisoner's Dilemma

		Player 2	
		Action A	**Action B**
Player 1	**Action A**	−1, −1	10, −10
	Action B	−10, 10	1, 1

Note: Represented as (Player 1 payoff, Player 2 payoff) for each possible pair of actions. Action A represents defection and Action B represents cooperation.

TABLE 3.2
Payoffs for Two Players in the Game of Chicken

		Player 2	
		Action A	**Action B**
Player 1	**Action A**	1, 1	−1, 10
	Action B	10, −1	−10, −10

Note: Represented as (Player 1 payoff, Player 2 payoff) for each possible pair of actions. Action A represents swerving off the road and Action B represents staying the course.

with the payoffs shown in Table 3.2. Participants did not know the number of trials, nor were these payoff tables given to participants who only had the repeated choice between two buttons labeled "Action A" and "Action B"; after making a choice they saw their own action, their opponent's action, each of their payoffs for the trial, and their own running total. This design allowed them to learn the game structure in a naturalistic way. After each game, participants completed a survey about the level of trust that had developed between them and their opponent. (They were not told that their opponent in the chicken game was the same as in the prisoner's dilemma, but just that they had been matched with a random other, so either inference was plausible.)

In addition to the games, participants completed a basic demographic questionnaire, a certainty equivalent for a gamble to measure risk tolerance, and several individual assessment scales aimed at determining their identities and cultural affiliations: the Self Report Altruism Scale (Rushton, Chrisjohn, and Fekken 1981), Ten-Item Personality Inventory (Gosling, Rentfrow, and Swann 2003), Cognitive Reflection Test (Frederick 2005), Brief Maximization Scale (Nenkov et al. 2008), Short Form of the Need for Cognition (Cacioppo, Petty, and Kao 1984), Social Value Orientation (Messick and McClintock 1968), and a hypothetical Ultimatum Game (Güth, Schmittberger, and Schwarze 1982).

In a preliminary analysis of the first 26 pairs of participants, we found that individual differences in cognitive style led to unexpected pair-wise outcomes due to the interaction between the two individuals. For instance, the interaction of two individuals' scores on the Cognitive Reflection Test (Frederick 2005) had a marginal positive effect on their proportion of mutual defection in the prisoner's dilemma ($p = .06$), although the independent effect of each of the two individual scores was in the negative direction ($p = .12$ and $p = .13$). This means that a single member of the pair with higher cognitive ability might be able to control the pattern of play and avoid mutual defection, yet both players having either low or high cognitive ability would make coordination more difficult. We expect social factors to be similarly compounded in our final analysis looking not only at the influence of each individual's cultural variables, but at the interaction between these factors for counterparts who are sequentially responding to one another.

TOP-DOWN APPROACH USING PEACEMAKER

Researchers have illustrated the importance of taking into account the task complexity in studies of dynamic decision, and have suggested microworlds as tools for manipulating these factors and assessing behavioral outcomes (Brehmer 1992; Brehmer and Allard 1991). For instance, Sterman (1989) used an investment task to investigate misperceptions of feedback due to temporal delays between cause and effect. In a classic inventory management task, Diehl and Sterman (1995) also found decreases in performance as causal relationships were made more ambiguous by side effects. In such situations, observing a participant's entire sequence of actions within a microworld allows for inferences about the underlying decision-making heuristics being employed. Furthermore, elements of complexity can be implemented to reflect just the key aspects of a real-world scenario that a researcher wishes to isolate, thus permitting greater realism than static laboratory tasks while retaining greater experimental control (and reduction of confounding variables) compared to traditional field studies.

The interactive game PeaceMaker was developed by Impact Games (2006) primarily as an educational tool for teaching players factual information and the subjective difficulties of reaching an agreement in the conflict between Israeli and Palestinian factions. It is a single-player computer game in which an individual takes on the role of either the Israeli prime minister or the Palestinian president and engages in a series of actions to work toward the goal of achieving a peaceful two-state solution that satisfies constituents on both sides of the conflict. A sustainable peace in the game is defined by the sufficient appeasement of constituents on *both* sides of the conflict, regardless of the role being played. This is, of course, a very difficult objective in the real world, which is captured in the game by ambiguous feedback and random inciting incidents that disrupt stability. Of greater interest is the manner in which a player works toward this end. The three main types of actions available on the menu are construction, political, and security. But within each of these categories, there are a variety of alternatives depending on the role being played. Both player actions and inciting incidents affect the approval of constituents (including factions within the region such as Fatah, as well as international parties such as the United Nations).

PeaceMaker can be played on three difficulty levels (calm, tense, and violent), which differ in the frequency of inciting incidents. In addition, the game environment changes dynamically due to inciting incidents and a player's own actions, so the player cannot predict exactly how various constituent groups will react to a particular action based on their reaction in the past. Throughout the game, a player views real storylines and news footage drawn from historical events with the input of experts. In this sense, PeaceMaker can be considered a microworld, or a representation of the actual conflict that immerses players in a rich context that may evoke similar cultural identifications as in the real world. See Figure 3.1 for a screenshot of the PeaceMaker interface.

The constituent views are aggregated into the two main scores of the game that we call *OwnScore* and *OtherScore*, representing the approval of stakeholders aligning with one's own role in the simulation and of stakeholders aligning with the other side, respectively. *OwnScore* and *OtherScore* each start at 0, and players are aware that both scores need to reach 100 in order to win the game, whereas they will

FIGURE 3.1 PeaceMaker Screenshot. The expandable menu of player actions appears on the middle left, the summary of OwnScore and OtherScore on the lower left, and the more detailed constituent approval ratings on the lower right of the screen. (Courtesy of Eric Brown, copyright © 2007. With permission.)

lose and the game will come to an end if either score drops below –50. We refer to *OwnScoreChange* and *OtherScoreChange* as the amounts by which each of these scores is adjusted as a result of a player action or an inciting incident. The formula that determines these score changes is defined within the game architecture based upon the input of experts on the historical conflict. Because the underlying formula is conditioned on the current state of affairs in the game, part of a player's challenge is predicting which actions will lead to increased satisfaction of constituents and thus an improvement in the corresponding score. Successful players will realize the need to take some actions that appease one side at the expense of approval from the other side to ensure that neither score drops too low. PeaceMaker thus encapsulates many interesting features of the conflict, such as the need to balance competing interests. Furthermore, players must deal with uncertain events beyond their control despite the lack of transparency about how differently constituents will respond to actions taken. In this sense, PeaceMaker allows us to look at strategic behavior in a highly complex environment (Gonzalez and Czlonka 2010; Gonzalez and Saner, in press).

We have used PeaceMaker to study a number of different dimensions of conflict resolution, such as the need to balance competing interests (Gonzalez, Saner, and Eisenberg 2011). Several key results have emerged from these studies thus far. First, individual identity variables and trust disposition correlate with scores obtained in the game (Gonzalez and Czlonka 2010). Second, personality type (assessed by the Myers-Briggs Type Indicator) predicts performance in the game, with those of "thinking" versus "feeling" personality winning the game more often; this suggests that those who are more assertive and impersonal, rather than affective and personal, are more successful in conflict resolution (Gonzalez and Saner, in press). Third, knowledge about the history of the conflict and repeated play of the game improves performance and leads to decreased correlation between performance and some individual variables (Gonzalez, Kampf, and Martin 2011). Fourth, exploration across a variety of action categories is positively related to performance in the game, with nonwinners being consistently less exploratory from beginning to end, and specifically, a disproportionate number of security actions predicting poor performance (Gonzalez, Kampf, and Martin 2011).

The various findings using PeaceMaker highlight many issues of social importance in dealing with conflict resolution situations. In the next section, we focus on a single experiment conducted across cultures, demonstrating the game's potential to test the effects of cultural identification in playing this game.

Example of the Top-Down Approach: Group Identification in PeaceMaker

There are many levels of group membership that shape an individual's cultural affiliations. Given the confluent influences of upbringing, schooling, community, religion, and nationality, any person's set of cultural affiliations could be considered unique, and yet at any point in time a particular subset of these identifications may be especially salient. It is one of the central tenets of the social identity theory that individuals define themselves by their group categorizations, which may be determined by minimal dimensions of intragroup similarity or intergroup difference (Tajfel et al. 1971; Turner 1975). An "us versus them" mentality may arise as individuals strive to receive a favorable reaction from their group, to maintain a satisfying connection to that group, and to attain congruence with group values (Kelman 2006). Because these motivations for group identification are so strong, it is no surprise that they play a major role in the development of self-concept and the formation of social stereotypes (Tajfel 1982). We wish to explore the way that the culture ascribed by group membership alters an individual's worldview and behavior in interpersonal interaction.

In-Group Favoritism

If individuals glean a sense of self from group membership, then members of a common group might even treat one another like extensions of themselves and engage in greater behavioral reciprocity (Brewer 1979). Game theory is one field that has demonstrated in-group favoritism, whereby decision makers bestow disproportionate rewards on members of their own group and/or disproportionate punishments on outsiders. In an intergroup prisoner's dilemma, for example, individuals were willing to punish another group if it benefitted their own group, especially when they were allowed to communicate and potentially establish in-group norms of reciprocity

(Halevy, Bornstein, and Sagiv 2008). Likewise, members of indigenous tribes in Papua New Guinea punished norm violators in a dictator game more severely when the violator belonged to another tribe and the victim belonged to the punisher's own tribe (Bernhard, Fischbacher, and Fehr 2006). Such "parochial altruism" has also been established in a prisoner's dilemma game played amongst members of randomly assigned Swiss Army platoons (Goette, Huffman, and Meier 2006). In these minimalist settings, in-group favoritism seems to be quite pervasive: People tend to protect in-group members, oftentimes at the expense of out-group members.

Out-Group Perspective-Taking

Group divisions may be further accentuated when they are reinforced by hierarchical power differentials, as discussed by social dominance theory (Sidanius and Pratto 1993). Some aspects of an individual's culture, however, may span various group allegiances and increase the ability to take the perspectives of others. System justification theory highlights the psychological mechanisms that cause subjugated individuals or groups to defend the status quo of a dominant social order, even if it is not in their best interest (Jost, Banaji, and Nosek 2004). Specifically, minority group members may be more likely to acknowledge cultural commonalities in addition to their distinct subgroup interests, and therefore prefer a system that legitimizes dual identification (Dovidio, Gaertner, and Saguy 2009). Similarly, members of high-status groups are more apt to sympathize with members of low-status groups if they feel that the power differential is unjust. Cognitive dissonance—or a discrepancy between self-perception and reality—may ensue (Halperin et al. 2010), and the former may experience guilt (Roccas, Klar, and Liviatan 2006) or anger, promoting the desire for compensatory political action (Leach, Iyer, and Pedersen 2006). Thus, although we expect individuals to exhibit a fundamental in-group bias, the degree of favoritism may be tempered by connections to overarching cultural norms and perspective-taking of other groups.

GROUP IDENTIFICATION IN A PEACEMAKER EXPERIMENT

As many researchers have shown, group identification can lead individuals to favor the interests of their own group members, sometimes at the expense of other competing interests (Bernhard, Fischbacher, and Fehr 2006; Goette, Huffman, and Meier 2006; Halevy, Bornstein, and Sagiv 2008). For the most part, we predicted in-group bias to be present in the game, with individuals having political or religious affiliation to one side taking more actions that favored that side's interests regardless of which role they were assigned to play. In the context of conflict resolution, however, an individual's desire to generate a peaceful solution may enhance the ability to take the other side's interests into account. We explore conditions under which out-group interests are internalized as one's own (Dovidio, Gaertner, and Saguy 2009; Jost, Banaji, and Nosek 2004), increasing motivations to balance in-group and out-group interests (Halperin et al. 2010; Leach, Iyer, and Pedersen 2006).

PeaceMaker is an ideal tool for the investigation of the in-group and out-group hypotheses described above. This is particularly relevant to participants with affiliations to the two parties represented in the game, but also for more general affiliations. In this specific experiment, differentials between the *OwnScore* and *OtherScore* variables (as well as differentials between actions tending to increase one at the expense of the other) were used to assess the bias of a player. Related to the discussion of in-group favoritism, we asked the political and religious affiliation of participants, and tested the effect of these cultural variables on the bias of actions taken in the game, measured using several composite metrics. The main hypothesis was that those who affiliated with the role they were playing would exhibit a greater proportion of actions in favor of that side, whereas those who affiliated with the opposite of the one they were playing would show a bias in favor of the other side. The Israeli–Palestinian conflict resolution scenario turned out to have induced more subtle motivations, sometimes favoring the side with which a participant affiliated, and sometimes the reverse.

Experiment Design

For this study, we recruited 142 participants (62 female, 80 male; age $M = 22.4$, $SD = 6.15$) from three computer labs in Israel (26 participants), Qatar (59 participants), and the United States (57 participants). The experimental protocol of the study was as similar as possible in all three locations. Each participant completed a tutorial of PeaceMaker that introduced them to the interface, and then played the game twice, once as the Israeli prime minister and once as the Palestinian president, in counterbalanced order such that half the participants played one role first and the other half played the other role first. A software patch allowed us to record the entire sequence of actions taken during the course of a game, their effects on *OwnScore* and *OtherScore*, and the game outcome (win or loss). Participants also completed a survey of demographic information, political affiliation, religious affiliation, background knowledge on the conflict, the social values orientation trust assessment, the Myers-Briggs type indicator personality assessment, and the number of gaming hours per week.

The main independent variables of interest were individual differences in religious and political affiliation because we wished to analyze whether affiliation with one or the other role induced bias toward that side. To characterize the degree to which a particular action favored the constituents of one's assigned role in the conflict relative to that of the other side, we created the metric *Bias* = *OwnScoreChange* – *OtherScoreChange*. *Bias* can be positive (indicating bias toward one's own assigned role) or negative (indicating bias toward the opposite side from one's own assigned role). A value of zero on average indicates no bias and a balance between the two positions of the game. Although the specific impact a particular action might have on each score depends on various factors in the game (including a stochastic element), this metric serves as a proxy for the intentions of a participant in satisfying one side versus the other by taking a causal action. *Bias* thus captures relative changes in the two scores due to a particular action, and is positive for actions that favor the role being played, negative for actions that favor the opposite side, and zero for actions that favor both sides equally. This measure is independent of the participant's own political or religious affiliation, so that we can compare bias of participants playing the same role with their own political and religious affiliations as predictor variables.

Results

Figures 3.2a and 3.2b (for the Israeli role and Palestinian role, respectively) show the relationship between *OwnScoreChange* and *OtherScoreChange* for all actions taken by at least one participant in the dataset, and a list of the actions with the most highly positive or negative values for these two variables. The negative correlation between

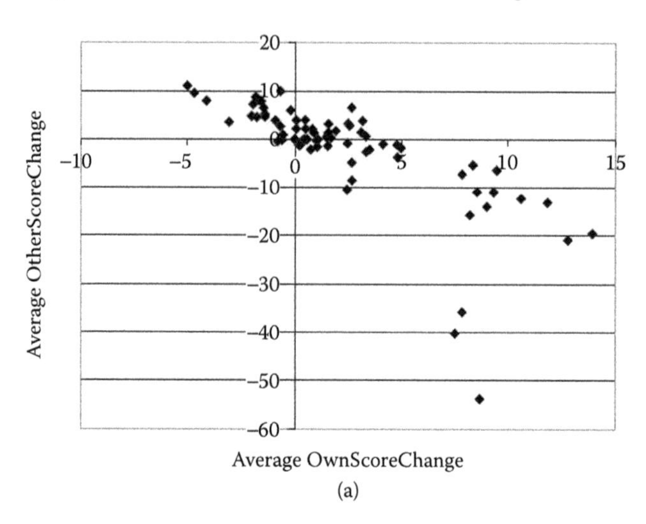

Actions with lowest average *OwnScoreChange*:

- Security >> ISRAEL RAISE CURFEW
- Security >> ISRAEL DECREASE CHECKPOINTS
- Security >> ISRAEL REMOVE IDF

Actions with highest average *OwnScoreChange*:

- Security >> ISRAEL ASSASSINATE COVERT OPERATION
- Security >> ISRAEL BULLDOZE PUNISH MILITANTS
- Construction >> ISRAEL BUILD WALL ON PALESTINIAN LAND

Actions with lowest averge *OtherScoreChange*:

- Construction >> ISRAEL BUILD NEW SETTLEMENT
- Security >> ISRAEL MISSILE PA POLICE
- Construction >> ISRAEL EXPAND SETTLEMENTS

Actions with highest average *OtherScoreChange*:

- Political >> ISRAEL INCREASE WORKER PERMITS
- Security >> ISRAEL DECREASE CHECKPOINTS
- Political >> ISRAEL YESHA ARREST LEADERS

FIGURE 3.2 (a) Relationship between OwnScoreChange and OtherScoreChange for all actions available in the role of Israeli Prime Minister (averaged across all occurrences of the action in the dataset). The correlation was −.74, $p < .0001$. (b) Relationship between OwnScoreChange and OtherScoreChange for all actions available in the role of Palestinian President (averaged across all occurrences of the action in the dataset). The correlation was −.15, $p = .03$.

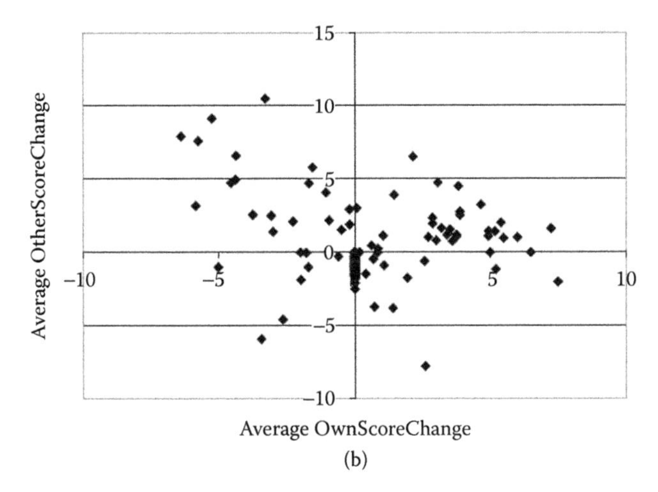

(b)

Actions with lowest average *OwnScoreChange*:

- Security >> POLICE PATROL FOR MILITANTS
- Political >> HOLY SITES PERMIT ACCESS FOREIGNERS
- Security >> POLICE ARREST KNOWN MILITANTS

Actions with highest average *OwnScoreChange*:

- Political >> SPEECH TO WORLD DEMANDING
- Political >> HOLY SITES RESTRICT ACCESS FOREIGNERS
- Security >> INTERNAL TRAVEL PUSH CONTROL

Actions with lowest average *OtherScoreChange*:

- Political >> SPEECH TO PALESTINIANS PRO-MILITANCY
- Political >> HAMAS ASK TO DISARM
- Political >> HAMAS ASK FOR CEASE-FIRE

Actions with highest average *OtherScoreChange*:

- Security >> POLICE ARREST KNOWN MILITANTS
- Security >> POLICE PATROL FOR MILITANTS
- Security >> POLICE SEIZE MILITANT ARMS

FIGURE 3.2 (*Continued*)

OwnScoreChange and *OtherScoreChange* (–.74 for the Israeli role and –.15 for the Palestinian role) confirms that participants often face a trade-off between satisfying constituents on one side or the other, although there are some actions that tend to either please or displease both sides.

The distributions in Figures 3.2a and 3.2b reveal that the nature of play is not identical across the two roles. In addition, higher approval scores for the Israeli side (reflected in generally positive role bias toward the role of Israeli prime minister and generally negative role bias toward the role of Palestinian president) suggest either an inherent bias in the game or a baseline favoritism of all participants for that side. Since we cannot disentangle the two, measures are compared across participants of different identities playing the same role. As such, any differences in dependent measures that we find can be attributed to individual differences in play rather than any artifact of the game architecture.

Participants were classified as having a political affiliation with Palestine (16 participants) or Israel (46 participants) if they listed one or the other. All others were characterized as having no political identification with either side and were excluded from this analysis for the sake of direct comparison between those who identified directly with one or the other side. Similarly, participants were classified as having a religious affiliation with the Israeli side (25 participants) if they reported their religion as Judaism, Judaism-Conservative, Judaism-Reform, or Judaism-Secular, and with the Palestinian side (30 participants) if they reported their religion as Islam. Only one participant reported Judaism-Conservative as their religion, so we were unable to test more specific differences depending on the strength of affiliation. There was a strong positive correlation between political and religious affiliation ($r = 0.67$, $p < .0001$ when they were each encoded as 1 = affiliated with the role being played, 0 = affiliated with neither role, and –1 = affiliated with the opposite role), yet we examine the two types of affiliation separately since the cases in which one affiliation happens to predict bias more strongly than the other may be of interest to readers (Martin and Gonzalez 2011).

By computing *Bias*, we were able to assess participant bias for any given action taken in the game. The relevant outcome variables—in bold on first appearance below—were intended to capture bias across the entire game in several different ways. ***AvgBias*** is defined to be the average of *Bias* across all actions in the game to represent overall preference for actions that increase *OwnScore* relative to *OtherScore*.

First, we tested the effects of political and religious affiliation on *AvgBias*. For the role of the Israeli prime minister, we discovered a marginally significant negative effect of political affiliation on *AvgBias* ($M = 1.52$ for those with Israeli affiliation and $M = 3.77$ for those with Palestinian affiliation, $t(60) = 1.9$, $p = .06$), and a significant negative effect of religious affiliation on *AvgBias* ($M = 1.41$ for Jewish participants and $M = 5.16$ for Muslim participants, $t(53) = 2.5$, $p = .02$) (Martin and Gonzalez 2011). This contradicts our hypothesis of in-group favoritism because Palestinian and Muslim participants took actions more biased on average toward Israeli constituents of their assigned role than did Israeli and Jewish participants. For the role of the Palestinian president, we found no effect on *AvgBias* of political affiliation with one's role ($M = -2.10$ for Palestine-identified participants and $M = -2.83$ for Israel-identified participants, $t(60) = 1.5$, $p = .14$) or religious affiliation with one's role ($M = -2.62$ for Muslims and $M = -2.57$ for Israelis, $t(53) = .14$, $p = .89$) (Martin and Gonzalez 2011).

SumBias is defined as the sum of *Bias* values across all actions in the game. This variable indicates the accumulated level of preference toward the participant's assigned role. Note that *SumBias* will be very close to 0 for the minority of participants who won the game, so the measure is a better indicator amongst those who lost of which side they favored in doing so. In the role of the Israeli prime minister, we found no effect on *SumBias* of political affiliation ($M = 8.28$ for Israelis and $M = 21.4$ for Palestinians, $t(60) = .84$, $p = .41$) and a negative effect of religious affiliation ($M = 5.72$ for Jewish participants and $M = 34.3$ for Muslim participants, $t(53) = 2.0$, $p < .05$) (Martin and Gonzalez 2011). Hence, the general trend of *SumBias* in the role of the Israeli prime minister was to favor constituents opposite one's own affiliations throughout the game. In the role of the Palestinian president though, we found support for our in-group favoritism hypothesis, with a positive effect on *SumBias* of both political affiliation ($M = -67.3$ for Palestinians and $M = -102$ for Israelis,

$t(60) = 2.1$, $p = .04$) and religious affiliation ($M = -69.2$ for Muslim and $M = -105$ for Jewish participants, $t(53) = 2.3$, $p = .02$) (Martin and Gonzalez 2011). In other words, when assuming the role of the Palestinian president, participants with Palestinian and Muslim identification tended to favor this side in the game—in sum across all game actions—more so than did participants of Israeli and Jewish identification. Alternatively, we could interpret this result as Israeli and Jewish participants favoring Israeli constituents more strongly even though they were playing the Palestinian role.

Finally, we measured a participant's tendency to take the most extreme actions in favor of the role to which they were assigned. We considered an action to be extremely in favor of one's own side if *Bias* was at least two standards deviations above the mean, across all actions taken in the dataset for that role. For the Israeli role, the distribution of *Bias* had a mean of 0.57 and standard deviation of 11.36, so the minimum cutoff for being considered extreme was $0.57 + 2 \times 11.36 = 23.29$. For the Palestinain role, *Bias* had a mean of −1.71 and standard deviation of 6.49, so the cutoff was $-1.71 + 2 \times 6.49 = 11.27$. The percentage of times that a participant took an action classified as extreme, relative to the total number of actions he/she took in the game, determined the value for ***ExtremeBias*** in that game. The complete distribution of *Bias* for the two roles is shown in Figures 3.3a and 3.3b.

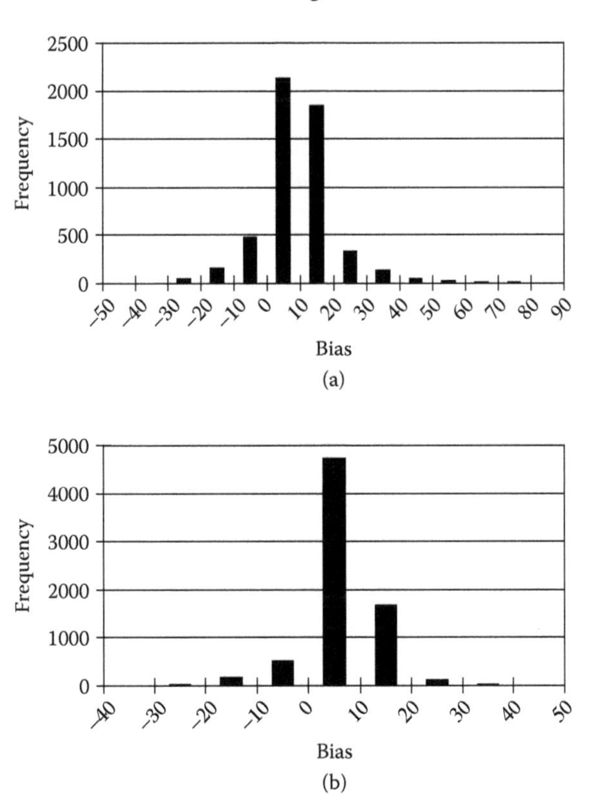

FIGURE 3.3 (a) Frequency of actions with different values of Bias for all actions taken in the role of Israeli Prime Minister. (b) Frequency of actions with different values of Bias for all actions taken in the role of Palestinian President.

We explored whether political and religious affiliation affected a participant's tendency to take more extreme actions in favor of one side or the other. In the role of the Israeli prime minister, a participant's Israeli (versus Palestinian) political affiliation negatively predicted *ExtremeBias* ($M = 6.10\%$ versus $M = 12.1\%$, $t(60) = 2.1$, $p = .04$), and a Jewish (versus Muslim) religious affiliation also had a marginal negative effect on *ExtremeBias* ($M = 7.11\%$ versus $M = 12.1\%$, $t(53) = 1.7$, $p = .09$) (Martin and Gonzalez 2011). This suggests that nonaffiliated participants took a higher proportion of actions that strongly favored an increase in *OtherScore* relative to *OwnScore* when playing the Israeli role. In contrast, for the role of the Palestinian president, we once again observe a more "standard" bias of extreme actions in favor of constituents with shared identification: *ExtremeBias* was positively predicted by both Palestinian (versus Israeli) political affiliation ($M = 3.31\%$ versus $M = 1.47\%$, $t(60) = 2.1$, $p = .04$) and Muslim (versus Jewish) religious affiliation ($M = 3.11\%$ versus $M = 1.30\%$, $t(53) = 2.1$, $p = .04$) (Martin and Gonzalez 2011). This indicates that when playing the Palestinian role, participants took a greater percentage of actions favoring constituents with whom they associate on these cultural dimensions.

Discussion

The previous analysis emphasizes the role of cultural affiliations in predicting strategic bias in PeaceMaker. The literature in social and political psychology generally supports the notion that individuals develop a sense of purpose in their lives from identification with groups, and will often defend the values of these groups as a means of self-enhancement. Surprisingly, our results on the average bias (*AvgBias*), aggregate bias (*SumBias*), and extreme bias (*ExtremeBias*) only partially support the hypothesis that participants who affiliate with their assigned role in the game will favor its constituents more strongly. This was the predominant finding for the role of the Palestinian president where Palestinian and Muslim participants indeed showed stronger relative favoritism toward their own side compared to Israeli and Jewish participants, respectively. However, we discover the opposite pattern for the role of the Israeli prime minister where Israeli and Jewish participants more strongly favored the opposite side in the conflict compared to Palestinian and Muslim participants, respectively.

From the standpoint of participants with Palestinian political affiliation or Muslim religious affiliation, we interpret this as a desire to assert a less powerful international position in the Palestinian role (Smooha 1978, 1992), but also identify with the Israeli role because of the more pervasive acceptance of those interests (Dovidio, Gaertner, and Saguy 2009). Perceived minority status by Palestinian-identified participants may have induced typical in-group bias in the role of the Palestinian president (Cehajic, Brown, and González 2009; Tajfel 1982), yet triggered adherence to majority group norms in the role of the Israeli prime minister (Jost, Banaji, and Nosek 2004). An alternative—and not incompatible—explanation of the different identification effect across the two roles is that Israeli and/or Jewish participants rationalized prevailing Israeli interests when they were assigned to the Palestinian role, but overcompensated in attempting to take the opposition's perspective in the Israeli role (Leach, Iyer, and Pedersen 2006). They may have assumed more extreme actions were necessary to appease the other side than was actually the case, which is compatible with evidence for "naïve realism" (Robinson et al. 1995), whereby partisan individuals overestimate the extremity of opposition views on contentious issues. On the whole, it appears that individuals may have a more visceral response to upholding their cultural identities and avoid behaviors that violate those interests when playing the Palestinian president, yet make a deliberate effort to put themselves in the shoes of the other side and avoid offending opposing interests in the role of the Israeli prime minister.

DISCUSSION OF TOP-DOWN AND BOTTOM-UP APPROACHES

In the combination of top-down and bottom-up approaches we have described, each offers some benefits to overcoming the limitations of the other in isolating the effects of culture on strategic interaction. The strength of using microworlds like PeaceMaker is that they provide participants with a complex simulation of the real world that can thus evoke many of the same cultural identifications that might not arise in more static or sterile lab experiments. We hope our experimental paradigm using PeaceMaker has illustrated the highly nuanced types of behavior that can be observed through the use of interactive computer simulations. Unlike traditional laboratory or field studies, empirical investigations with microworlds permit some control and measurement of independent variables while still supplying participants with engaging, context-rich stimuli.

The strength of using abstract games like the 2×2 software developed is to observe individual decision making under very minimal conditions so that it is easier to isolate the effect of individual differences and other manipulations. The biggest challenge is unifying these two approaches so that the findings from one can inform the design of future studies in the other. For example, results from PeaceMaker indicate that in-group favoritism may be more pervasive amongst individuals assigned to a less mainstream position, which is an aspect of the context that we can manipulate more specifically in our 2×2 games via power framing or payoff differentials. As results are gathered from the 2×2 games, they will also supply insights about the cultural factors that can be built into more realistic microworlds. If there is one result common across the two approaches, it is that culture cannot be simply neglected in the study of dynamic decision making in complex social interaction.

ACKNOWLEDGMENTS

This research is supported by the Defense Threat Reduction Agency (DTRA), grant number HDTRA1-09-1-0053 to Cleotilde Gonzalez and Christian Lebiere.

REFERENCES

Anderson, J. R., and C. Lebiere. 1998. *The Atomic Components of Thought*. Hillsdale, NJ: Lawrence Erlbaum Associates.

Aumann, R. J. 1959. Acceptable points in general cooperative *n*-person game. In *Contributions to the Theory of Games IV. Annals of Mathematics Studies*, ed. A. W. Tucker and R. C. Luce, Vol. 40, 287–324. Princeton, NJ: Princeton University Press.

Axelrod, R. 1984. *The Evolution of Cooperation*. New York: Basic Books.
Bernhard, H., U. Fischbacher, and E. Fehr. 2006. Parochial altruism in humans. *Nature* 442:912–5.
Boellstorff, T. 2006. A ludicrous discipline? Ethnography and game studies. *Games Cult* 1:29–35.
Bohnet, I., and B. S. Frey. 1999. The sound of silence in prisoner's dilemma and dictator games. *J Econ Behav Organ* 38:43–57.
Bohnet, I., and R. Zeckhauser. 2004. Trust, risk, and betrayal. *J Econ Behav Organ* 55:467–84.
Brehmer, B. 1992. Dynamic decision making: Human control of complex systems. *Acta Psychol* 81:211–41.
Brehmer, B., and R. Allard. 1991. Dynamic decision making: The effects of task complexity and feedback delay. In *Distributed Decision Making: Cognitive Models of Cooperative Work*, ed. J. Rasmussen, B. Brehmer, and J. Leplat, 319–34. Chichester, UK: Wiley.
Brewer, M. B. 1979. In-group bias in the minimal intergroup situation: A cognitive-motivational analysis. *Psychol Bull* 86:307–24.
Cacioppo, J. T., R. E. Petty, and C. F. Kao. 1984. The efficient assessment of need for cognition. *J Pers Assess* 48:306–7.
Camerer, C., and R. H. Thaler. 1995. Anomalies: Ultimatums, dictators and manners. *J Econ Perspect* 9:209–19.
Cehajic, S., R. Brown, and R. González. 2009. What do I care? Perceived ingroup responsibility and dehumanization as predictors of empathy felt for the victim group. *Group Process Intergroup Relat* 12:715–29.
Charness, G., and U. Gneezy. 2000. *What's In a Name? Anonymity and Social Distance in Dictator and Ultimatum Games*. Unpublished manuscript.
Diehl, E., and J. D. Sterman. 1995. Effects of feedback complexity on dynamic decision making. *Organ Behav Hum Decis Process* 62:198–215.
Dovidio, J. F., S. L. Gaertner, and T. Saguy. 2009. Commonality and the complexity of 'we': Social attitudes and social change. *Pers Soc Psychol Rev* 13:3–20.
Frederick, S. 2005. Cognitive reflection and decision making. *J Econ Perspect* 19:25–42.
Goette, L., D. Huffman, and S. Meier. 2006. The impact of group membership on cooperation and norm enforcement: Evidence using random assignment to real social groups. *Am Econ Rev* 96:212–6.
Gonzalez, C., and L. Czlonka. 2010. Games for peace: Empirical investigations with PeaceMaker. In *Serious Game Design and Development: Technologies for Training and Learning*, ed. J. Cannon-Bowers and C. Bowers. Hershey, PA: IGI Global.
Gonzalez, C., and V. Dutt. 2011. *Making Instance-Based Learning Theory Usable and Understandable: The Instance-Based Learning Tool*. Unpublished manuscript under review.
Gonzalez, C., R. Kampf, and J. Martin. 2011. Action diversity of Israeli students in a simulation of the Israeli-Palestinian conflict. Unpublished manuscript under review.
Gonzalez, C., and C. Lebiere. 2005. Instance-based cognitive models of decision making. In *Transfer of Knowledge in Economic Decision-Making*, ed. D. Zizzo and A. Courakis, 148–65. New York: Macmillan (Palgrave Macmillan).
Gonzalez, C., C. Lebiere, J. M. Martin, and I. Juvina. 2011. Dynamic decision making games and conflict resolution. In *Advances in Cross-Cultural Decision Making,* ed. D. Schmorrow and D. Nicholson, 312–21. Boca Raton, FL: CRC Press.
Gonzalez, C., J. F. Lerch, and C. Lebiere. 2003. Instance-based learning in dynamic decision making. *Cogn Sci* 27:591–635.
Gonzalez, C., and L. D. Saner. In press. Thinking or feeling? Effects of decision making personality in conflict resolution. In *Emotional Gaming*, ed. J. V. Brinken, H. Konietzny, and M. Meadows.

Gonzalez, C., L. D. Saner, and L. Eisenberg. 2011. *Learning to Stand in the Other's Shoes: A Computer Video Game Experience of the Israeli-Palestinian Conflict*. Unpublished manuscript under review.

Gonzalez, C., P. Vanyukov, and M. K. Martin. 2005. The use of microworlds to study dynamic decision making. *Comput Human Behav* 21:273–86.

Gosling, S. D., P. J. Rentfrow, and W. B. Swann. 2003. A very brief measure of the big-five personality domains. *J Res Pers* 37:504–28.

Güth, W., R. Schmittberger, and B. Schwarze. 1982. An experimental analysis of ultimatum bargaining. *J Econ Behav Organ* 3:367–88.

Habyarimana, J., M. Humphreys, D. N. Posner, and J. M. Weinstein. 2007. Why does ethnic diversity undermine public goods provision? *Am Polit Sci Rev* 101:709–25.

Halevy, N., G. Bornstein, and L. Sagiv. 2008. "In-group love" and "out-group hate" as motives for individual participation in intergroup conflict: A new game paradigm. *Psychol Sci* 19:405–11.

Halperin, E., D. Bar-Tal, K. Sharvit, N. Rosler, and A. Raviv. 2010. Socio-psychological implications for an occupying society: The case of Israel. *J Peace Res* 47:59–70.

Hoffman, E., K. A. McCabe, and V. Smith. 1998. Behavioral foundations of reciprocity: Experimental economics and evolutionary psychology. *Econ Inquiry* 36:335–52.

Impact Games. 2006. *PeaceMaker Computer Software*. Pittsburgh, PA: ImpactGames, http://www.peacemakergame.com. (accessed January 13, 2010).

Jost, J. T., M. R. Banaji, and B. A. Nosek. 2004. A decade of system justification theory: Accumulated evidence of conscious and unconscious bolstering of the status quo. *Polit Psychol* 25:881–919.

Kelman, H. C. 2006. Interests, relationships, identities: Three central issues for individuals and groups in negotiating their social environment. *Annu Rev Psychol* 57:1–26.

Kelman, H. C. 2008. A social-psychological approach to conflict analysis and resolution. In *Handbook of Conflict Analysis and Resolution*, ed. D. Sandole, S. Byrne, I. Sandole-Staroste, and J. Senehi, 170–83. New York: Routledge.

Leach, C. W., A. Iyer, and A. Pedersen. 2006. Anger and guilt about ingroup advantage explain the willingness for political action. *Pers Soc Psychol Bull* 32:1232–45.

Lejarraga, T., V. Dutt, and C. Gonzalez. In press. Instance-based learning: A general model of repeated binary choice. *J Behav Dec Mak*.

Martin, J. M., and C. Gonzalez. 2011. *The Cultural Determinants of Strategic Bias: A Study of Conflict Resolution in an Interactive Computer Game*. In *Proceedings of the 3rd ACM International Conference on Intercultural Collaboration*, ed. P. Hinds, A. M. Søderberg, and R. Vatrapu, 151–60. Copenhagen, Denmark: ACM.

Messick, D. M., and C. G. McClintock. 1968. Motivational bases of choice in experimental games. *J Exp Soc Psychol* 4:1–25.

Nash, J. 1950. Equilibrium points in n-person games. *Proc Natl Acad Sci U S A* 36:48–9.

Nenkov, G. Y., M. Morrin, A. Ward, B. Schwartz, and J. Hulland. 2008. A short form of the maximization scale: Factor structure, reliability and validity studies. *Judgm Decis Mak* 3:371–88.

Poundstone, W. 1993. *Prisoner's Dilemma*. New York: Doubleday.

Rabin, M. 1993. Incorporating fairness into game theory and economics. *Am Econ Rev* 83:1281–302.

Rapoport, A., and A. M. Chammah. 1965. *Prisoner's Dilemma: A Study in Conflict and Cooperation*. Ann Arbor: University of Michigan Press.

Rapoport, A., M. J. Guyer, and D. G. Gordon. 1976. *The 2 × 2 Game*. Ann Arbor: University of Michigan Press.

Robinson, R. J., D. Keltner, A. Ward, and L. Ross. 1995. Actual versus assumed differences in construal: "Naive realism" in intergroup perception and conflict. *J Pers Soc Psychol* 68:404–17.

Roccas, S., Y. Klar, and I. Liviatan. 2006. The paradox of group-based guilt: Modes of national identification, conflict vehemence, and reactions to the in-group's moral violations. *J Pers Soc Psychol* 91:698–711.
Rushton, J. P., R. D. Chrisjohn, and G. C. Fekken. 1981. The altruistic personality and the self-report altruism scale. *Pers Individ Dif* 2:293–302.
Sidanius, J., and F. Pratto. 1993. The inevitability of oppression and the dynamics of social dominance. In *Prejudice, Politics, and the American Dilemma*, ed. P. M. Sniderman, P. E. Tetlock, and E. G. Carmines, 173–211. Stanford, CA: Stanford University Press.
Smooha, S. 1978. *Israel: Pluralism and Conflict*. London: Routledge & Kegan Paul.
Smooha, S. 1992. *Arabs and Jews in Israel: Change and Continuity in Mutual Intolerance*. Boulder, CO: Westview.
Sterman, J. 1989. Misperceptions of feedback in dynamic decision making. *Organ Behav Hum Decis Process* 43:301–35.
Tajfel, H. 1982. Social psychology of intergroup relations. *Annu Rev Psychol* 33:1–39.
Tajfel, H., M. G. Billig, R. P. Bundy, and C. Flament. 1971. Social categorization and intergroup behaviours. *Eur J Soc Psychol* 1:149–78.
Turner, J. C. 1975. Social comparison and social identity: Some prospects for intergroup behaviour. *Eur J Soc Psychol* 1:5–34.

4 Why Do People Think Culturally When Making Decisions? Theory and Evidence

Yuk-Yue Tong and Chi-Yue Chiu

CONTENTS

INTRODUCTION

Cross-cultural comparisons of the relative likelihood of exhibiting a certain judgment style have become the gold standard test for psychologists who are interested in effects of culture on judgment (Lehman, Chiu, and Schaller 2004). For example, an investigator interested in the effect of individualism-collectivism on the likelihood of committing the fundamental attribution error (attribution of the cause of an act to the dispositions of the actor despite the presence of an obvious situational explanation of the act) would measure the relative likelihood of committing the fundamental attribution error in individualist versus collectivist cultures (Miller 1984).

Comparative studies of judgment and decision-making styles in the past two decades have revealed marked differences in the cognitive habits between cultural groups. For instance, compared to individuals from Eastern cultures, individuals from Western cultures are less inclined to attend to the focal objects (e.g., the actor in a social situation) as opposed to the background (e.g., the social context of the situation) (Chua, Boland, and Nisbett 2005), spontaneously infer the presence of internal dispositions in the focal object (Newman 1993), group objects on the basis of their inferred internal dispositions (Ji, Zhang, and Nisbett 2004), and use these dispositions to explain the behaviors of the focal objects (Morris and Peng 1994). In contrast, due to the dynamic nature of the social environments in East

Asia (see Weber and Morris 2010), individuals from Eastern cultures tend to exhibit less choice-congruent behavior and compliance with their initial decisions than do American students (Petrova, Cialdini, and Sills 2007).

These striking differences seem to support the cultural relativist thesis that the ways in which individuals come to understand and interpret the world around them differs from one culture to the next (see Chiu, Kim, and Wan 2008). According to this thesis, each culture is characterized by a certain "indigenous" tradition, consisting of ideas and practices that are often untranslatable and sometimes even nonexistent in another culture. For example, some cultural researchers (Peng and Nisbett 1999) have attributed Westerners' greater preference for dispositionist thinking to the distinctive emphasis on analytical thinking (a methodical approach to solving complex problems by breaking them into their constituents' parts, and identifying the cause and effect patterns of the constituent parts) in Western philosophical traditions. Likewise, Easterners' greater preference for contextualized thinking has been explained in terms of the distinctive emphasis on holism (the idea that the properties of a given system cannot be determined or explained by its component parts alone, but the system as a whole determines how the parts behave) in Eastern philosophical traditions. The assumption that all cultures understand the world and form judgments in their own unique way is antithetical to the notion of *psychic unity*, the assumption that all people when operating under similar circumstances will think and behave in similar ways (see Chiu and Hong 2005). Cultural relativism also implies that behavioral scientists need to develop multiple indigenous systems of knowledge to understand the psychological processes of different cultural groups.

In the present chapter, we advocate a probabilistic approach to understanding cultural differences. We contend that the presence of cultural differences in social judgment and decision making does not entail incommensurability of psychological processes across cultures. A key assumption in our approach is that different cognitive practices exist in different cultural groups, although their relative popularity differs across cultures.

Specifically, a distinction should be made between mean differences between the countries on the outcome variable one wishes to understand (level effect) and the process leading to the outcome (linkage effect) (Bond 2007). Level effect and linkage effect are orthogonal cultural effects. Two cultures may differ in both the overall likelihood of displaying a certain judgment or decision-making pattern as well as the processes that mediate this pattern. For example, compared to individuals in American culture, people in Asian cultures are more likely to practice collective responsibility attribution, blaming the collective or its members for a negative event caused by another member of the collective. Furthermore, the reasons for practicing collective responsibility attribution also differ between American and Asian cultures. Americans practice collective responsibility attribution for the sake of maintaining group harmony; to Americans, responsibility sharing may protect the individual wrongdoer from the potential threats of public denouncement and social isolation and promote group solidarity and intimacy. In contrast, Asians practice collective responsibility attribution both for the purpose of maintaining group harmony and for deterring individual wrongdoers through social monitoring. Asians think that group members should be punished not because they are deemed collectively responsible for the wrongdoing but simply because they are in an advantageous position to identify, monitor, and control responsible individuals, and can be motivated by the threat of sanctions to do so (Chao, Zhang, and Chiu 2008).

However, two cultures may differ in the overall likelihood of displaying a certain judgment or decision-making practice only. In the present chapter, we illustrate this possibility with our work on spontaneous trait inferences—the tendency to spontaneously infer traits from behaviors (Uleman, Newman, and Moskowtiz 1996). As mentioned, the overall likelihood of making spontaneous trait inferences is higher in American than in Chinese culture (Newman 1993). However, we review results showing that spontaneous trait inferences are made to support the same mental practice of making dispositional attributions in both American and Chinese cultures. Because dispositional attribution is more widely practiced in American than Chinese culture, spontaneous trait inference is also more prevalent in American than Chinese culture. Returning to the debate on the commensurability of cultures, our results show that despite the presence of striking cultural differences in the overall likelihood of making spontaneous trait inferences, the psychological process contributing to spontaneous trait inferences may be the same in American and Chinese cultures.

In short, in the probabilistic perspective to cultural differences, not all Americans spontaneously make dispositional attributions, and some Chinese also make spontaneous dispositional attributions. Thus, the challenge of cultural research is to account not only for the origin of cognitive variations between cultural groups but also individual and situational variations within a cultural group (see also Chapter 1). Specifically, three important questions from the probabilistic perspective are: Why are some mental practices more popular in some cultural groups? Who in the culture are more likely than others to adopt the popular mental practice in the group? When would people in a cultural group adopt the popular mental practice (versus other practices) in the group? We seek to answer these questions in the second part of the chapter. Specifically, we argue that a mental practice would become popular in a group if it can facilitate attainment of individual goals under a set of physical and human-made constraints and opportunities. Because a popular mental practice provides epistemic and existential security to the individual, it is likely to be adopted among those who have a strong need for epistemic and existential security and when the individual experiences a strong need for such security in the situation.

RELATIVE POPULARITY OF MENTAL PRACTICES

We assume that cultures differ from one another primarily in the relative popularity of different mental practices. This assumption is consistent with some recent theorizations of culture and its evolution (Tay et al. 2010). A cultural tradition consists of a constellation of loosely organized ideas and practices that are shared (albeit imperfectly) among a collection of interdependent individuals and transmitted across generations for the purpose of coordinating individual goal pursuits in collective living (Chiu, Leung, and Hong 2011). This definition implies that a cultural practice is created and retained because it is useful for coordinating individual activities in the group. In the course of cultural evolution, a cultural tool that works well may be replaced with one that works slightly better, although frequently the culture retains both (Triandis 2004). Thus, several mental practices can coexist in the same culture. Through a process of random variation and selective retention, the positively selected variants become most popular in the culture.

This idea has been applied to explain cultural differences in self-construal. According to Triandis (1989), people are active agents who selectively appropriate symbolic ideas about the self from their cultural environment to further valued goals. Triandis has identified three major kinds of the self: the private self (knowledge about a person's traits, states, or behaviors), the public self (knowledge about the generalized other's view of the self), and the collective self (knowledge about some collective's view of the self). Every person possesses these three kinds of the self, although people in different cultural groups sample these three kinds of self with different probabilities.

How likely the private self and the collective self will be sampled depends in part on the relative emphasis on personal versus collective goals in the cultural context. In some cultural contexts (e.g., individualist contexts), pursuit of personal goals is widely accepted and highly valued (Triandis et al. 1988). The private self (self-reliance, independence, self-esteem, self-concept clarity, and self-realization) is likely to be sampled in such contexts.

In other cultural contexts (collectivist contexts), the widely accepted view is that people should avoid pitting their personal goals against the collective goals. If a conflict between personal and collective goals is inevitable, people should subordinate their personal goals to the collective goals (Triandis et al. 1988). In such cultural contexts, higher rates of sampling the collective self are expected, and there are greater emphases on group agency, social acceptance, intergroup distinction, and competition.

Accordingly, although most people in an individualist culture value personal goals, some people in an individualist culture (the allocentrics) emphasize collective goals. Similarly, although the prevailing norms in a collectivist culture emphasize collective goals, some people in a collectivist culture (the idiocentrics) give high priorities to personal goals (Triandis et al. 1988). Triandis (1989) also recognizes the role of context in determining the probability of sampling a particular kind of self. For example, the probability of sampling the collective self would increase as the salience of ingroup–outgroup boundaries increases.

Judging Social Events

A robust finding in cross-cultural difference is that when comprehending social events, compared to Easterners, Westerners have a greater tendency to make spontaneous trait inferences (Newman 1993). Spontaneous trait inference (inferring dispositional traits in the actors from the actor's trait-relevant behaviors) is a cognitive procedure that supports dispositional attribution (explanation of behaviors in terms of the actor's internal dispositions), because identification of the trait implications of behaviors is a prerequisite for dispositional attributions. That is, spontaneous trait inference is a means or cognitive procedure in the service of dispositional attribution (Uleman, Newman, and Moskowitz 1996).

Nonetheless, perceivers do not always attribute social events to dispositional causes (Chiu, Hong, and Dweck 1997). Sometimes, perceivers will attribute the cause of a social event to the actor's situational goal or intention (an unstable internal cause) or to the demand of the situation (an external cause). For reasons we will explain later, cultures may differ from one another in terms of the relative popularity of dispositional, intention, or situational attributions. To illustrate cross-cultural differences in the propensity to make different kinds of attributions, we have designed a task to tap the relative readiness to make dispositional, intentional, or situational attributions among a sample of European Americans ($N = 70$) and a sample of Singaporean Chinese ($N = 83$). In this task, the participants were shown the following four concepts on the computer screens: traits, situation, intention, and behavior. The participants were asked to draw arrows between these concepts to indicate the perceived interrelations among the concepts. If the participants thought that one concept would affect another concept, they would draw an arrow from the first concept to the second one. If they thought that the two concepts could affect each other, they would draw two arrows, one from the first to the second concept and one from the second concept to the first concept. Participants could draw as many or as few arrows as they liked. To tap the first causal link that came to the participants' minds, we classified the participants into four groups based on the first arrow drawn by the participants into four groups: (1) participants whose first arrow was drawn from "trait" to "behavior" (dispositionists); (2) participants whose first arrow was drawn from "intention" to "behavior" (intentionists); (3) participants whose first arrow was drawn from "situation" to "behavior" (situationists); and (4) others.

Sixty-nine American participants and 61 Singaporean Chinese participants fell into one of the first three categories—these participants readily attributed behaviors to actor trait, actor intention, or the situation. As shown in Figure 4.1, for most Americans ($N = 39$, or 56.5% of the 69 Americans who fell into one of the first three categories), the behavioral cause that came to mind first was actor trait. In contrast,

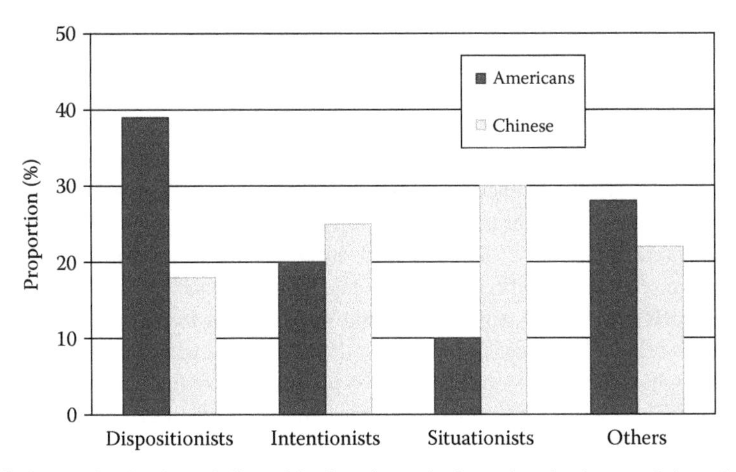

FIGURE 4.1 Distribution of dispositionists, intentionists, situationists, and American and Chinese participants.

among the 61 Singaporean Chinese who fell into one of the first three categories, only 15 (24.6%) drew the first arrow from trait to behavior; actor trait, actor intention, and situational demand were equally likely to be the first behavioral cause that came to the mind of these Chinese participants. This result suggests that (1) all three social explanation practices exist in both American and Singaporean Chinese cultural contexts, and (2) Americans and Singaporean Chinese differ in the relative popularity of the three attribution practices.

To demonstrate that adopting a spontaneous dispositional attribution practice is linked to the same cognitive procedures in American and Singaporean Chinese culture, we also measured in this study the tendency to make spontaneous trait inference from behaviors. Recall that spontaneous trait inference is a cognitive procedure that supports dispositional attribution. Therefore, we hypothesized that for both Americans and Chinese participants, the tendency to spontaneously make dispositional attribution would be accompanied by a greater tendency to make spontaneous trait inferences from behaviors.

We used the false memory paradigm (Uleman, Hon, et al. 1996) to measure spontaneous trait inference. The test consists of a learning phase and a recognition phase. In each of the 36 trials in the learning phase, the participants saw a photo of a different person and read a behavior performed by this person (e.g., "Jeff picked out all the good pieces of chocolate for himself before his guests arrived"). In each trial of the testing phase, the participants saw the photo of the target person in the learning phase together with a trait word (e.g., "selfish," "conscientious") and judged whether the trait word was in the behavioral description of the target person. If during the learning phase, the participants had spontaneously inferred a personality trait from the target person's behavior (e.g., inferred that Jeff was a selfish person because of his behavior), they might falsely remember later that the trait word (e.g., "selfish") was in the behavior displayed by the target person (e.g., Jeff). That is, when they saw the picture of the target again in the recognition phase, they would be more likely to have the false memory of seeing a word that referenced the inferred trait (e.g., "selfish") than a word that referenced an irrelevant trait (e.g., "conscientious"). Accordingly, we used the difference in the proportion of false memories of relevant versus irrelevant traits to form a measure of spontaneous trait inference.

Consistent with past findings (Newman 1993; Zarate, Uleman, and Voils 2001), American (versus Singaporean Chinese) participants displayed a greater tendency to make spontaneous trait inferences from behaviors ($M_{\text{Americans}} = 12.85$ and $M_{\text{Chinese}} = 8.93$). Nonetheless, upon closer examination, as shown in Figure 4.2, among both Americans and the Singaporean Chinese, only the dispositionists showed a significant tendency to make spontaneous trait inferences. This result indicates that the previously obtained East–West difference in the tendency to make spontaneous trait inferences is a consequence of the relatively greater prevalence of dispositional attribution in American contexts. Although dispositional attribution is less popular among the Singaporean Chinese, Singaporean Chinese dispositionists, like their American counterparts, also make trait inferences from behaviors spontaneously.

Taken together, the results cast doubt on the incommensurability of cultures. Although an *average* Singaporean Chinese shows a lesser tendency to make spontaneous trait inference, this inferential practice is present in Singaporean Chinese

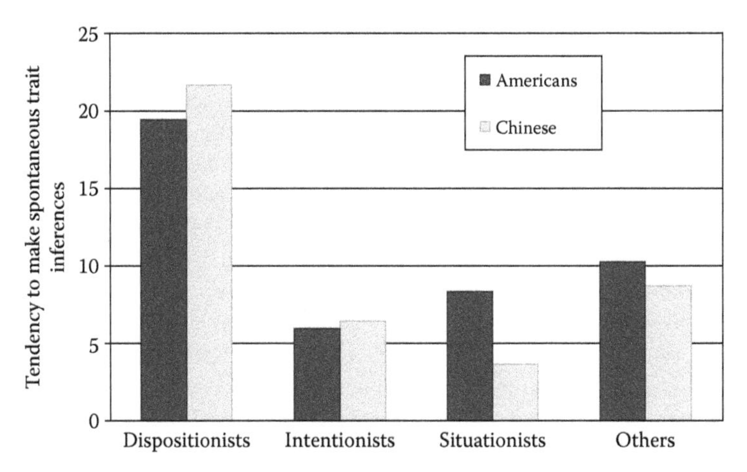

FIGURE 4.2 Connection of dispositional attribution to spontaneous trait inference among American and Chinese participants.

culture, although it is not a popular practice. More important, this practice supports dispositional attribution in both American and Singaporean Chinese cultures, suggesting that spontaneous trait inference has the same psychological significance in both cultures. The fact that cultural differences in cognition are probabilistic rather than absolute highlights the importance of considering both cross-cultural differences and intra-cultural variations in cultural analysis of cognitive processes.

SOCIETAL CONSTRAINTS AND INFERENTIAL HABITS

Why is dispositional attribution a more popular inferential practice in American than Singaporean Chinese culture? We contend that a mental practice would become popular in a cultural group if it can facilitate attainment of personal goals under a set of physical and human-made constraints and opportunities. All societies must allow and support pursuit of individual goals. However, unregulated pursuits of individual goals can also hurt the interests of the collective through cut-throat competitions, exploitation of shared resources, and neglect of the public good (Heylighen and Campbell 1995). Thus, it is the imperative of every society to keep the pursuit of selfish goals within bounds.

Societies can regulate individual behaviors through institutional control. Some common forms of institutional control are social reputation control (e.g., mutual monitoring, gossip) (Kniffin and Wilson 2005) and social sanctioning (e.g., social ostracism, collective punishment) (Chao, Zhang, and Chiu 2008; Spoor and Williams 2007).

Societies differ in the amount of institutional control they impose on the pursuit of selfish goals. Research has shown that it is relatively easy to implement institutional control in societies that lack job mobility (Chen, Chiu, and Chan 2009). For example, the United States has very high job mobility (Fuller 2008; Wheeler 2008). According to the Organization for Economic Cooperation and Development, in 1997, the average length of tenure (job duration) in the United States was 7.4 years, compared to

11.3 years in Japan (which is also a wealthy, industrialized country). In the same year, 34.5% of American employees had stayed in their current job for under 2 years, whereas only 9.0% of working Americans had stayed in their current job for over 20 years. In contrast, in the same year, only 23.6% of the employees in Japan had stayed in their current job for under 2 years, whereas 21.4% of the employees in Japan had stayed in their current job for over 20 years (Knight and Yueh 2004). The open nature of American societies renders institutional control of selfish goal pursuits difficult. Thus, these societies tend to rely on internalization of prosocial values to regulate individual interests: As long as other people's rights are not violated, individuals are not required to suppress expression of their personal qualities. Rather, they are encouraged to make choices based on their own preferences (Kitayama et al. 2004). Thus, there is relatively strong correspondence between personal dispositions and behaviors in such high job mobility societies as the United States (Chen, Chiu, and Chan 2009). In contrast, in low job mobility societies, people feel the need to change their personal qualities to meet the role expectations in their professions. Because behavioral choices are often driven by strategic conformance to role expectations, oftentimes, behaviors do not signal the actors' personal qualities (Chen, Chiu, and Chan 2009). Accordingly, perceivers in high (versus low) job mobility societies can more confidently infer personal dispositions from the actors' behavioral choice.

WITHIN-CULTURE VARIATIONS IN COGNITIVE PRACTICES

Although cultures differ in the prevalence of dispositional attributions, there are also considerable within-culture variations in attribution practices. How do we account for within-culture variations in the prevalence of dispositional attributions? To answer this question, we first review the psychological functions of culture.

Why do people connect their self to their culture? One answer to this question is that culture confers psychological benefits to the self. Recall that culture refers to a constellation of ideas and practices that are shared among a collection of interdependent individuals and transmitted across generations. Two defining characteristics of a cultural knowledge tradition are its sharedness and continuity (Chiu and Liu 2011). A unique family tradition is one that has a history but is not widely shared in the community. A fad is a fashion, notion, or manner of conduct followed enthusiastically by a large group, but its popularity is temporary. An individual's eccentric belief is not shared by others and would not likely be passed down through history. As illustrated in Table 4.1, unlike a unique family tradition, a fad, or an eccentric belief, a cultural tradition is both shared among many people and has a history.

TABLE 4.1
Defining Characteristics of Culture

	Low Continuity	High Continuity
Low sharedness	Eccentric idea	Family tradition
High sharedness	Fad	Culture

By virtue of its sharedness and consensual validity, culture provides its followers a sense of epistemic security (Chiu et al. 2000; Fu et al. 2007). Widely shared cultural knowledge provides individuals with a consensually validated framework to interpret otherwise ambiguous experiences. It informs individuals in the society what ideas or practices are generally considered to be true, important, and appropriate. Thus, it protects individuals from the epistemic terror of uncertainty and unpredictability. Consistent with this idea, research shows that individuals are more likely to follow consensual norms and expectations when the need for epistemic security is salient. Indeed, many documented cultural differences in cognition and behavior are enlarged when the need for epistemic security is high. For instance, compared to each other, when explaining an ambiguous event, collectivists are more likely to reference dispositions of a group and individualists are more likely to reference dispositions of the individual (Menon et al. 1999). These differences are found only among those who have a high need for firm answer (i.e., a high need for cognitive closure) or those in high time pressure situations, in which the need for firm answer is pronounced (Chiu et al. 2000). Similar moderation effects of the need for firm answers were reported in subsequent studies on cultural differences in reward allocation and conflict resolution (Chao, Zhang, and Chiu 2010; Fu et al. 2007).

By virtue of its continuity, culture provides its followers a sense of existential security, protecting the individual from the terror of recognizing one's mortality (Kesebir 2011). Despite the finitude of an individual's life, the cultural tradition one belongs to will be passed down through history. Thus, connecting the self to a seemingly immortal cultural tradition can help assuage existential terror.

Whereas adherence to cultural norms confers epistemic security, connecting the self to an endurable cultural tradition confers existential security (Chiu et al. 2011). Thus, when the need for existential security is salient, individuals will be particularly motivated to follow the dominant mental practice in their culture (Kesebir 2011). For example, mortality-salient Americans are particularly likely to make dispositional attributions.

CONCLUSION

Why do people think culturally when making judgments and decisions? A common answer to this question is that culture shapes people's worldview and the way they form judgments and decisions. For example, Kitayama and Markus (1999) submitted that culture gets into the head of individuals in a cultural group through a set of mutually constitutive relations between culture and the person. According to them,

> Everyone is born into a culture consisting of a set of practices and meanings ... To engage in culturally patterned relationships and practices and to become mature, well-functioning adults in the society, new members must come to coordinate their responses to their particular social milieu. That is, people must come to think, feel, and act with reference to local practices, relationships, institutions and artifacts; to do so they must use the local cultural models, which consequently become an integral part of their psychological systems. (p. 250)

As an alternative to this absolutist view of culture, we presented in this chapter a probabilistic view of cultural influence on social judgments and decision making. Using person perception as an example, we showed that American and Chinese cultures differ from each other not because all Americans are dispositionists and all Chinese are situationists. Rather, the two groups differ in how widely spread the practice of dispositional attribution is in the group. Because dispositional attribution is more widely practiced in American culture than in Chinese culture, and because dispositional attribution is linked to the tendency to make spontaneous trait inferences from behaviors, Americans (versus Chinese) show a greater overall tendency to make spontaneous trait inferences. However, as our data show, a sizable proportion of Americans also spontaneously attribute behaviors to the actor's intention or the situation. Similarly, a sizable proportion of Chinese make dispositional attribution spontaneously.

The probabilistic view of cultural differences invites cultural researchers to address the origins of cultural differences as well as within-culture variations in social inferences. In this chapter, drawing on our previous research, we argued that differences between cultures in the relative popularity of dispositional attribution may result partly from such ecological factors as job mobility. We also argued that individual and situational variations in dispositional attribution may result from the chronic and momentary needs for epistemic and existential security.

Consistent with our view, Weber and Morris (2010) make the following conclusions after reviewing the cross-cultural literature on judgment and decision making:

> Cultural influences on individual judgment and decision making are increasingly understood in terms of dynamic constructive processing and the structures in social environments that shape distinct processing styles ... These structures include the society's observable patterns of normative actions and responses, its prevalent forms of interpersonal interaction, the typical size and density of social networks, the ideational frames represented publically in texts and institutions, and so forth. (p. 410)

Berger and Heath (2005) also believe that culturally preferred problem-solving strategies might have evolved in response to environmental cues that prime people to think about an idea and cause them to believe that it may be relevant to pass along. These convergent views invite researchers to consider the ecological embeddedness of cultural differences in judgment and decision making.

Finally, the cultural similarity in the link between dispositional attribution and spontaneous trait inferences suggests that despite cultural differences in the average level of spontaneous trait inferences, the functional connectedness of dispositional attribution and spontaneous trait inferences may be the same across cultures. Thus, the presence of cultural differences in the likelihood of displaying a certain inferential practices does not always imply cultural relativity in basic cognitive processes.

In conclusion, recent cultural research has uncovered marked cultural differences in the likelihood of engaging in many cognitive practices (Lehman et al. 2004). Such differences have led researchers to focus on cultural differences in fundamental cognitive processes. The probabilistic perspective to cultural differences advanced in this chapter acknowledges the presence of cultural differences in the likelihood of displaying a certain cognitive practice, while at the same time inviting attention to within-culture variations in the practice and possible universal linkages between cognitive practices.

REFERENCES

Berger, J. A., and C. Heath. 2005. Idea habitats: How the prevalence of environmental cues influences the success of ideas. *Cogn Sci* 29:195–221.

Bond, M. H. 2007. Fashioning a new psychology of the Chinese people: Insights from developments in cross-cultural psychology. In *Interpersonal Perspectives on Chinese Communication*, ed. S. J. Kulich and M. H. Prosser, 233–51. Shanghai, China: Shanghai Foreign Language Education Press.

Chao, M. M., Z.-X. Zhang, and C.-Y. Chiu. 2008. Personal and collective culpability judgment: A functional analysis of East Asian-North American differences. *J. Cross Cult Psychol* 39:730–44.

Chao, M. M., Z.-X. Zhang, and C.-Y. Chiu. 2010. Adherence to perceived norms across cultural boundaries: The role of need for cognitive closure and ingroup identification. *Group Process Intergroup Relat* 13:69–89.

Chen, J., C.-Y. Chiu, and F. S.-F. Chan. 2009. The cultural effects of job mobility and the belief in a fixed world: Evidence from performance forecast. *J Pers Soc Psychol* 97:851–65.

Chiu, C.-Y., and Y. Hong. 2005. Cultural competence: Dynamic processes. In *Handbook of Motivation and Competence*, ed. A. Elliot and C. S. Dweck, 489–505. New York: Guilford.

Chiu, C.-Y., Y. Hong, and C. S. Dweck. 1997. Lay dispositionism and implicit theories of personality. *J Pers Soc Psychol* 73:19–30.

Chiu, C.-Y., Y.-H. Kim, and W. N. Wan. 2008. Personality: Cross-cultural perspectives. In *The Sage Handbook of Personality Theory and Assessment. Vol. 1: Personality Theory and Testing*, ed. G. J. Boyle, G. Matthews, and D. H. Salofske, 124–44. London: Sage.

Chiu, C.-Y., A. K.-Y. Leung, and Y.-Y. Hong. 2011. Cultural processes: An overview. In *Cultural Processes: A Social Psychological Perspective*, ed. A. K.-Y. Leung, C.-Y. Chiu, and Y.-Y. Hong. New York: Cambridge University Press.

Chiu, C.-Y., and Z. Liu. 2011. Culture. In *Routledge Companion to Race & Ethnicity*, ed. S. M. Caliendo and C. D. McIlwain. New York: Routledge.

Chiu, C.-Y., M. Morris, Y. Hong, and T. Menon. 2000. Motivated cultural cognition: The impact of implicit cultural theories on dispositional attribution varies as a function of need for closure. *J Pers Soc Psychol* 78:247–59.

Chiu, C.-Y., C. Wan, Y.-Y. Cheng, Y.-H. Kim, and Y.-J. Yang. 2011. Cultural perspectives on self-enhancement and self-protection. In *The Handbook of Self-Enhancement and Self-Protection*, ed. M. Alicke and C. Sedikides. New York: Guilford.

Chua, H. F., J. E. Boland, and R. E. Nisbett. 2005. Cultural variation in eye movements during scene perception. *Proc Natl Acad Sci U S A* 102:12629–33.

Fu, H.-Y., M. W. Morris, S.-L. Lee, M.-C. Chao, C.-Y. Chiu, and Y.-Y. Hong. 2007. Epistemic motives and cultural conformity: Need for closure, culture, and context as determinants of conflict judgments. *J Pers Soc Psychol* 92:191–207.

Fuller, S. 2008. Job mobility and wage trajectories for men and women in the United States. *Am Sociol Rev* 73:158–83.

Heylighen, F., and D. T. Campbell. 1995. Selection of organization at the societal level: Obstacles and facilitators of metasystem transitions. *World Futures J. Gen Evol* 45:181–212.

Ji, L.-J., Z. Zhang, and R. E. Nisbett. 2004. Is it culture or is it language? Examination of language effects in cross-cultural research on categorization. *J Pers Soc Psychol* 87:57–65.

Kesebir, P. 2011. Existential functions of culture: The monumental immortality project. In *Cultural Processes: A Social Psychological Perspective*, ed. A. K.-Y. Leung, C.-Y. Chiu, and Y.-Y. Hong. New York: Cambridge University Press.

Kitayama, S., and H. R. Markus. 1999. Yin and yang of the Japanese self: The cultural psychology of personality coherence. In *The Coherence of Personality: Social Cognitive Bases of Personality Consistency, Variability, and Organization*, ed. D. Cervone and Y. Shoda, 242–302. New York: Guilford Press.

Kitayama, S., A. C. Snibbe, H. R. Markus, and T. Suzuki. 2004. Is there any "free" choice? Self and dissonance in two cultures. *Psychol Sci* 15:527–33.

Kniffin, K. M., and D. S. Wilson. 2005. Utilities of gossip across organizational levels: Multilevel selection, free-riders, and teams. *Hum Nat* 16:278–92.

Knight, J., and L. Yueh. 2004. Job mobility of residents and migrants in urban China. *J Comp Econ* 32:637–60.

Lehman, D., C.-Y. Chiu, and M. Schaller. 2004. Culture and psychology. *Annu Rev Psychol* 55:689–714.

Menon, T., M. Morris, C.-Y. Chiu, and Y. Hong. 1999. Culture and the construal of agency: Attribution to individual versus group dispositions. *J Pers Soc Psychol* 76:701–17.

Miller, J. G. 1984. Culture and the development of everyday social explanation. *J Pers Soc Psychol* 46:961–78.

Morris, M. W., and K. Peng. 1994. Culture and cause: American and Chinese attributions for social and physical events. *J Pers Soc Psychol* 67:949–71.

Newman, L. S. 1993. How individualists interpret behavior: Idiocentrism and spontaneous trait inference. *Soc Cogn* 11:243–69.

Peng, K., and R. E. Nisbett. 1999. Culture, dialectics, and reasoning about contradiction. *Am Psychol* 54:741–54.

Petrova, P. K., R. B. Cialdini, and S. J. Sills. 2007. Consistency based compliance across cultures. *J Exp Soc Psychol* 43:104–11.

Spoor, J., and K. D. Williams. 2007. The evolution of an ostracism detection system. In *Evolution and the Social Mind: Evolutionary Psychology and Social Cognition*, ed. J. P. Forgas, M. G. Haselton, and W. von Hippel, 279–92. New York: Psychology Press.

Tay, L., S. E. Woo, J. Klafehn, and C.-Y. Chiu. 2010. Conceptualizing and measuring culture: Problems and solutions. In *The Sage Handbook of Measurement*, ed. G. Walford, E. Tucker, and M. Viswanathan, 177–202. London: Sage.

Triandis, H. C. 1989. The self and social behavior in differing cultural contexts. *Psychol Rev* 96:506–20.

Triandis, H. C. 2004. Dimensions of culture beyond Hofstede. In *Comparing Cultures: Dimensions of Culture in a Comparative Perspective*, ed. H. Vinken, J. Soeters, and P. Ester, 28–42. Leiden, The Netherlands: Brill Publishers.

Triandis, H. C., R. Bontempo, M. J. Villareal, and M. L. Asai. 1988. Individualism and collectivism: Cross-cultural perspective on self-group relationships. *J Pers Soc Psychol* 54:323–38.

Uleman, J. S., A. Hon, R. J. Roman, and G. B. Moskowitz. 1996. On-line evidence for spontaneous trait inferences at encoding. *Pers Soc Psychol Bull* 22:377–94.

Uleman, J. S., L. S. Newman, and G. B. Moskowitz. 1996. People as flexible interpreters: Evidence and issues from spontaneous trait inference. *Adv Exp Soc Psychol* 28:211–79.

Weber, E. U., and M. W. Morris. 2010. Culture and judgment and decision making: The constructivist turn. *Pers Psychol Sci* 5:410–19.

Wheeler, C. H. 2008. Local market scale and the pattern of job changes among young men. *Reg Sci Urban Econ* 38:101–18.

Zarate, M. A., J. S. Uleman, and C. I. Voils. 2001. Effects of culture and processing goals on the activation and binding of trait concepts. *Soc Cogn* 19:295–323.

5 Cross-Cultural Decision Making

Impact of Values and Beliefs on Decision Choices

Parasuram Balasubramanian

CONTENTS

INTRODUCTION

Mobile telephony represents the fastest-growing industry in most countries of the world today. It is not widely known that the cell phone service providers have chosen diametrically opposite charging mechanisms for phone use in the Western and Eastern hemispheres. Whereas the phone charges are shared equally between the caller and the recipient in the West, it is mostly the caller who bears the entire charge in the East. Why would a rule seen as fair and just in one society be seen as unfair in another?

Vendors extend credit to buyers in the corporate world. Sophisticated buyers in the West use systems to track all payments and sequence the release of funds to optimize the credit terms extended. The pioneering firms in the East, on the other hand, have chosen to pay up the bills well ahead of time so that they can demand high-quality service and better price from vendors, not to mention the intent to build strong long-term relationships. Why do they differ in their decisions, one preferring an *arms length* approach while the other seeks *tight coupling*?

Newer ideas with unproven technologies or even unknown market potential are often funded by venture capitalists (VCs). One of the mechanisms employed by VCs to assess a proposal for funding is to evaluate the exit strategy. This would enable the VC to judge the duration of funding needs as well as to estimate likely return for their investment. The related frameworks are very well developed in the West. Yet in earlier years, more often than not, the VC would find the exit strategy plan missing from the proposals from the East, even when the proposal is comprehensive in all other aspects. Is there a missing link here?

There is a wide gap observable in the decision options and the decision processes of the oriental and occidental worlds. Where do such differences in thought and action come from? Can we attribute them to context variations (such as supply demand gaps and regulations) or to factors beyond? Could they arise from cultural differences? My intent is to explore this issue in-depth in this chapter.

Management science in general and decision theory in particular have provided the framework for dealing with multiple and often conflicting objectives in a system arising out of differing goals of various stakeholders. Further, the uncertainty prevailing in the environment can result in unreliability or unavailability of data, incomplete or inaccurate linking of cause and effects, and unpredictable outcomes. Government regulations or societal concerns can vary from one place to another. Hence, context is defined as the enveloping environment within which the decision content lies. Differences in context will give rise to differing decisions even if content remains invariant. The term cultural intelligence (CQ) refers to the ability to make sense of unfamiliar contexts and then to blend in (Early and Mosakowski 2004).

IMPACT OF CULTURE ON DECISION CHOICES

Culture is best defined as the attitude and behavior of a community as derived from their underlying values and beliefs. It has been revealed time and again that even when the context and the content are similar, the decisions made differ significantly if they are made by persons from different cultures. Culture can be considered not as yet another layer on context but as a matter that permeates across both the context and the content of any decision. Context and content defined to represent a given scenario in one culture are understood and interpreted differently in different cultures.

Although anecdotal references to negotiation strategies adopted in the occidental versus oriental cultures are many, systematic studies comparing, say, Chinese and Japanese buyer behaviors, or German and French preferences for communication modes, also show differences. Multinational firms have struggled with the issue of globalization for decades. Prahalad and Lieberthal (1998) have asked that the multinationals, in their quest to enter developing markets, shun the imperialistic mind-set and move beyond developing and displaying cultural sensitivity. "Global mind set and local implementation" was the mantra of multinationals for a long time. Even such strategies have failed, though, as evident from the struggles of Wal-Mart, Vodafone, and McDonald's corporations to penetrate the Indian market (Bloomberg Business Week 2009; Parker 2010).

Literature Review

Culture has been defined in many ways. In fact, Kroeber and Kluckhohn (1952) compiled 164 variations in its definition almost half a century ago. Culture is treated akin to the collective body of knowledge that resides in a society, as well as the shared beliefs, attitudes, values, and behaviors of the constituent members.

A cross-cultural scenario involves taking one's products and services from a native country, society, or community to another place with a significant variation in culture. Thanks to transportation and telecommunication technology advances in the last century, we can now be instantly connected with anyone across the globe, and the Internet era has even enabled the delivery of digital goods across borders within seconds. Yet it has not been smooth sailing for the international firms to succeed in global reach. One corporation after another has found it a challenging issue to replicate its success in one country in other countries. Cultural factors seem to be the biggest hurdle that has been realized by many in this process.

Buyer behavior and the influence of local culture on purchasing habits have been studied extensively in both China and India (de Mooij 2003; Tamminga 2009). Hansen (2000) interviewed a sample of buyers in supermarkets from four different cities in China. This study found significant differences even within these four cities. Urban buyers were willing to accept more self-service counters compared to their counterparts in smaller cities, and their focus was on product certification, while others in smaller cities gave more weight to known brands. Kwok (2006) analyzed consumer behavior in urban China to identify brand loyalty, market segmentation, and factors that influence buying behavior. Based on three decades of consumer research in India, Bijapurkar (2007) stressed that Indian buyers across all segments exhibit price consciousness, yet they are tolerant of ambiguity in communications.

Another aspect of corporate interaction across cultures that has received wide attention is that of negotiating postures and approaches. The transaction orientation of U.S. executives is in contrast to the long-term relationship focus of the Orientals (Chinese, Japanese, and Koreans in particular); the collective wisdom of the latter differing from the assertions of individuals of the former is documented extensively in literature (Berten 1998; Hupert 2009).

Cultural differences are also visible in delivery of services between the East and the West. Andrews and Boyle (2008) reflect on cultural differences in healthcare delivery and note the preference for a tight schedule in the West, whereas a flexible schedule is the choice of the patients from the eastern hemisphere. Further, Andrews and Boyle highlight significant differences in information sharing and communication preferences between the two, particularly with family members. The differences in healthcare delivery mechanism even among the countries in the Western world are articulated by Reid (2009).

Banks and other financial institutions have embraced the Internet, landlines, and mobile phone technologies to reach out to their customers in a cost-effective manner. This is a global phenomenon. Yet almost all banks have been forced to adopt a mixed mode distribution channel approach due to customer preferences. Cultural differences are evident in this case as well. Indian customers, particularly from rural communities, have a preference for face-to-face meetings with the banker in spite of it being both time and cost inefficient (eHow.com 2010).

Hofstede (2001) has probed deeper to identify underlying differences between cultures, identifying five dimensions on which they differ. Tolerance within a society for a wide distribution of authority and power, the role of individuals versus a group, gender-based role assignment, comfort level with uncertainty, and long-term focus are the dimensions used in his analysis to compare and contrast cultures.

Decision Manifestations

A decision maker uses words and actions to communicate his or her thoughts. We have to infer the decision maker's intentions by studying the words and actions; we rarely get an opportunity to know her or his thoughts precisely. The impact of culture on decisions has to be derived similarly by observing and studying human words and actions.

Liu (2010) made a seminal contribution to this area through her illustrations of cultural differences between the Orient and the Occident. She explored and explained how people from the former choose an oblong path to communicate, whereas those from the latter prefer a straight path of communication. Orientals' preference for collective wisdom, the individual's willingness to subordinate his or her position to the group leader, tolerance for ambiguity, and preference for sidestepping an issue rather than confronting it are in contrast to the positions preferred by the Westerner, and are documented by Liu. It should be noted that all of her findings on cultural differences are derived from words communicated or actions performed. In fact, Kaplan (1966) is to be credited for identifying the differences in communication patterns across cultures. His diagram of patterns of communication is widely referred to in the literature (Campbell 1998).

Decision Manifestation Model

Decisions made by any individual or group are the subject of study of decision theorists and analysts. Their interest lies in understanding the decision choices (implied or explicit) generated and the decision process by which one of the alternatives is chosen. The need to demarcate decision content from the decision context and to study their interplay has been described in the section on Impact of Culture on Decision Choices. Such an approach is internal to the group with the decision problem.

Entities external to the group may not have access to the decision choices considered or the decision process of the responsible group. They can surmise or infer based on available outputs. The spoken or written words and carried-out action of the deciding group are the outputs available as evidence to the external entities. The underlying thought processes of the decision maker need to be inferred or derived from the observed actions and received communications. The decision manifestation process is represented in Figure 5.1. I wish to stress that the representation in Figure 5.1 is a conceptual model. In practice, errors of observation, deduction, and judgment can occur while viewing, studying, and interpreting actions and words. Hence what is concluded, as the underlying thoughts of the decision maker (DM) are prone to be erroneous, also.

It is often perceived that factors such as art, literature, food habits, and social etiquette are representative of the culture prevailing in a society. These are to be treated as the "stock" factors, meaning stored or accumulated representations. Actions and words, on the other hand, are the "transactional" factors representing a culture. We often tend to use either of the factors to infer or deduce the underlying culture.

Missing Link

That studying the stock and transaction factors is not sufficient for understanding the impact of culture on decisions is borne out by empirical evidence. Models utilizing these factors have at best given a partial view of the decision choices and processes so far. When implemented, particularly in cross-cultural settings, they have failed to achieve the intended consequences. Sebenius (2002) notes that the decision-making process differs from culture to culture and alludes to core beliefs of the culture impacting heavily on the decision process.

For example, the maxim "Think globally but act locally" was derived after extensive study of cross-cultural factors and their impact on decision choices in communities. Numerous corporations adopted this approach to penetrate other cultures, particularly from Occidental to Oriental and vice versa. Most of these initiatives have failed to yield the spectacular success achieved by these corporations in their parent cultures (Bloomberg Business Week 2009; Parker 2010). Hence, there seems

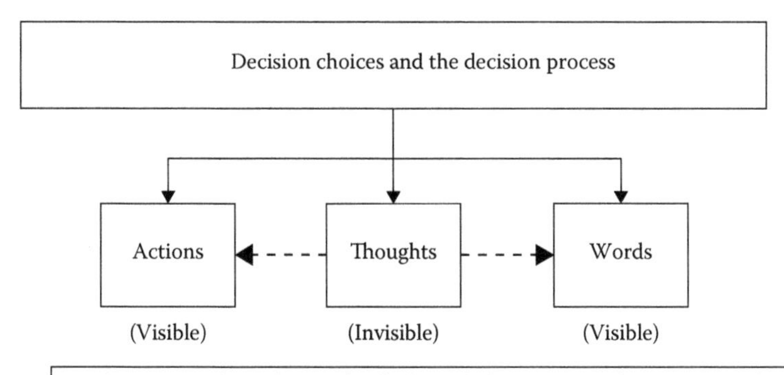

FIGURE 5.1 Conceptual model of the decision process.

to be a missing link in existing models that relate cultural factors to decision choices, and in what follows, I endeavor to identify this missing link.

VALUES AND BELIEFS

The culture that is native to a community is essentially derived from its long-held and cherished values and beliefs. It is certain that these values and beliefs have been questioned often and have remained accepted over time by the collective affirmative experience of the community. For example, consider the "don't hold grudges" view taken by the Amish community when five girl students of an Amish school were shot dead by a local milk vendor in Pennsylvania in October, 2006 (Kraybill, Nolt, and Weaver-Zercher 2007). Faith plays as equally a critical role as science at any given time period, as values and beliefs are shaped. The latter can be viewed as at the root of the culture tree. Hence, a study and in-depth analysis of underlying values and beliefs of a society can help to differentiate its culture from another (see Kotelnikov 2010, and Difference Between 2010 for the distinction between beliefs and values).

Differences in values and beliefs across cultures can occur even when both cultures profess to accept certain basic axioms. Consider the example shown in Table 5.1.

TABLE 5.1
Values that are Common or Specific to Oriental and Occidental Cultures

Common to Both Cultures	Specific to Oriental Culture	Specific to Occidental Culture
End does not justify the means	Journey is the destiny; enjoy the journey; do your duty without focusing on results	Focus on the end and drive to achieve it
Arrogance is to be shunned	Humility at all times	Self assertion is a virtue
Set aspirational goals	Goals have to be ascetic; selfless	Goals can include material possession
Individual coexists with the society	Public good overrides individual benefit	Let every individual strive to attain his/her best
Wisdom is derived from experience	Age is a proxy to experience and hence to wisdom	Age is not correlated to wisdom
Women are to be empowered	Women are best suited to play certain roles	Women are equal to men in all respects
Individuals and families need to coexist	Every individual belongs to a family	May or may not be part of a family; individuality can be asserted
Contents and contexts are both important	Decision context has to be well understood (implying that the context cannot be easily altered)	Decision content needs to be optimized in a given context (context can be altered)
Happiness = (material possession)/(desire)	Spartan living is a virtue	Aspire for a comfortable life
Ecological balance is to be preserved	Mankind must coexist with other living beings and nature	Mankind is superior; it should aspire to dominate and overpower natural hurdles

It can be noted that there can be substantive differences in values and beliefs across differing cultures while the underlying tenets can be common. In reality, societies live in a complex and hybrid environment. Some tenets seem to be common across a society, but many tenets are substantially different, also.

FOUR MAJOR FACTORS

Understanding cultures from the root level has been attempted previously. Hofstede's five cultural dimensions model is a step in this direction. They help to explain the fundamental factors by which cultures differ. The factors can be used to compare cultures and determine what to stress when interacting with members of a particular culture. Yet these dimensional factors are derivable from observation and study of actions and words. They do not attempt to probe the thought process itself.

Thoughts govern both actions and words, and influencing thoughts is tantamount to controlling attitudes and behavior. It has already been noted that the values and beliefs of a society are the major influencers of thought, along with the context and content of a decision scenario. This leads to the proposition that cultural differences are best understood not by studying words and actions alone, nor by identifying the cultural dimensions in which they are manifested, but by getting to the root by means of an in-depth study of values and beliefs. Values and beliefs must be analyzed to identify the key factors that have a major impact on decisions. By focusing on the decision dimension and deriving the factors, we can then construct usable models. These models can be used to redesign systems and processes, reconfigure products and delivery channels, and refocus marketing messages to appeal to a given culture. The models also may reveal the unsuitability of an existing product or service to a new culture.

Values and beliefs prevalent in a society relate to their accumulated wisdom and emergent attitude regarding happiness, wealth accumulation, austerity, self-worth, humility, passion, ego, ecological concerns, and gender-role expectations. They define the rules of governance whenever collective interest emerges on top of individual preferences. As societies advanced over years, the values and beliefs even prescribed protocols for dealing with adversaries and emissaries.

Four major factors of values and beliefs can be identified that have an impact on decisions. They have been extracted by analyzing many known cases of success and failures recorded in the media. Two major goals were kept in mind while deciding on these four factors: They must be mutually exclusive as far as possible, and they can be converted into action items regarding decisions. The four factors are (1) age versus wisdom; (2) gender and role expectations; (3) collective versus individual ownership of assets; and (4) belief in end versus means.

AGE VERSUS WISDOM

The parental role as a protector and nurturer of the young as well as the teacher of livelihood has been the seed for a loving and caring relationship between the child and the parents in all societies. Children were expected to be grateful and show due respect to their parents. This has also translated into a corresponding obligation to take care of the parents in their old age. This chapter, however, is exploring a different dimension of the relationship between them as well as with other elders in society. Given that our study focuses on decision choices and how they are impacted by culture, the current concern is not filial obligations but the equation that emerges between the young and the old from a decision perspective.

When confronted with the unknown, societies have turned to the man (or woman) who can best explain what has happened, how, and why. The explanation, of course, has to be logical and convincing so that it can be validated against multiple occurrences over time and at different places. The three queries of what, how, and why require increasingly difficult responses in the same order as they appear. Theosophy came to mankind's rescue whenever available scientific knowledge fell short of providing the required explanation. Further, early civilizations depended on oral and person-to-person transmission and sharing of information. Under such contexts it is no wonder that an older person was revered as a repository of knowledge and wisdom more than a younger one. It is more likely that he or she would have had more encounters with events and opportunities to exchange thoughts with other people to develop a better understanding of occurrences. Hence, age stood as the strong proxy variable to wisdom in all societies for a long time.

Sternberg (1990, 2010) makes the case that wisdom is one of a triad of mental traits, along with intelligence and creativity. He differentiates education focused on knowledge and creativity from one whose core is wisdom and asks for schools to shift focus to the latter. A study by University of Alberta (2008) has found a correlation between age and wisdom, and there is considerable literature on Indian culture that calls for respect for elders (Jranks Marriage and Family Encyclopedia 2010), as they are the embodiment of knowledge and wisdom.

There are two events that changed the world order regarding this link of age and wisdom or knowledge. The first phenomenon was the invention of the printing press by Gutenberg in 1436 AD. It revolutionized the manner in which information was shared within and across societies. More information could be spread faster across many people in a shorter time span. For example, nearly 8 million manuscripts were printed and circulated within 50 years of this machine being invented. Hence, dependency on older people to acquire information and knowledge has decreased gradually over the centuries, particularly in the Western world.

The second phenomenon is the disintegration of joint families into nuclear families from the seventeenth century onwards. The nuclear family became an economically viable unit due to industrialization in the Western world. Mass migration to the new colonies also meant the movement of younger and more able-bodied persons to the new continents, thus further splitting the families into smaller units.

These two phenomena occurred early in the Western world and impacted their populace sooner. Consequently, accumulated wisdom came to be questioned with increasing skepticism, and reliance on scientific scrutiny gained common ground. With the breaking up of the joint family system, reliance on the elderly for guidance and advice also diminished.

Scientific advancement and economic development are recent entrants into Oriental communities. They have made an impact on these societies only in the late twentieth century. Consequently, reliance on elders has remained intact until recent decades in these societies. The joint family system, where extended families live together or in close proximity, is prevalent even today in rural India. Religious and social practices have endorsed respect for elders in these societies for a long time. Consequently, age-based seniority has garnered enough authority on its own in many corporate firms in the Eastern world. An informal mentorship system flourishes within the society and the firms, whereby younger members seek the guidance of senior people from time to time. The respect for the senior in role in the office environment is a natural consequence of this phenomenon. Respect for the higher role is expected and given in most places.

Closely tied to this concept is the need to be humble. Since one is expected from a young age to learn continuously from elders, humility, rather than self-assertion, is taught as a virtue. "I will try my best" is the best assurance one can get from an Oriental even when she or he is in a position to guarantee results. The relative emphasis placed on understanding context (in decision choices) can be grasped when knowledge is sought from external entities and with humility. Content resolution happens only when the context is thoroughly examined.

The media invasion through print and television, the penetration of mobile phones, computers, and Internet, and growing urbanization has impacted this equation of age with wisdom in Asian countries. Nuclear families are more common than joint families. Much of the information needed is obtained from technology sources. Access to the elderly for face to face interaction is diminishing. Yet, it is premature to say that the strong correlation established between age and wisdom is under threat in these societies.

Gender and Role Expectation

Industrialization, migration to the new world, and the two world wars have played significant roles in empowering women, giving them access to education and employment opportunities and providing them with equal status to men in societal affairs in the Western hemisphere. Certain tasks that depended on physical strength were considered to be the domain of men two centuries ago. The introduction of machine power in lieu of manpower and the automation of systems to eliminate the need for human inputs altogether have altered this scenario. Today, there is hardly any task or role that a man can perform that a woman cannot handle. The attitude and behavior of the society has been shaped accordingly in the past two centuries.

Oriental societies, on the other hand, are slow to adopt the view of equal status for women. Even though in Vedic Indian Society, before 2000 BC, women were treated as equals (Prakash 2003), were given roles outside homes, and were allowed to marry a person of their choice at the right age, women in India have played a subordinate role to men during the past several centuries. Child marriage, prohibitions of widow remarriage, the practice of Sati (in which the wife jumps into the funeral fire of her deceased spouse), and so on has been prevalent over the past two millennia. Women are held in esteem, and the Hindu religion accords equal respect to female gods as well, yet gender-based role assignment is still the norm (Wells 2011). Women were denied opportunities for education, better health care, and in some cases, voting rights, until a hundred years ago. A woman's role has been confined to the home as the homemaker, but rural India has accepted her as a farm hand too.

There has been remarkable change since India's independence in 1947. India has had a female prime minister, president, numerous state chief ministers, and has passed a law recently to reserve one-third of the seats in Parliament and other legislative bodies for women. Enrollment of women in schools and colleges has grown, and it is commonplace to see women in the workplace, especially in newer service industries. Women are becoming socially aware and economically independent. Indian society, however, continues to maintain values and beliefs assigning gender-biased roles to women. In print and film, in billboards and advertisements, in editorials and public fora, one cannot miss such values and attitudes. How would this be relevant in the context of decision choices? If men have the leadership roles by assignment at home and by de facto at the workplace, what effect does a woman's opinion have in the decision made or in the process of decision making?

There are ample differences noticed in the manner in which buyer behavior is being influenced through the media in India versus the Western world. For example, women are still shown in the home setting, caring for family members, but making decisions based on facts and figures. They are treated as value-conscious buyers in settings that are heavily culture biased.

Collective versus Individual Ownership of Assets

Triandis (1995) has articulated the concept of individualism versus collectivism as it exists in Western and Eastern societies. He has also shown how culture shapes the way we think and has offered evidence for culture-specific influence on thoughts and actions. The issue of collective versus individual ownership of assets erupted with tremendous internal strife in Western society, post industrialization. Capitalism stood head to head with communism and socialism, with Adam Smith and Carl Marx as leaders at either end. After fierce debates and societal experimentation, it is widely believed now that the benefit of individual ownership of certain assets far outweighs the danger of creating an unequal society. The laws of most countries in the Occidental world have been structured to protect an individual's right to own assets of various forms, and the individual can derive economic advantage through them within the framework of legal provisions. The assets considered include both tangible and intangible assets. Land and buildings, plants and machinery, and goods of all types can be owned by the individual. He or she can even own natural assets such as coal and diamond mines, a pond or a river, and cattle. Inventions and discoveries in the form of patents and artwork held as copyrights can also be his or her possessions.

Consequently, society has chosen to encourage individual enterprise and to structure a reward system that promotes drive and excellence in any sphere of life. That the collective welfare of the society is enhanced through individual initiatives is a strong belief in this framework. The new world has had abundant natural resources to support the ambition and aspiration of the populace in this regard. Cultural divergence of the East from the West is nowhere more significant than in this issue. From Vedic scriptures in India to Jainism and Buddhism to the Confucius philosophy of China, the basic tenet practiced is one of spartan living and virtuous thinking.

Greed is said to be the cause for the unhappiness of mankind, and hence one is asked to limit desire and live in modest means rather than aspire to possess enormous wealth. It is likely that one may end up with vast possessions due to circumstances. Then, he or she is to assume the role of a trustee or guardian and not the owner of such assets and to utilize them for the larger benefit of society.

This concept is held true even for knowledge and intellectual assets. Knowledge is treated as a community asset and is meant to be shared. It was perceived that the optimal way to enrich the body of knowledge is to share it with others. Residual knowledge was tested against observations, and the one who could best explain the observed results through a conceptual framework was revered as "guru."

In today's context, this view of knowledge has created an ambivalent environment in the East. Although the need to protect new intellectual property (IP) created through scientific studies is acknowledged, traditional thoughts prevail as well. Any incremental knowledge gained by building on top of existing knowledge for centuries is still treated as common property. This dichotomy has been the key source of contention between the East and the West in the World Intellectual Property Organization (WIPO) and the World Trade Organization (WTO) (Watal 2001). Negotiations over the past decades have resulted in creation of a new type of IP known as geographical indicators (GI). No individual can claim proprietary rights over GI. For example, Basmati rice, Mysore silk, and the antiseptic properties of turmeric are knowledge that has been resident in India, in certain parts or all over the country, for many centuries. They all come under the description of GI. Anyone coming up with a new strain of Basmati rice cannot hope to get a patent for Basmati rice in his or her name. Nondisclosure agreements (NDA), indemnity against third party liabilities, IP rights, and so on, are concepts deeply embedded in the commercial contracts of the West. They all become serious issues when East meets West for business purposes.

Belief in End versus Means

Table 5.1 highlights the common belief across all societies that the end does not justify the means except in special circumstances. However, the relative emphasis placed on end and means has an unbridgeable chasm between the cultures. In the West, setting a goal as an end point and driving towards the goal, purposefully and relentlessly, is encouraged and success is rewarded. Competitive spirits are spurred among individuals, and the one first to reach the goal post is amply rewarded. In general, it is a "winner takes all" attitude. That there could be multiple paths to reach the goal, and one should take the most optimal path, provided it is within the law, is a theme that is endorsed. The concept of six sigma, as articulated by the Juran Institute (de Feo and Barnard 2005), and its implementation in a corporate setting to drive employee focus on quality and productivity at General Electric company (Welch and Byrne 2001), signify the goal-oriented behavior of the Western world.

Eastern cultures take an entirely different view on this issue. The means to attain a goal are treated as equally important as the goal itself. The purpose of any mission is to not be the first to reach the goal but to enjoy the journey, as well. If there are multiple paths to reach a destination, then the one that is ethically acceptable and sustainable is considered the optimal path. This is further accentuated by the belief in India that one must do one's just duty without being focused on what the outcome could be (International Gita Society 2001).

The desire of the Westerner to communicate to the point and to negotiate swiftly to reach a conclusion is a direct result of his or her frame of mind. Similarly, the approach of the Easterner to establish a trusting relationship with a potential alliance partner prior to the execution of business also emanates from his or her frame of mind.

The context versus content view that differs between the East and the West also can be tracked to the differences in their beliefs on end versus means. While content is the overwhelming focus of the West, the Easterner, with his concern for the means, cannot help but assign significant weight to the context. Context is considered to be controllable and manageable by the West. It can also be influenced or modified whenever required to accomplish a greater goal. The Orientals believe otherwise. Their belief is that context cannot be fully controlled but needs to be managed. Since context is treated equally with content, as a natural corollary, it is believed that a proper balance must exist between the two. Many times context includes other stakeholders, who could be human beings, animals, plants, or water bodies. The Eastern approach calls for identifying a solution to a given problem by finding the right balance between all entities sharing a common resource. Hence, the need to coexist with nature and its other creatures is important. A Westerner tends to be happy to solve a given problem optimally and fix any adverse contextual impact appropriately. The Easterner on the other hand, will insist on finding a solution that has no adverse impact on the environment from the beginning.

REAL-WORLD ISSUES

While challenges encountered in implementing a solution can arise due to any of the above four dimensions, it is likely that some issues cut across these dimensions. Hence, an issue can fall in a hybrid domain of two or more of these dimensions. An approach to solve the issues arising in cross-cultural environments is to question the efficacy and implementability of the solution. These are explored through each of the four specified dimensions. System, product, and process design are reviewed to detect cultural divergence and remedial action. Both design and implementation activities need to be subject to this cultural divergence test. However, fundamental design flaws cannot be overcome by superior implementation. Hence, design review is as mission critical as ensuring implementation effectiveness.

Attacking and addressing cross-cultural issues from the values and beliefs perspective is akin to confronting them at the root cause level. The traditional approach is to treat at the level of manifestation of cultural differences. Such an approach fails to recognize the permeation and diffusion of culture issues into both the context and content layers of decisions; instead, these layers are treated merely as an add on. We trust that the proposed approach will yield better results due to seamlessly dispersing the cross-cultural issue solutions into the architecture of the decision models.

I demonstrate this approach through two examples, the first of which deals with the help desk process and the second of which concerns healthcare delivery in all its aspects in two different cultures. An existing process design is subject to validating culture fit in the help desk example. A different approach is taken with respect to healthcare delivery. The existing delivery mechanism is described in detail, highlighting the contextual as well as cultural variations. Hence, a system designer can treat this section as a high-level specification prior to design.

EXAMPLE 1: HELP DESK PROCESS

Figure 5.2 depicts the help desk process as it exists in many firms in the West. The authenticity of the customer is verified first, followed by customer authentication. If no such customer (by name or code) exists in the system or the authentication process is not a success, then "regret" for the inability to proceed further is conveyed and the call is terminated. The time spent on servicing a "noncustomer" is minimized.

The next level of verification is about the customer's service eligibility. The maintenance or service agreement may not be up-to-date. A quick assessment of the above facilitates fast termination of the call (and hence time savings for the service provider). It is only at this stage that the service agent prefers to listen to the customer regarding the issue. Issue discussion can be time-consuming and is ridden with many interactive queries and responses from both ends. It is quite likely that this could be the stage at which the service provider discovers that the current agreement does not provide coverage against the current issue. Hence, service denial can occur at this stage, also.

Once the service eligibility is determined with regard to the current issue, then the resolution phase begins. If the issue is known to be common, then the system database is likely to identify the solution quickly. (Similar issues encountered earlier and their resolution logs are available in the system.) The solution is

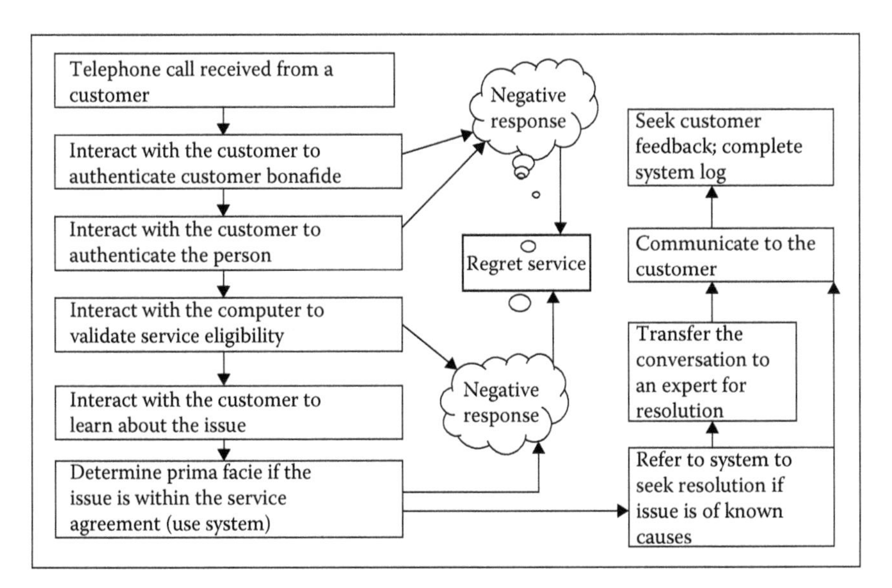

FIGURE 5.2 Help desk process as designed in the Western world. (With kind permission of Springer Science + Business Media: *Handbook of Automation*, ed. S. Y. Nof, Automating information and technology services industries, 2009, p. 1271, Figure 71.2, P. Balasubramanian.)

conveyed to the customer. The interaction from the service provider side, until this time, can be manual or automated. Most firms in the Western world prefer the latter with the inclusion of an interactive voice recognition system.

If the issue is uncommon and intricate, then the intervention of an expert may be required. The expert typically is a person and hence an expensive resource. He or she listens to the customer, understands the peculiar nature of it and suggests remedial action. After the process is completed, the customer is requested to provide feedback about the entire process so that any process improvement can be considered for the future.

Many design features of this system can be noted. It recognizes the need to be time efficient from both the customer and service provider perspectives. It is a fair system of service for agreed value. It recognizes the need for division of labor. It creates separate queues based on the severity of the problem. And it treats the customer as a valuable resource for system improvement. Another notable feature of this system is the complete segregation of marketing and selling of the service from the servicing process. Nowhere in the process is a customer given the choice to sign up for the relevant service and seek issue resolution simultaneously. Service is extended only to currently valid and legitimate customers. The ineligible customers are expected to revert to another part of the system to sign up for the service and then come back to seek issue resolution. This aspect is also justifiable from the time efficiency and division of labor concepts.

Let us review this process for its success if implemented, on an as-is basis, in a totally different cultural setting, for example, in India. It seems that the process has been defined for high throughput and efficiency, and the cross-cultural issues are not visible initially. However, the context itself could be dramatically different. This is how.

Customers prefer to talk to a person at the other end rather than to a robot and an interactive voice recognition system. Establishing a rapport with a person is a prerequisite to sharing an issue, besides the discomfort in communicating with a synthetic voice. It is the customer's expectation that the service agent will listen to him or her first, hear out the issue completely, and then move to the next step. Hence, queries regarding customer authentication and verification, preceding discussion of the issue, can mar the customer relationship severely (the end versus means dimension of beliefs serves to identify this difference). The original system is age and gender neutral. It has a strict adherence to link with the customer and no one else. But this neutrality could be the cause for certain additional issues.

It is common in India that the service agreement is registered in the name of the husband or the head of the household, while any member of the family, for example, spouse, child, or others, may seek the service. Hence, a service call can come from any of the family members (most likely the wife), and the service provider is expected to oblige. This can be done only by design modification and not by altering the implementation process alone. A well-designed system would provide the option to seek service by designated members of the family, provide for registering their names, and permit a common verification process such as date of birth of the head of the family or the residential address. (The gender and role expectation dimension as well as the collective versus individual ownership dimension of beliefs serves to identify this difference.)

It is prudent to combine the sales and service opportunities in the Indian context. Many a well-listened-to customer can easily be convinced to sign up for additional services and renew expired contracts. It would be considered not an interruption of the service process but a continuation of it. Manpower costs rarely exceed 15% of the total costs of any service in India. Hence, it is worthwhile to add additional service agents and train them on multitasking rather than to seek strict division of labor. A general expectation is that all sales personnel are well-versed in the field-level issues arising during maintenance. Hence, knowledge sharing is encouraged and job enrichment is preferred. (The collective versus individual ownership dimension of beliefs with particular reference to attitude on knowledge sharing serves to identify this difference.)

The issue resolution process needs a minor adjustment as well. Since the issue has been described in detail to start with, transferring the call to an expert when necessary must be done with adequate explanation to the customer. In other words, the context for switching the call to another service provider who is an expert has to be conveyed. A general belief is that the expert has had ample opportunity to experience diverse situations and hence is in a position to resolve the matter quickly. (The age versus wisdom dimension with the spotlight on the link between experience and knowledge helps to identify this difference.)

This example demonstrates how the values and beliefs dimensions of cultural differences can be utilized to design systems and processes that work best in a given culture. It is also feasible to design these dimension specific filters and use them in a systematic way to redesign existing processes.

EXAMPLE 2: HEALTHCARE DELIVERY SYSTEM

Cross-cultural issues impact heavily on every aspect of design of a healthcare delivery system in one society versus another. The four dimensions of values and beliefs lead to a bipolar scenario resulting in the need for the total redesign of delivery systems. Yet this aspect is not well-understood or reflected while implementing solutions.

The context of stakeholders is divergent between the West and the East. It can be understood with reference to Figure 5.3, which depicts healthcare delivery in the Western world. Health care is delivered by a bundle of interconnected services, yet each one is provided by a distinct service provider. The physician/surgeon is at the core of it, while the peripheral services are comprised of a laboratory for diagnostics, a clinic or hospital for inpatient and outpatient services, a pharmacy for drug dispensing, and a host of other auxiliary services such as physiotherapy and home care. The patient is in need of these services, while the payer for the services can be either the patient him- or herself or his or her employer and the government. The system functions efficiently due to the presence of an insurer.

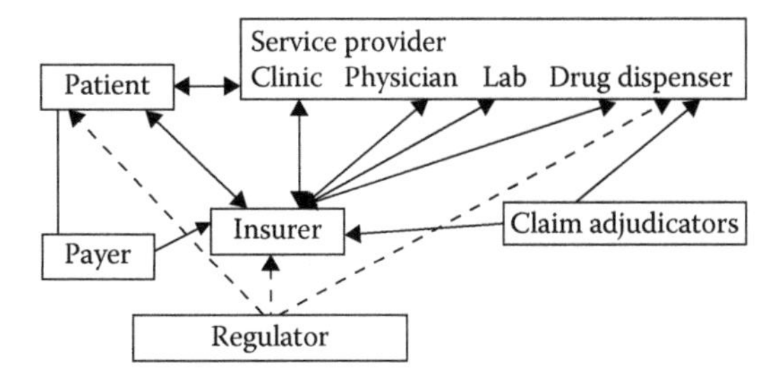

FIGURE 5.3 Health care delivery in the Western world: Context diagram.

The payer contracts for a set of services meant for the patient with the insurer. The insurer in turn has a contract with various service providers to reimburse the costs on an agreed basis. Often an intermediary (i.e., a claim adjudicator) is required to assess the fair cost of the service that needs to be reimbursed. Hence, the system is quite complex. The complexity of the system is further compounded by the malpractice insurance requirement of all service providers. Each one of them runs the business risk of being sued when a patient suffers unduly. Hence, service providers need to protect their professional practice and service with malpractice insurance as well as against the risk of loss of revenue.

Needless to say, government plays an overseeing role to ensure equity in the system. In addition, recent legislative changes seek to protect patient privacy and confidentiality while simultaneously facilitating portability of health records from one service provider to another.

Figure 5.4 shows the contrasting context diagram of healthcare delivery in India. The key differences are

- The role of the insurer is weak from the patient perspective. The majority of the population has no insurance coverage. They pay from their own resources for many services.
- There is a wide variation in the quality and cost of services provided by a multitude of service providers. Government-run hospitals charge the least, while the multispecialty hospitals run by the private sector provide world-class services at a high cost. There are many in-between mode clinics and hospitals run on a full or partial charity basis.
- The link between the service providers and the insurers (from the malpractice perspective) is almost nonexistent.
- The regulators play a limited role. There is hardly any enforceable patient bill of rights. The insurers and the service providers are subject to certain regulations but nothing of the intensity as in the West.
- There is no regulation governing health information portability or patient privacy and confidentiality.

Consequently the billing, receivables, and record-keeping systems tend to be simple and localized. Most of the medical records tend to be on paper.

Yet another difference at the context level pertains to the wide gap between supply and demand prevalent widely in society. The physicians, paramedic

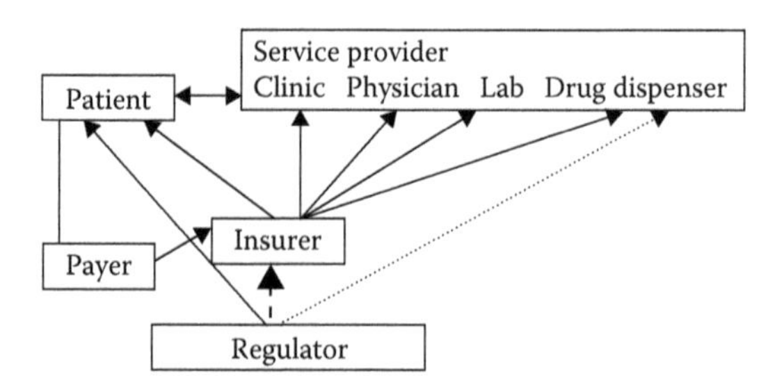

FIGURE 5.4 Health care delivery in India: Context diagram.

staff, and government-run hospital facilities are stretched beyond capacity. The difference in healthcare quality is even more significant between the urban and rural India.

Impact of Culture on Healthcare Delivery

Patient Registration and Administration

A hybrid system of preregistration and acceptance of walk-in patients prevails in most places. The capability of the patient to pay for the likely services is often not checked in advance (except in labs and pharmacies). An advance may be demanded in private hospitals if there is a likelihood of inpatient admission. It is common for most patients to wait long even for registration. The quality of service varies widely across the spectrum and the populace waits, albeit impatiently.

The appointment system to see a practitioner is hybrid, too. Patients can call in advance and fix a convenient time for consultation. However, most clinics, hospitals, and general practitioners allow walk-ins. The physician accommodates all patients within the day by stretching their work hours. This is a distinguishing feature of the Indian system. Rarely is a patient turned away. While the appointment system is honored, the walk-in patients are seen during gaps or with priority if emergency prevails.

Admission as an inpatient in a hospital is not so orderly, particularly in government-run hospitals. Queue priority is modified dynamically to accommodate preferred patients, much to the annoyance of those waiting. It is not uncommon for patients to wait for a few days before they are provided a room. Many inpatients are accompanied by relatives. They usually stay in and around hospital premises, thus crowding the place. The presence of solicitors in hospital premises is not rare.

The private hospitals are managed better, with a higher level of discipline. Yet they too suffer the queue-jumping phenomenon due to societal pressures. Although the quality of care provided is better, the hospitals' charges are much higher too. Hence, it is unlikely that the poor patients can afford to utilize their services.

How does a system designer understand the contextual differences described above and then provide for the cultural variations as well? How does culture play a role in all of these? Which of the cultural dimensions described in the section The Four Major Factors impact significantly on the patient registration and appointment processes? First and foremost, a system of this kind, where there is a significant gap between supply and demand, can work only by stretching each resource's utilization. The physicians and other paramedic personnel work much longer hours in India than their counterparts in the Western world. For many, the workday can end anywhere from 6 p.m. to 10 p.m. They take fewer holidays and are not afraid to change queue discipline if needed. As a consequence the quality of service varies widely. In the absence of a strong malpractice insurance protection, the service providers could face a serious practice risk. However, rarely do they get sued for delay in admission or even for incomplete examination to start with. Culture plays a subtle yet strong role in ensuring this.

Most patients appreciate that the physician is stretched and is obliging them by working long hours and seeing many patients in a day. They understand the constraints and limitations within which the physician works. As discussed in Age versus Wisdom Dimension, age is treated as a proxy to experience and knowledge: the elderly are respected in most walks of life, and that respect is carried forward to anyone in a superior role, with humility as a virtue rather than self assertion. Physicians are held in high esteem in society due to their knowledge, and yet the public understands that knowledge also has its limitations. Hence, the decision to sue a physician for deficiencies in the service does not occur as a choice or option to most patients.

As per the prevailing culture, women have been the housekeepers and men have had the responsibility to be the bread winners. But globalization and growing exposure to mass media has softened this stand, and women are considered to be the ideal choice as employees in certain roles. Healthcare administration is one of them. That a woman as a caregiver can extend her role from homes to hospitals is a realization that has occurred in both urban and rural environments (gender and role expectation dimension). Clinics and hospitals employ women as receptionists, billing clerks, and admission assistants.

The impact of the end versus means dimension is felt across many of the activities that occur while admitting a patient. Requests to start the operation of a patient at an auspicious hour (even for a minor cataract surgery) is most common. (What begins well will end well, is the belief.) Stoicism is at the top of all thoughts and concerns as the patient is being treated. The physician is held responsible for the care given and rarely blamed if the end result is unfavorable. This is yet another reason as to why the caregiver is not under the malpractice threat. In general, Indian society is not a litigious society and often seeks to give the benefit of doubt to a knowledgeable caregiver.

Diagnosis and Treatment

Medical science has advanced by leaps and bounds in the twentieth century. Gone are the days of treating a patient based on symptoms alone. Numerous machines and testing devices are lined up to help the diagnostic process so that the root cause of the ailment can be identified and treatment can be started as soon as possible. The costs too vary from the insignificant to the exorbitant.

The physician's dilemma is to decide which tests to order and in what sequence. When symptoms are common to many diseases, the tests become necessary for early diagnosis. Any delay in performing the appropriate test will prolong patient suffering and is a definite source of displeasure. Many patients in the Western world have chosen to sue the physician or clinic on account of this delay. The physician or clinic would not be held liable if a patient chose not to undergo the test because he or she was unable to afford it (with or without insurance coverage).

In the Indian context, the severity of resource constraints both at the supply and demand sides translates into inadequate availability of high end CT scanners, MRIs, and so on, in both the system for those who can afford to pay and also those who are unable to pay, even for medium-cost testing procedures for the poor patients. Hence, the physicians (instead of the patients) become the decision makers and order the required tests and their sequence. The process of informing the patient of the pros and cons of a test and consequences of delay is well developed in the West. It is rarely adopted in the East, and even if adopted is likely to receive lukewarm reception.

The cultural dimensions of age versus wisdom and end versus means are relevant here too. (The physician as a superior knowledge source is respected and the process of caring is as critical as the cure itself.)

A multitier pricing system is common for the use of such resources, including the physician's services. The majority of the physicians are willing to charge less or even nothing if the patient's case genuinely warrants it. Along with many clinics, they play a dominant role in providing the testing services at a low cost to deserving patients. The system of charity coexists seamlessly with commercial prudence. This is yet another demonstration of the caregiving process considered in a larger context and as more critical than the cure. However, the unfortunate aspect of women receiving lesser care than men, particularly when it means paying for expensive diagnostic tests, is a societal truism. It can be directly traced to the prevalent values and beliefs on gender and role expectations.

The void created by the absence of a strong insurance system is filled by all stakeholders in the society. While the physician and the hospitals are willing to be flexible in the charging mechanism, many also play an active role in identifying and linking patients with charity organizations. Even before this, the extended family usually rallied around to pool resources to pay for the expensive services. The cultural dimension of collective versus individual ownership is quite visible in such an event.

Drug Discovery and Clinical Trials

Indian history through the first two millennia AD contains many references to healthcare-related discoveries from India. The medicinal properties of neem, amla, thulasi, and turmeric are fully documented, as are the discoveries and descriptions of human anatomy. There were many protocols of treatment with varying beliefs, such as ayurveda, homeopathy, and unani, that existed prior to colonization by the British. The allopathic system of treatment was introduced by them in the seventeenth century and is the dominant method at present. Still, it coexists with all other forms of native systems.

The greatest source of contention between the Western world and India has occurred with respect to drug discovery and associated IP. Until a decade ago, product patents were not recognized, but the processes were. This system has been adopted partially due to the cultural belief that knowledge is a societal property and also due to the need to make modern medicine available at a low cost to the masses. As a signatory to the new WIPO regime, India recently has accepted product patent rights.

Surprisingly, innovations in devices and procedures (as opposed to drugs) continue to flourish in India. Surgeons pool their knowledge with engineers and constantly innovate to develop new devices that are locally relevant and inexpensive. Process variations while performing surgeries are common. In the absence of a litigious culture and with the strong belief in collective ownership of knowledge, information and knowledge is shared between relevant persons in the process of innovation. This has also benefited many patients when the need for consultation between say an endocrinologist, urologist, and nephrologist exists. They share information easily and apply their minds collectively to solve a problem.

India has become one of the preferred destinations for outsourcing clinical trials in the last decade. A combination of factors such as ever-tightening regulations and the unwillingness of patients to go through a trial for an unknown level of improvement has lead to scarcity of pool of patients available for clinical trials in the United States. India, on the other hand, has millions of patients, many of whom look forward to a high-end treatment at no cost. Many make a conscious choice that the attendant risks of an unknown drug or process are well compensated by cost considerations. This system works due to their trust in the physician as one who would not recommend a process of known adversity. These values and beliefs prevalent in the culture act as a strong catalyst for this growing opportunity for India.

FURTHER RESEARCH AND THE WAY FORWARD

This chapter has sought to link the values and beliefs aspect of culture to the decision choices of people and has proposed that such an approach is better than just identifying cultural factors and attending to them while implementing systems in cross-cultural environments. Like all new approaches, the approach advocated here must also be subject to the process of scientific scrutiny and validated.

In addition, the four dimensions identified are to be assessed for comprehensiveness and mutual exclusivity. By continuing the research and considering many different real-world circumstances, we can aim to complete this task. Finally, a set of filters can be developed using these dimensions for use as strainers in testing and verifying systems for cultural fit. These strainers can be used with both existing and proposed systems.

REFERENCES

Andrews, M. M., and J. S. Boyle, eds. 2008. *Transcultural Concepts in Nursing Care*. 5th ed. Philadelphia, PA: Wolters Kluwer Health/Lippincott Williams & Wilkins.

Balasubramanian, P. 2009. Automating information and technology services industries. In *Handbook of Automation*, ed. S. Y. Nof, 1265–84. New York: Springer.

Berten, P. 1998. How unique is Japanese negotiating behavior? *Jpn Rev* (10):151–61.

Bijapurkar, R. 2007. *We Are Like That Only: Understanding the Logic of Consumer India*. New Delhi, India: Penguin Books India.

Bloomberg Business Week. 2009. Big retailers still struggle in India. http://www.businessweek.com/globalbiz/content/oct2009/gb20091016_385819.htm (accessed March 15, 2010).

Campbell, C. P. 1998. Beyond language: Cultural predispositions in business correspondence. Paper presented at the Region 5 STC Conference, Fort Worth, TX. http://infohost.nmt.edu/~cpc/internationalethos.htm (accessed March 15, 2010).

de Feo, J. A., and W. W. Barnard. 2005. *Juran Institute's Six Sigma Breakthrough and Beyond*. New Delhi, India: Tata McGraw-Hill.

de Mooij, M. K. 2003. *Consumer Behavior and Culture: Consequences for Global Marketing and Advertising*. 2nd ed. Thousand Oaks, CA: Sage Publications.

Difference Between. 2010. Difference between values and beliefs. http://www.differencebetween.net/language/difference-between-values-and-beliefs (accessed March 15, 2010).

Early, P. C., and E. Mosakowski. 2004. Cultural intelligence. *Harv Bus Rev* 82(October):139–46.

eHow.com. 2010. Problems related to net banking in India. http://www.ehow.com/about_5147496_problems-related-net-banking-india.html (accessed July 28, 2010).

Hansen, K. 2000. Buying behavior in Chinese supermarkets—A comparison across four major cities. In *Proceedings of ANZMAC 2000 Visionary Marketing for the 21st Century: Facing the Challenge*, ed. Aron O' Cass. 473–477. Gold Coast, Australia: Australia and New Zealand Marketing Academy.

Hofstede, G. 2001. *Culture's Consequences: Comparing Values, Behaviors, Institutions and Organizations Across Nations*. 2nd ed. Thousand Oaks, CA: Sage Publications.

Hupert, A. 2009. Europeans versus American negotiating styles in China. http://www.chinese-negotiation.com/2009/04/europeans-vs-american-negotiating-styles-in-china (accessed March 15, 2010).

International Gita Society. 2001. The Bhagavad Gita, Chapter 2: Transcendental Knowledge. http://www.gita-society.com/section2/2_chap_02.htm (accessed March 15, 2010).

Jranks Marriage and Family Encyclopedia. 2010. India—family life and family values. http://family.jrank.org/pages/859/India-Family-Life-Family-Values.html (accessed July 25, 2010).

Kaplan, R. B. 1966. Cultural thought patterns in inter-cultural education. *J Lang Learn* 16(1–2):1–20.

Kotelnikov, V. 2010. Personal beliefs, values, basic assumptions and attitudes. http://www.1000ventures.com/business_guide/crosscuttings/character_beliefs-values.html (accessed August 6, 2010).

Kraybill, D. B., S. M. Nolt, and D. L. Weaver-Zercher. 2007. *Amish Grace: How Forgiveness Transcended Tragedy*. Hoboken, NJ: John Wiley & Sons.

Kroeber, A., and C. Kluckhohn. 1952. *Culture: A Critical Review of Concepts and Definitions*. New York: Vintage Books.

Kwok, S. 2006. A systematic analysis of consumer buyer behavior in urban China. Thesis, University of New South Wales School of Marketing, Australian School of Business. http://unsworks.unsw.edu.au/vital/access/manager/Repository/unsworks:931 (accessed March 15, 2010).

Liu, Y. 2010. The Design of Cultural Difference. http://blog.sina.com.cn/s/blog 5d20abc20100hkwy.html (accessed October 24, 2010).

Parker, A. 2010. Vodafone struggles in Indian market. www.ft.com/cms/s/0/4eb74d68-6111-11df-9bf0-00144feab49a.html#axzz1PkOv0r1L (accessed June 11, 2011).

Prahalad, C. K., and K. Lieberthal. 1998. The end of corporate imperialism. *Harv Bus Rev* 76(July-Aug):69–79.

Prakash, N. 2003. Status of women in Indian society—issues and challenges in processes of empowerment. *Proceedings of the 11th International GASAT Conference, Mauritius*, 249–60. http://www.gasat-international.org/conferences/G11Mauritius/proceedings/proceedings%205.pdf (accessed March 15, 2010).

Reid, T. R. 2009. *The Healing of America: A Global Quest for Better, Cheaper and Fairer Health Care*. New York: Penguin Press.

Sebenius, J. K. 2002. The hidden challenge of cross border negotiations. *Harv Bus Rev* 80(March):76–84.

Sternberg, R. 1990. Understanding wisdom. In *Wisdom: Its Nature, Origins, and Development*, ed. R. J. Sternberg, 3–12. New York: Cambridge University Press.

Sternberg, R. 2010. Teaching for wisdom in our schools. Center for Development and Learning, Los Angeles. http://www.cdl.org/resource-library/articles/teaching_wisdom.php (accessed July 25, 2010).

Tamminga, B. 2009. Why consumers buy: The cultural factors that affect buyer behavior. http://ezinearticles.com/?Why-Consumers-Buy---The-Cultural-Factors-That-Affect-Buyer-Behavior&id=1850442 (accessed March 15, 2010).

Triandis, H. C. 1995. *Individualism and Collectivism*. Boulder, CO: Westview Press.

University of Alberta Faculty of Medicine & Dentistry. 2008. Wisdom comes with age, at least when it comes to emotions. *Science Daily*. Retrieved from http://www.sciencedaily.com/releases/2008/06/080612185428.htm.

Watal, J. 2001. Chapter VIII. Distinctive signs: Trademarks and geographical indicators. In *Intellectual Property Rights in the WTO and Developing Countries*, ed. J. Watal, 243–75. The Hague: Kluwer Law International.

Welch, J., and J. A. Byrne. 2001. Chapter 21: Six Sigma and beyond. In *Straight from the gut*, 325–40. New York: Warner Books.

Wells, D. 2011. The dichotomies in Indian women's lives. The Alicia Patterson Foundation. Retrieved from http://64.17.135.19/APF2003/Wells/Wells.html (accessed June 14, 2011).

Section III

International Partnership and Collaboration

6 Cultural Factors, Technology, and Operations in Developing Countries

Two Case Studies

David M. Upton

CONTENTS

INTRODUCTION

The development of information technology and operations strategy has been widely recognized as an important potential source of wealth building and social benefit in developing countries (Negroponte 1998). The potential to reach and enrich the poorest of the poor is a laudable goal, and one that the operations and information technology communities are seeing as an increasingly important element of their fields. Bridging the digital divide, for example, and advancing from exploitative craft production are merely small examples of the potential impact (Center for Ethics of Science and Technology 2009; Scrase 2003).

Many authors have decried the blunt use of purely financial foreign aid, while programs that have injected technology directly without meticulous regard for the cultural milieu of their targets have suffered numerous failures (Dichter 2003). The successful, broad improvement of the standard of living through technology remains enigmatic. While there are important initiatives in microfinance and education (Ledgerwoord and White 2006), there remains vast potential to create widespread impact.

The two case studies presented in this chapter should not be considered examples of best practice. Rather, they are teaching and discussion vehicles that examine the decision-making principles that guide two very successful initiatives in India. Both are aimed at improving the lot of the poorest of the poor, and both are home grown.

Part of their success derives from that homegrown development. They provide a deep understanding of the cultural factors that shape decision making that, in turn, ultimately form a strategy. Without knowledge of the local culture, both of the initiatives would have failed. Even within India, very few middle-class urbanites have a good understanding of the life of an impoverished farmer (Sainath 1999). ITC Ltd. made great efforts to use their extant knowledge of tobacco farming, and the trust the farmers had in them, to provide great benefit to both.

ITC'S eCHOUPAL

The innovators behind ITC's eChoupal developed a number of principles for making decisions as part of their strategy (see Case Study 1). While many are technical and financial in nature, the most important are those decisions that are contingent on the local culture. For example, to smooth the transition from an ancient model of farming to one more profitable for farmers, the decision was made to build systems that mirrored existing social structures. The *choupal*—a nightly village meeting place for farmers—provided an analogue for the meetings that subsequently took place around a computer using information from the Internet. Informal meetings were still held, congruent with the traditional *choupal*. This built a conceptual bridge helping the often illiterate farmers to use information technology (IT) to their great advantage. Farm productivity increased by 20%. They were able to avoid being cheated of their profits. They were also able to beat the market inefficiencies resulting from asymmetric information (such as those described by Akerlof 1970).

There are many other examples in the case of such culturally apposite decision making: the requirement that the *sanchalak* (the trustee of the equipment and information) not be a religious or political leader, for example. He or she also took an oath to tell the truth and use the technology to the benefit of the village. This decision was underpinned by an understanding that one's oath is truly meaningful in an isolated village in which ostracism would mean destitution. Without doubt, deft technology deployment was important in the success of the project. But the dominant force behind its success was the culturally informed decision making.

AKSHAYA PATRA

Akshaya Patra, described in the second case, is the largest school meals program in the world, serving over 1.2 million children every day. As a charity, the organization has followed a slightly different path to improve the lot of the disadvantaged. Their genesis sprang from the moment when his Divine Grace A. C. Bhaktivedanta Swami Prabhupada saw a group of children fighting with street dogs over scraps of food. From this simple, yet heart-breaking incident, came the determination that no one within a ten mile radius of their religious center should go hungry—a principle that derived from fundamental religious beliefs. Even if one remains agnostic concerning the motivation for the project, the effect has been much deeper than originally intended. First, it provides adequate nutrition to young brains and bodies, which, at the margins of starvation, has a dramatic effect on intellectual and physical development. Second, and probably as important, is that it makes children go to school. It provides a reason for parents to have their children educated, rather than beg on the street or perform menial labor. Often, lunch is the only meal they will eat in a day—without school, they would go without.

Operationally, the system is very elaborate. Centralized production is carried out each morning by the monks just outside Bengaluru. A brilliantly effective distribution system gets the hot meals to the schools. On its own, it is a fine example of a centralized operations strategy: large-scale production being meted out across a well-organized, star-shaped distribution system. The organization quickly realized, however, that those in rural areas were being left out of the program, and they were anxious to reach out given their initial success. Rural India however, is a very different kettle of fish: distances are large, and roads are often impassable or nonexistent, making centralized production an infeasible solution. A different set of decisions was made to build the operations strategy for rural areas. Local production, using small kitchens and local, village labor (usually women) became the primary method of production. Cultural factors were the primary force shaping this new distributed strategy. Poorly educated, many of the women had a meager understanding of the link between cleanliness and the health of the children. The decision was made, therefore, that the first meal produced be offered to God: to whom one would never offer something unclean. This assured the quality and cleanliness of the subsequent meals. As the case describes, the dramatically different culture in the rural areas drove decision making, and that sensitivity has improved the lot of tens of thousands of people—not only now, but for the generation the children will become.

CONCLUSION

It might be simply a truism that technology and operations should take into account the culture in which they will be embedded. Empirical evidence suggests that they rarely do (Madu 1989). Countless decisions are made in all countries that use only a broad brush to match culture and technology. The idea needs more than a Powerpoint slide vaunting its importance. The devil, as ever, is in the details. Each of a myriad of decisions must be made, like any strategy, with due regard to its situation. A deep understanding of the culture in which decisions are made concerning operations and technology can have a powerful impact on the world. The cases presented here may not embody perfection, but they shine a light on a path forward to benefit the many people who remain untouched by the potential advantages of the modern world, without abandoning their extant cultural traditions.

REFERENCES

Akerlof, G. A. 1970. The market for "lemons": Quality uncertainty and the market mechanism. *Q J Econ* 84:488–500.

Center for Ethics of Science and Technology. 2009. *Meaningful Broadband Report 2.0.* Bangkok, Thailand: Digital Divide Institute.

Dichter, T. W. 2003. *Despite Good Intentions: Why Development Assistance to the Third World Has Failed.* Amherst, MA: University of Massachusetts Press.

Ledgerwoord, J., and V. White. 2006. *Transforming Microfinance Institutions: Providing Full Financial Services to the Poor.* Washington, DC: The World Bank.

Madu, C. N. 1989. Transferring technology to developing countries—Critical factors for success. *Long Range Plann* 22(4):115–24.
Negroponte, N. 1998. The third shall be first. *Wired* 6(1):96.
Sainath, P. 1999. *Everybody Loves a Good Drought: Stories from India's Poorest Districts.* London: Headline Review.
Scrase, T. J. 2003. Precarious production: Globalisation and artisan labour in the third world. *Third World Quarterly* 24(3):449–61.

CASE STUDY 1 The ITC eChoupal Initiative

David M. Upton and Virginia A. Fuller

On the challenge of inclusive growth:

"It is now universally acknowledged that no long-term economic growth agenda for India can be feasible without including in its fold the agricultural sector, which is home to 72% of the population and 60% of the nation's workforce. The challenge lies in sustaining high rates of economic growth with equity over many years in order to convert the world's largest pool of economically disadvantaged people into viable consumers, thereby translating development into economic freedom".

On ITC's purpose:

"ITC consciously exercises the strategic choice of contributing to and securing the competitiveness of the entire value chain of which it is a part. This philosophy has shaped the vision for your company, the vision I have referred to in earlier years as "A Commitment Beyond the Market." Creative use of information technology through the eChoupal initiative has enabled your company to bring together diverse agencies, each with specialized competencies, in a bid to empower the Indian farmer".

Excerpts from a speech by Chairman Shri Y. C. Deveshwar, ITC Annual General Meeting, 2003.

CHOUPAL: A TIME-HONORED TRADITION

The village of Dahod appeared to be an unlikely setting for a technological revolution. Located 25 kilometers south of Bhopal in India's central state of Madhya Pradesh, Dahod was dominated by soybean farmers who made their living as their ancestors did, harvesting their crop and selling it in the local market yard. Kamal Chand Jain was one such soybean farmer. Jain had spent 40 years cultivating a reputation as a trustworthy unofficial leader in this quiet community of 3000 people. He lived in a simple concrete home that opened onto a dusty crossroads, providing both a physical and social center to the village. For years, his fellow villagers had gathered in the cool cement front room of Jain's home on their way in from the fields or a trip to town to chat, gossip, or share stories and news from the day. This evening gathering was a traditional staple of Indian farm life, not only in Bhopal but all over Madhya Pradesh. In Hindi, the word for this meeting place was *choupal*. The *choupal* constituted an informal assembly, a forum that villagers could call their own, a place where knowledge could be shared and captured.

Meanwhile, in the corporate offices of ITC Limited's International Business Division (IBD) in Hyderabad, Chief Executive S. Sivakumar pondered the *choupal* concept. IBD was the agricultural commodities export division of ITC, and, by March 1999, it was clear that it was lagging behind the other divisions of the company. In 1998, IBD had grossed Rs. 450 crore* ($100 million) in agricultural commodities sales, a marginal addition to the total Rs. 7701 crore ($2 billion) in sales generated by ITC's other divisions, which included tobacco, paperboard, retail, hospitality, and foods, among others.

The soybean and its derivatives comprised two-thirds of ITC's agricultural export business.[†] ITC sourced soybeans from farmers located throughout rural Madhya Pradesh. Madhya Pradesh had been dubbed India's "soyabowl," as its farmers contributed 4 million of India's 5 million tons of soybean crop. ITC had had a 100-year relationship with farmers (based originally on the tobacco industry)[‡] that gave it an integrated presence along the entire value chain, from procuring soybeans from farmers and processing the beans in exclusively hired processing plants, to exporting the processed soymeal via vessel loads and container shipments. When soybeans were processed, about 80% of the crushed bean was turned into soymeal, a high-protein extract that was added to poultry and cattle feed. ITC exported soymeal to countries such as China, Pakistan, Bangladesh, and the United Arab Emirates, as well as other parts of Southeast Asia.[§] The remaining 20% of the soybean material became edible oil, highly valued for its nutritional content and a very popular cooking medium in the domestic market.

ITC had been successful in selling soybean oil domestically and processed soymeal internationally, but both the input and output sides of the agricultural supply chain in India were still far from efficient. The limited technological resources in India had constrained the dissemination of know-how in rural farming communities. Farmers did not have access to quality inputs, such as sowing seeds, herbicides, and pesticides, or information, such as accurate weather reports, that would help them improve their crop quality as well as the process of bringing it to market. They did not reap financial benefits from any profits made off the valuable soybean-derived materials. In fact, farmers were losing 60–70% of the potential value of their crop, with agricultural yields only a third to a quarter of global standards. Similarly, on the output side, middlemen clogged the supply chain, reducing profit margins for both farmers and buyers such as ITC. Unfair practices affected the way the farmers were paid, the weighing of the

---

* One crore equals 10 million and one lakh equals 100,000. Case uses exchange rate of US$1 = 45.23INR (2003).

[†] Soybeans represent IBD's oil seeds, grains, and pulses (OGP) group and comprise approximately 65% of the company's gross agricultural exports. IBD's other product groups include coffee and spices, seafood, and value-added horticultural products.

[‡] See http://www.itcportal.com for more information about ITC Limited.

[§] http://www.itcibd.com/feedhis.asp

produce, and the amount of time taken by the process. This drastically increased transaction costs, slashing potential profits for the farmer.

Both farmers and soybean processors were locked in an unproductive cycle. Farmers had limited capacity for risk and therefore tended to minimize their investment in crops, lest inclement weather or pests destroy their investment. This, however, meant a lower-value crop, which translated into slim margins for both the processor and the farmer. With such risk aversion, farmers were also loath to experiment with new farming methods. Since this meant that few new sources of value were found, the cycle continued unabated.

A SEED IS PLANTED

In March 1999, Sivakumar was challenged by ITC Chairman Y. C. Deveshwar to generate a new business plan for the IDB that would bring it up to speed, both in ITC's realm as well as in the global commodities exporting market. But Sivakumar knew that a host of factors—fragmented farms, overdependence on monsoons, and lack of sophisticated inputs and farming practices—undermined the competitiveness of Indian agriculture and, in turn, ITC.

Noting that many of the challenges for both ITC and farmers arose from the ineffective supply chain for agricultural goods, Sivakumar pondered Deveshwar's credo: "What can we do to secure the competitiveness of the entire value chain, so that this business achieves its full potential?" On the other hand, how could a small business think of investing large sums towards such a goal, he thought. Deveshwar suggested exploring the digital technologies that were changing so many of the companies around them. Sivakumar began to rethink the soybean supply chain. He studied the farmers' villages and market yards to identify pieces of the supply chain that could be improved, so that IBD might reach its goal of Rs. 2000 crore ($442.6 million) in revenue by the year 2005.

STUNTED GROWTH: FROM FIELD TO FACTORY

Farmers in Madhya Pradesh made their living in much the same style as their predecessors 50 years earlier. The process of getting crops to market began with farmers harvesting the soybeans and loading them onto tractors and bullock carts. Farms varied in size from under five acres for a small farmer to greater than 12 acres for a large farmer.* An average farmer, with about nine acres of farmland, could expect an annual net income of approximately Rs. 20,000 ($443) from soybeans and wheat together.† After the harvest, farmers hauled their loads of produce 30–50 kilometers to the closest *mandi*‡ and then waited for the crop to be auctioned. The auction began when a government-appointed bidder valued

* For comparison, the average American farm is about 450 acres.

† Farmers typically raised more than one type of crop to take advantage of the varying seasons. The soybean season, for example, was from June to September. From October to May, most soybean farmers grew wheat.

‡ Hindi word meaning market yard. Madhya Pradesh had 308 *mandis*, 175 of which were soybean dominated.

the produce and set the initial bid. From here, government-licensed buyers called commission agents (CAs) bid upwards until the crop was sold.

ITC contracted with a specific CA in each *mandi* to bid on behalf of the company. Prices were authorized by ITC's office in Bhopal, Madhya Pradesh. Here ITC employed a team of traders who followed the global market. Although the CA knew what price ITC would pay, nothing prevented him from buying from the farmer at a much lower price, selling to ITC at market price, and pocketing the difference.

Once a CA won an auction, the farmer brought his tractor to that CA's shop in the *mandi* and waited for the produce to be weighed on a manually operated balance scale that accommodated only small increments of the lot. The actual weight of the crop was often manipulated at this point because of the inaccuracy of the crude beam scales. For example, if the farmer brought 20 quintals* of loose soybeans to the *mandi*, he could expect to lose about 10 kilograms total during the transactions, or 0.5% of his original lot. This translated to a loss of about 100 Rs. ($2.22) per lot. After the weighing process, the product was bagged and the farmer was paid. According to the law, CAs were supposed to pay the farmer immediately, but, in smaller *mandis*,† farmers were often paid after an unofficial credit period. The CA would simply tell the farmer to return after a few days for the money.

On any given day, at least 1000 farmers‡ could be found trying to file into the market to sell their produce. Some had to wait for two or three days just to get into the crowded marketplace. Once inside the *mandi*, the farmer was faced with further challenges of the chaos and pressure that characterized the market yard. The Bhopal *mandi*, hosting an average of 1700 farmers a day and the sole destination for farmers in Dahod, was a dusty yard with a perimeter of booths belonging to the various CAs. It teemed with adolescent boys who ran through the crowd, kicking up dust and eating beans off the farmers' carts. Laughing, joking men loitered and watched the auctioneers.

Farmers suffered as a result of the time it took to sell produce in the *mandi*, for they were dependent on timely cash flow for subsistence. Thus, when harvest time arrived, they all descended upon the *mandi* at once. The crop had to go to market immediately, and, more importantly, it had to be sold. Farmers were stuck in the position of not being able to turn down a CA's offer; in many cases it had taken him all day to reach the *mandi* from his village, and to return with a full cart of unsold produce would be a waste of time and money. Farmers rarely had access to adequate storage facilities in which to hold the crop if it was not sold. If a farmer were able to store the soybeans, and sell before or after harvest, without the time pressures associated with a perishable product,§ he would have

* A quintal = 100 kilograms. Twenty quintals (two metric tons) was a typical lot size.

† Containing two to three commission agents. Larger *mandis* contained 25–30 CAs.

‡ More than 4000 per day visit larger *mandis*, 1500 visit medium-size *mandis*, and 1000 visit smaller *mandis*.

§ Soybeans would perish if allowed to get wet after harvest; thus, ill-timed rains could ruin a stockpiled product.

more leverage over their value. This was impossible, however, under the prevailing system, where the farmer did not have other options.

Once a transaction had taken place, the CA brought the produce to an ITC processing facility. There, ITC paid him for the cost of the soybeans. This was effectively a reimbursement, since the CA had paid the farmer in the *mandi* from his own resources at the time of the sale. The farmers' isolation from one another and lack of telecommunications meant they had no way of knowing ahead of time what price would be offered the day they arrived at the *mandi* other than word of mouth. As a result, price discovery occurred only at the end of their growing and selling process.

The Seed Is Cultivated

In May 1999, Sivakumar anchored a brainstorming session of the ITC management team in Patancheru, near Hyderabad. The team knew that, in order to reduce costs and inefficiencies incurred along the current supply chain, the "village A → *mandi* B → factory C" cycle had to be broken. Deveshwar's idea of digital technologies came in handy. Indeed, the team worked to develop a business model that incorporated "e" into the age-old tradition of village *choupals* to facilitate a reorganization of the channel. Sivakumar believed that the team had to work with the cultural infrastructure that had evolved in the villages rather than owning or controlling the entire value chain from top to bottom.

Knowledge shared and captured in the traditional *choupal* could be extraordinarily useful to farmers, but it had traditionally been limited to verbal communication. In the absence of telecommunications, and even electricity in some places, news from the closest city could take days to reach an outlying farming village. The uncertainty surrounding cash flow prevented the farmers from creating a sound financial base; instead, they had become locked into subsistence living. As D. V. R. Kumar, manager of trading in ITC's Bhopal office, said, "We know we can't predict the market, and no one expects to, but some guidance would certainly serve the farmers well. Under the 'old' system, they remained completely in the dark with no reference points for pricing other than word-of-mouth reports of yesterday's numbers."

Sivakumar and his team knew that the price trends of soybeans and their derivative products could be forecasted. Prices of Indian soybeans generally followed the agriculture futures market on the Chicago Board of Trade and the Kuala Lumpur Commodity Exchange. Given the volatility of the spot market, and the fact that the value of agricultural commodities was based on largely uncontrollable factors such as weather, disease, and pest infestation, farmers needed to be aware of market activity. They needed to understand their product in its global context, so that they could plan their activities with more confidence.

eChoupal

At the May 1999 meeting, Sivakumar and his team conceived ITC's eChoupal initiative. The eChoupal was based on the knowledge sharing found in the traditional *choupal* model, but took the concept one step further. ITC supplied a computer kit to each village with the following components:

- A PC with a Windows/Intel platform, multimedia kit, and connectivity interface
- Connection lines, either telephone (with bit rate between 28.8 and 36 Kbps) or, more commonly, VSAT* (in 75% of eChoupals; average 2003 usage 64 Kbps inbound, 1 Mbps outbound)
- A power supply consisting of UPS† and solar-powered battery backupA dot-matrix printer

The total setup cost to ITC was Rs. 170,000 ($3762) per *choupal.* Another Rs. 100,000 ($2213) was spent on people, travel, communication, software, and training.‡ With the arrival of these components, nightly *choupals* at the home of Kamal Chand Jain were no longer limited to stories and gossip of the village. Farmers were instead accessing the World Wide Web through a site dedicated specifically to them, ITC's http://www.soyachoupal.com. This website was updated by the ITC Bhopal office. The data uplink (which provided the source information for the site), however, took place in Bangalore, home of ITC Infotech India Ltd., ITC's own information technology subsidiary responsible for developing the software. The site contained much useful information that was previously unavailable to farmers in Madhya Pradesh. The site opened up by welcoming farmers into the "community" of the eChoupal. On the left side of the screen, there were eight links to the areas of key information that comprised the eChoupal: weather, best practices, crop information, market information, FAQs, news, feedback, and information about ITC. The feature set had been developed progressively with full involvement of the farmers using the system.

Weather Page

India had no private weather service; the government was the only provider of weather information. Before the eChoupal, weather forecasts were rarely communicated to the remote villages of the rural farmers. When they did reach the villages, they were too generalized and did not accurately cover the 30.75 million hectares of Madhya Pradesh. Farmers needed to know their regional weather in order to accurately expect the rains. ITC negotiated with the Indian Meteorological Department, the national weather service, to get localized forecasts for district pockets of 70–80 square kilometers within Madhya Pradesh.

* A VSAT is a very small aperture terminal. Traditionally, VSATs had a few disadvantages; VSAT bandwidth was not very high and restricted to a few hundrend Kbps. There was also a certain amount of latency (the time between initiating a request for data and the beginning of the actual data transfer) between nodes. But these limitations have been overcome to a large extent due to advancement in technology. VSAT providers in India offered up to 52.5 Mbps outroute (from hub to VSAT) and 307.2 Kbps inroute (from VSAT to hub) data rates, with 270 millisecond latency (Network Magazine India).

† UPS stands for uninterruptible power supply.

‡ The company believed it would be able to recover the cost and make a profit within three years of the initial eChoupal rollout.

The result was that, on the soyachoupal.com weather page, a farmer could click on his home district to see his localized forecast. This knowledge made a lot of difference in the timing of various farm operations such as the application of herbicides and fertilizers.

Furthermore, the difference between harvesting before or after a big rainfall drastically affected the quality of the crop. Ill-timed rains reduced the value of a soybean crop irrespective of the amount of money that was put into it at sowing time. "Farmer A," for example, might spend 8000 Rs. ($177)/hectare on inputs in the hopes of getting 15,000 Rs. ($332) in return. Similarly, "Farmer B" could put in 5000 Rs. ($110)/hectare and get about 10,000 Rs. ($221) in return. Poor rains, however, would reduce the value of both A's and B's crops to a 6000 Rs. ($133) return. Without reasonable knowledge of weather trends, there was reduced incentive for a farmer to spend additional money to produce a higher-quality crop from higher-quality seed. Without an accurate weather forecast, farmers tended to err on the side of frugality. The ability to predict rain patterns would therefore make a difference to the quality of soybeans sown by the farmers. The soyachoupal.com site served to reduce farmers' weather-based risks and took the guesswork out of determining the best time to harvest.

Best Practices Page

Here, a farmer could find out what other farmers of similar land area and crop volume were *actually* doing and compare these "actual" practices to the "ideal" practices described on the page. This was done in the simple local Hindi vernacular, not in obscure academic lingo. This way, the farmer could immediately identify the gaps between what he was doing and what he should have been doing. Such practices included how to prepare the soil before sowing and how to space the seeds as they were sown. For example, 18-inch spacing was considered "best practice"; many farmers, however, had been spacing their seed rows nine inches apart, which meant that their crops did not receive proper ventilation or light. Such conditions led to an undernourished crop.

Crop Information Page

This section contained instructional material such as "How to take a good soil sample" and information as to why soil testing was required. It also provided suggestions for further actions to be taken based on soil test results.

Market Information

Four links on this page gave the farmer the options of exploring world demand, world production, *mandi* trading volume, and *mandi* price lists. This way the farmer became involved in the context of his livelihood. He was given knowledge about the *mandi*: the prices (lows and highs), as well as the number of bags that had arrived at the *mandi* to date, and the estimated daily arrivals (usually about 40,000 tons/day in the peak season). The farmer could thus assess the demand for his produce at a particular *mandi*. This information used to be available only from research institutions or corporations. Now it was being provided directly to the farmer. The site also contained a link to the Chicago Board of

Trade, where farmers could find a 7-to-10-day market outlook and track global soybean price trends. One of ITC's strengths was its ability to communicate with the global markets daily. Sivakumar reasoned, "If we had access to this information, why not translate this into another context and share it with the farmers?"

Q & A Forum (FAQs)

Through this interactive feature, the farmer could pose a question, and it would be answered by an appropriate "panel" of experts. Individual farmers with relevant experience also had the chance to answer based on the question's category. Weather-related questions were routed to the meteorological department; crop questions went to four or five agriscientists on the panel. One farmer asked, for example, "Should we do soil testing before soy harvest or after?" The answer was posted to the forum soon after: "You can do soil testing any time but preferably before the rains." (To find out when the rains were due, the farmer could check the weather page on the soyachoupal.com site.) Another farmer wanted to know why the U.S. soybean crop was "so big" in the world. The answer was a photograph of an American soybean field, with special notation on how much space was provided between plants. This way, farmers in India could compare their methods with those of other countries.

The questions from all of the currently operating eChoupals were stored in a central database so farmers in other locations would be able to access and use the information. The computer's storage ability presented a significant advantage over television or radio, both of which had been considered as other options for this kind of knowledge dissemination.

News Page

This contained excerpts of relevant news items, such as the government's decisions on subsidies or minimum support prices (MSPs), and innovations in other countries' farming systems. If a farmer did something that was particularly successful or innovative, that was posted to the news section as well. This recognition provided incentive for the farmers, who would otherwise not have the chance to be heard, to try new things.

Lastly, www.soyachoupal.com had a place for suggestions. One of the advantages of the system was that the soyachoupal.com site could be continually tailored to the needs of the farmers. ITC had relied on farmers' input since the start of the project, and, in an effort to keep the site's content dynamic and relevant, it was important that the farmers could continue to be involved in its improvement.

The eChoupal served different purposes over the changing seasons. At harvest time, for example, farmers were more concerned about prices, at sowing time they were more concerned about weather forecasts. The eChoupal initiative was based on the belief that the farmer needed an alternative to the *mandi* system. By participating in the eChoupal network, farmers were offered new channels through which they could sell directly to ITC, thus eliminating the cost inflation and cheating that occurred through the middlemen. ITC selected a lead farmer in the village to become the caretaker of this equipment and a liaison between ITC and the farmers. This lead farmer, designated *sanchalak*, was ideally someone like Kamal Chand Jain, someone who was well-known in the village and whose home had already been a natural *choupal* platform for years. The sanchalak had to be someone whom people felt comfortable visiting and someone who, in turn, welcomed such gatherings in his or her home.

Sanchalak

In the evening, 15–20 people at a time showed up at the sanchalak's home for the usual *choupal* gathering. The ITC computer system that had been put in place offered new impetus to the discussions. In addition to the regular chatter, the sanchalak would use his* assigned user name and password to access www.soyachoupal.com and share with his neighbors the interactive features of the site. The sanchalak had received some basic IT training from ITC, as well as instruction in effective methods of communication. This qualified him to open the site to other farmers, who could then navigate the site themselves. The log-in feature was designed with the idea of offering customized content based on the log-in location; in the future, if the content were to evolve into a more personalized form, individual farmers would log in themselves. Until then, however, the sanchalak, distinguished for his literacy and communication skills, served as the liaison between ITC and the farmers.

Weather forecasts and *mandi* transactions were printed out and posted on a notice board in the sanchalak's house. This way, the sanchalak did not need to open the website with the arrival of every visitor. Farmers could stop in and read the printed information at any time throughout the day. If a farmer was unable to read, he only had to ask the sanchalak's advice. It was in the sanchalak's best interest to advise the farmer *correctly*, for better-quality produce from each farmer would fetch a higher price from ITC, and this meant a greater commission for the sanchalak, as well as supporting his reputation as an honest broker.

Farmers brought samples of their soybean crop to the sanchalak's home, where he was equipped by ITC with moisture meters and other tools used to assess the quality of the beans. The company provided "control samples" with which the sanchalak could perform a quality comparison. Each sanchalak was trained (as part of orientation) in quality assessment of his particular crop, so he would be qualified to judge the material based on damage or foreign matter. The soyachoupal.com website provided the "best material price" for "best-quality" beans, so when farmers brought their crop sample to the sanchalak, he priced the material based on its degree of variance from that "best-quality" example. The sanchalak then determined whether or not each farmer's sample matched what he had learned to identify as the best material. Using the samples, he could physically show the farmers: If you grow it like *this*, you will get a better price.

REORGANIZING THE SUPPLY CHAIN

The physical setup of the eChoupal kiosks facilitated a new kind of supply chain, of which technology was at the crux. Trading outside the *mandi*, for example, was very difficult before the eChoupal. First, the *mandi* provided the only means

* While most sanchalaks were men, there were also women who had taken on the role.

for price discovery, and farmers reasonably assumed they would fare best in open auction. Second, transactions outside the *mandi* were officially prohibited by the Agricultural Produce Marketing Act. The government had confined agricultural transactions to the *mandis* to protect the farmers from exploitation by unscrupulous buyers. Open auctions were considered the best safeguard against this. At the conception of the eChoupal, however, ITC was able to convince the government of the potential benefits to the farmers and the economy, and the government amended the act to legalize purchases of beans (and other agricultural commodities) outside the *mandi.* The transparency of the eChoupal—the fact that the website was accessible to anyone, including the government, to cross-check ITC's prices at any time—facilitated the government's acceptance of the initiative.

The web technology brought price discovery to the village level. This changed the way farmers did business. First, empowered with the knowledge of what price he would get at an ITC hub, as well as the reports on prices at nearby *mandis*, the farmer was able to make an informed decision about where to go to sell his beans. This knowledge was important, given the costs associated with traveling to the *mandi* with the beans. By learning about prices in the village itself, the farmer could determine how his revenue would compare to the cost of transportation. If he felt he could get a better deal at the *mandi* through the open auction process, he could choose to go there. But given the uncertainty of the *mandi* versus the set published prices offered by ITC's hubs, in addition to the perk of being reimbursed for transport cost, farmers began regularly defecting from the *mandis* and choosing ITC. Second, by following the real-time prices on the website, the farmers could decide *when* to sell. Knowing the price in advance meant that the farmer could go to an ITC hub (assuming he was happy with ITC's price) on his own schedule, even if there were no other reasonable bids on the beans at the *mandi.*

A third feature distinguishing the eChoupal was its transparency. It is arguable that prices could be communicated to farmers by other means, such as telephone or radio broadcast. These methods, however, still relied upon spoken word. The ability to actually see prices being offered, in writing, on the computer screen (in spite of the illiteracy of some of the farmers), was instrumental in establishing the trustworthiness that made the eChoupal effective. The web model was also more scalable, since one kiosk could be used by hundreds of farmers.

Without price discovery via the web portal in the villages, selling to ITC hubs directly might be little different than going to the *mandi.* In fact, the *mandi* might be more attractive because the farmers would have the opportunity to sell in an open auction. But with the eChoupal, the farmer was able to make his own informed choice. ITC worked to make its hubs attractive destinations for farmers. In addition to competitive pricing, the hubs contained multiple amenities that were not available to farmers in the *mandis.*

ITC Hubs

ITC had five processing units in Madhya Pradesh and 39 warehouses, making a total of 44 points to which a farmer could bring his soybeans. This compared to 51 large soybean-based *mandis* in the state. The farmers traveled an average of 20 kilometers to reach a hub, the range being five to 30 kilometers. ITC had 1695 eChoupals in Madhya Pradesh, covering 8400 villages and reaching 80% of soybean- and wheat-growing areas in the state. The physical architecture of the eChoupal model called for a web kiosk within walking distance (less than 5 kilometers) and a hub within driving distance (less than 30 kilometers) of every targeted farmer. To make sure this was fulfilled, ITC added three processing hubs and 36 warehouses in Madhya Pradesh after the eChoupal project got under way. The distance between farms and hubs was about the same as what a farmer would travel to get to a *mandi*.

Once the farmer arrived at one of ITC's hubs, his beans were weighed on a computerized weighbridge, and the weight was multiplied by ITC's published price. The farmer then received cash on delivery. ITC maintained enough cash in a secure kiosk at the processing plant so that the farmer was fairly and immediately paid. In addition, the farmer was reimbursed for the cost incurred transporting his material to the factory. Depending on how far the farmer had traveled, ITC repaid him based on fixed freight-cost parameters, and that sum was added to the payment for the produce.

Simple amenities at the ITC processing plant made the experience considerably more pleasant than at the *mandi* alternative. After the soybeans had been weighed, a tented seating area provided the farmer with a shaded spot for him to sit and await payment. Restroom facilities were available. None of these existed at the traditional *mandi*. In addition, there were 15-liter jugs of soybean oil available for purchase. ITC made a point of saying, "This is your oil; this was made from your beans." When the farmer bought soybean oil directly from ITC, he skipped four or five people in *that* supply chain, keeping his own purchasing costs to a minimum. The oil was pure and unadulterated because it came directly from the ITC factory.

As an added convenience, the processing facility included a soil-testing lab on the premises, where scientists offered recommendations for fertilizers or additives based on the chemical composition of the farmer's sample. This took three days, while the alternative—going to a government lab—would take longer. Scientists employed by ITC made recommendations on the nutrient dosages that the soil needed based on its properties. They did not recommend any specific brand of fertilizer; they just gave the farmer the soil properties, and then the farmer could choose his own brand of fertilizer based on his soil composition. Freedom of choice was an important principle of the eChoupal concept. Trading manager Kumar said,

> Visiting the eChoupal is not an extra job; this is part of [the farmer's] routine. Their routine is their agriculture. Before, if they wanted information, they had to go to town and ask somebody. Now, we are bringing the information into the village, into the home. It's natural for them.

For his involvement in the ITC procurement process, ITC paid the Sanchalak a 0.5% commission on the sale of soybeans. He was, after all, effectively doing ITC's buying—buying that would otherwise have taken place at the *mandi*.

The eChoupal system effectively turned the sanchalak into an entrepreneur, for he also had the opportunity to earn a 2–3% commission on orders placed for input items, such as herbicides, sowing seeds, and fertilizers, provided through ITC. Instant-glow gas lanterns and edible oil were also popular sells. With the help of commission from edible soybean oil sales, Dahod Sanchalak Jain earned about Rs. 35,000 ($775) over the three to four months of the 2002 soybean season. "That's a good amount of money for him," said Kumar.

Samyojak

ITC benefited because of the increase in turnover resulting from the sanchalaks organizing and mobilizing farmers to sell to the company. The farmers were happy to have a better-defined channel through which to sell. But what of the middlemen, the commission agents of *mandi* life? All of ITC's CAs were kept, albeit with a new title: *samyojak*. In this role, former CAs were given additional money-making opportunities within the eChoupal system. ITC mandated that its CAs become samyojaks if an eChoupal was being set up in their geographic area. In most cases the transformation was achieved by convincing CAs of the potential revenue to be gained through the transactions in the eChoupal. The samyojak role comprised three major areas of responsibility: 1) setting up the eChoupals; 2) facilitating ITC's purchasing transactions; and 3) helping with ITC's selling transactions.

In the establishment of new eChoupals, the samyojak assisted ITC teams in village surveys to lay the groundwork. This meant assisting in the selection of the sanchalak, acting as a liaison between villagers and ITC, and helping villagers understand the potential for the new system to be more efficient and profitable for all parties. Samyojaks also managed warehousing hubs attached to the processing facilities that stored bought soybeans. They assisted in the logistics of the cash disbursements to farmers arriving at the processing facility. They also helped facilitate transportation links for farmers who could not reach the ITC processing facility themselves.

ITC's selling transactions comprised the "one-stop-shop" feature of the eChoupal. Farmers could buy herbicides, sowing seeds, gas lanterns, fertilizers, and soybean oil, among other sundries, directly from the company. While the sanchalak had the responsibility of aggregating the orders in his village, the samyojak would assist in actually moving the goods from ITC's manufacturing units to the eChoupals and/or warehousing hubs.

Three systems enabled ITC to sell and deliver goods to farmers through the eChoupal. The first was at the village level, where the sanchalak aggregated demand for products through orders placed by his fellow farmers. This was done during the traditional *choupal* time. The sanchalak then e-mailed the order to ITC, and the items were either a) picked up by the sanchalak at the ITC warehousing hub, or b) delivered by the samyojak to the villages. In either case the sanchalak collected cash payments from his neighbors and remitted them to ITC. Seeds and fertilizers were sold in this way.

The second system did not involve any prior orders. Instead, the sanchalak bought products based on estimated demand and stocked them in his home.

Products sold this way were again procured by either sanchalak pickup or samyojak delivery. This system was most effective for consumer goods such as salt, matchboxes, soybean oil, and confectionary items. These were all ITC products, and both the company and the sanchalak earned a fee from any sales.

The third system for selling additional goods through the eChoupal was "shopping" for the products at the ITC processing facility. When sanchalaks and farmers visited the ITC facility to sell their produce, they also had the opportunity to peruse the warehousing hubs for items on which they might like to spend their freshly earned cash. Samyojaks managed these warehouses, assisting ITC in creating retail storefronts in the setup for 2003. This interactive feature of the eChoupal system created an opportunity for ITC, the sanchalak, and the samyojak to turn profits that were simply not possible under the traditional system. The sanchalak's job of arranging and mobilizing the farmers to take their soybeans directly to the ITC processing facility meant greater revenue for the company and commission for the sanchalak. It was estimated that ITC saved $5/ton on freight cost; from those savings, ITC would reimburse the farmer for the time it took to travel to the ITC facility. The farmer, in turn, earned an average increase of $8/ton. Chairman Y. C. Deveshwar said,

> By creatively reorganizing the roles of traditional intermediaries who deliver critical value in tasks like logistic management at very low costs in a weak infrastructure economy like India, the eChoupal ploughs back a larger share of consumer price to the farmer. Besides providing an alternative marketing channel, this model engenders efficiency in the functioning of *mandis* through competition and serves to conserve public resources that would otherwise be needed to upgrade the *mandi* infrastructure to handle higher volumes of agri output.

For the samyojak, income potential in the eChoupal system exceeded that of a CA in a *mandi*. ITC paid 1% of the transaction value to the CA when buying through the *mandi*. The typical CA turned his cash three times in a month during October to December (the time of peak activity), meaning that his gross return came to 3% per month for three months. Thereafter, money lay idle for the most part, except during May and June, when wheat and pulses were harvested. For an ITC samyojak, there were more frequent opportunities to earn commission off farm inputs and other products being sold through the eChoupal. Commissions ranged from 2% to 5% per transaction, depending on the product. Samyojaks who assisted ITC in cash disbursements at processing units earned a fee of 1% of the transaction. The advantage of this was that the samyojak now had the potential to earn from his working capital year-round. The eChoupal leveraged the physical capabilities of the current middlemen while removing them from the flow of information and market signals. CAs-turned-samyojaks had good terrain knowledge as well as long-standing relationships with villagers, which ITC also hoped to leverage. Samyojaks continued to operate out of *mandis* as hub points, allowing the eChoupal system to coexist with *mandis*.

In fact, ITC continued to buy from the *mandis* even as the eChoupal gathered steam. By 2003, ITC was procuring about 50% of its soybeans from the *mandis* and 50% from the eChoupal. The company hoped to shift the ratio to about 20%

mandi, 80% eChoupal, as new hubs were added to the eChoupal system. The hubs had limited capacity, and, at certain points in the season, when market prices for soybeans were particularly low, ITC benefited from buying from the *mandi* as well as the eChoupal in order to maximize procurement. Furthermore, the *mandi* provided a source of "market intelligence" for ITC, as samyojaks continued to operate there as well. Samyojaks still got a significant amount of income from their work in *mandis*; they could not yet afford to work solely for ITC.

PRINCIPLES OF THE eCHOUPAL

The eChoupal depended strongly on trust. The website was simply a medium for the more important human element, the interaction. Interaction was, after all, the initial driver of the *choupal* concept. No contract bound farmers to sell to ITC once they used the website. ITC did not ask for commitment; the farmer was free to do as he wished. In fact, the farmer could use all of the eChoupal's facilities, soak in all the information, and still choose to take his crop to the *mandi*.

The bet was that, once given these tools, farmers would realize on their own that selling directly to ITC was the best alternative to the *mandi*. Rajnikant Rai, vice president of trading in IBD Hyderabad, commented:

> We feel that this is how we can win people in the long run, by giving them the tools. Communication and information are developing; mobile phones are here, the Internet is here, we must use these things for education, and let there be no question of hiding information. We let the farmers understand and let them decide who is best in an open, competitive market scenario. Let them decide. They are the judges.

To ensure the sanchalak's integrity in the process, when he received the computer equipment he took an oath that connected him to ITC. At this oath-taking event, the sanchalak pledged, before the entire village, to uphold ITC's high standards, not to use the computer for "wrongful" purposes, and to maintain the ethics, image, and concept that ITC has created through the soyachoupal.com site. In addition, the ITC logo was painted on the front of the sanchalak's home. This vibrant green and yellow mural stretched across the wall from floor to ceiling, identifying the sanchalak as a liaison of ITC. Through these steps, ITC created a sense of pride and responsibility that infused the sanchalak nominations. The sanchalak became a highly visible figure after the ITC overhaul; as a result, potentially unscrupulous sanchalaks were deterred from taking advantage of the system. If a sanchalak were to act dishonestly at any point, it would create uproar throughout the village, and ITC would be able to take immediate action. Because the role was viewed as an honor, however, sanchalaks usually treated their duties with the utmost pride and seriousness.

There were many benefits realized by ITC when the company found a way to buy directly from farmers. First, ITC had more control over the quality of the product it sourced. Direct contact with the farmer enabled knowledge sharing in terms of best practice for sowing, irrigation, and harvesting; therefore, ITC had some additional leverage over what was available to them for purchase. Higher-quality produce, of course, enabled more competitive pricing in the international market. Furthermore, buying directly from the farmer reduced the chance of the produce being adulterated with impurities (which often happened with middlemen). Deveshwar said, "ITC has demonstrated that it is possible, nay, most crucial to combine the need for creating shareholder value with the superordinate goal of creating national value."

HUNTING FOR GROWTH

The imperative for growth promulgated by Deveshwar had naturally prompted Sivakumar to seek additional applications of the eChoupal concept. Sivakumar was anxious to see the success of the soybean eChoupal model leveraged into IBD's three other product groups: coffee and spices, aqua food, and value-added horticultural products. Sivakumar thought hard about other commodities. While the basic character of agriculture was the same across India, value chains of different crops had their own intricate dynamics, as did the socioeconomic conditions of each region.

Soon after soybeans started showing promise, ITC had set up pilot eChoupals in three other crops in three different regions of India that were as diverse from one another as possible and representative of all crops in ITC's product portfolio: coffee in Karnataka, aqua (seafood) in Andhra Pradesh, and wheat in Uttar Pradesh. Lessons from these pilots were expected to help the company scale up on a national level. Each of these projects shared a common management approach with respect to their scale and scope: First, pilot test the concept in a small number of villages; second, make changes based on the learning from the pilot phase and validate them in a larger number of villages; and third, grow the project to reach as many villages as possible and saturate the region. ITC called this approach "Roll Out, Fix It, Scale Up."

Other Commodities

As seen with soybeans, margins could be generated in many other commodities through logistics cost savings between the farm and the factory, where non-value-added activities were eliminated. While these savings could be instantly realized, Sivakumar wondered if they were sustainable over a period of time, as the savings were benchmarked against the current inefficient market. And the market was bound to become more efficient. He reasoned that ITC could potentially generate value via three other primary mechanisms: traceability (i.e., accountability for the quality of the product vis-à-vis its source), ability to match farmer production to consumer demand, and facilitation of an electronic marketplace. The three new eChoupal models were essentially a validation of these mechanisms.

With aqua products, or seafood, traceability provided an opportunity for ITC to generate value (and additional revenue) through the eChoupal. Global consumers of India's seafood would pay premium prices for shrimp that could be traced back to its source. If ITC were able to tell customers not only where a given commodity came from but also how it was produced (e.g., with antibiotics or not), significant gains would be possible. By controlling the source, ITC could guarantee the safety and sanitation of the product and thus receive higher prices.

Similarly, when ITC had greater knowledge of just *what* it was purchasing when procuring crops from rural farmers, value increased for both the company and the farmer. The wheat market was an example of this. Wheat varied greatly in both chemical composition and physical appearance, and, through the eChoupal, farmers learned to recognize which physical characteristics represented certain chemical qualities, such as gluten, protein, or starch content. Customers placed orders based on these chemical qualities. When ITC became able to analyze the crop *before* purchase, at the farm level, and then purchase and store it by chemical-composition category, there was an opportunity for cost savings, as it was expensive to separate wheat after it had been purchased and aggregated at the *mandi*. The farmer, too, benefited from this education. By identifying his high- and low-quality wheat, he could then price the varieties appropriately. He could command a higher price for the high-quality wheat and offer low-quality wheat at a reasonable price to customers who might need it, say, for animal feed. This way, he did not have to charge one *low* price for an amalgamation of wheat of disparate quality.

Coffee presented a new challenge to ITC. Coffee was an estate crop, grown by a large number of small-scale farmers. ITC had a deep knowledge of coffee farm practices; much research had already been done on the industry. The price volatility of coffee was high; variance from base price could reach 40% (compared with 20% volatility in the soybean market), and buyers would routinely renege on contracts if prices altered beyond tolerance. Market participants were savvy speculators. The importance of an agent in coffee transactions was paramount, and effective price discovery was often the critical part of a deal. With its electronic-trading platform, called Tradersnet, ITC improved real-time price discovery by hosting anonymous trades and letting the prevailing selling prices be known. Information sharing carried over to ITC's customers as well. "The task of adapting the eChoupal concept for different crops and regions continues to test ITC's entrepreneurial capabilities," said Chairman Deveshwar.

eCHOUPAL AS A MARKETING CHANNEL

As Sivakumar pondered the relative potential for eChoupal in each of the four commodities, he also turned his eye to the long-term future of a wired rural India. As he clicked through the constantly changing pages of the eChoupal's website, he wondered if marketing and distribution to the 60% of India's workforce living in rural areas might be the real growth engine for ITC.

ITC's vision for marketing via the eChoupal involved three features: superior product and distinctive functional benefits, process benefits (simplified transactions between buyer and seller), and relationship benefits (farmers' willingness to identify themselves and reveal their purchasing behavior). ITC had conceived ideas for various input items that could be developed for new business given this framework. The company believed that these products could be made available to farmers through the eChoupal, thus increasing the value of the farmer's product as well as generating additional revenue for ITC. Deveshwar called this philosophy a "commitment beyond the market."

Fertilizers

Farmers spent an average of Rs. 26,000 crore ($5.7 billion) annually on urea, diammonium phosphate (DAP), and muriate of potash (MOP). Still, farmers could not easily access the fertilizers they needed. Thirty-five percent of DAP and 100% of MOP were imported. Logistics had proved to be complicated for most companies. They were unable to access many rural markets because of fragmented, or nonexistent, distribution channels.

Agrichemicals

Farmers also spent Rs. 3500 crore ($774.5 million)/year on insecticides, herbicides, and fungicides. The agrichemical market was highly fragmented and consolidated by multinational corporations such as Dupont, Novartis, and Cyanamid. New chemicals were introduced frequently; however, their life cycles in the market were only two or three years. Given the short product cycle, big companies needed immediate market access. Farmers, too, suffered when they could not access these products. High costs of labor, for example, hurt soybean and wheat farmers whose fields could have been covered with herbicides, instead of weeding workers, at a lower cost.

Seeds

This was a relatively small, fragmented market of Rs. 3000 crore ($663.9 million)/year, but only 4% of farms used commercial seed. Government-promoted seed corporations made different types of seeds available though cooperatives, and large multinational companies had entered the market with better-quality material. Still, there were lead times of up to three years to make seed varieties available to rural farmers.

Insurance

At the time, Indians were collectively paying Rs. 50,000 crore ($11 billion) in yearly life insurance premiums, and that market was expected to reach Rs. 150,000 crore ($33 billion) by 2010.* Life Insurance Corporation (LIC) was a government-run insurance provider that had already taken a shine to rural markets. Even in the relatively poor states of West Bengal and Bihar, 6 million rural farmers had taken out policies. In 2000, private insurance companies were allowed to enter the market, and, as a result, at least 12 new companies were seeking to expand their business into rural India to compete with LIC. By 2003, rural business comprised 16% of LIC's portfolio, but only 9% of private companies' portfolios. These markets were largely untapped because of a lack of trustworthy intermediaries. ITC believed that it could create a relationship of trust and help farmers understand the rules and benefits of insurance plans. Eventually, ITC envisioned Sanchalaks being able to offer the eChoupal infrastructure to LIC agents for a fee or to set up its own insurance brokerage company.

* Company information.

An opportunity also existed for other types of insurance policies, covering fire, marine, motor, and workmen's compensation. Insurers, however, had been biased toward larger accounts, leaving less prosperous farmers unable to participate. Insurers lacked quality data on risks and parameters of farm life and were hesitant to insure rural customers. With ITC as a liaison, data on rural farmers could be delivered to insurance companies, thus demystifying and uncovering the rural market.

Credit

A national survey in 2001 showed that Indians were saving about 30% of their annual income, though not through financial institutions. Both private- and public-sector banks lacked a customer-friendly approach and were often avoided by rural farmers. One of the main reasons that farmers avoided saving through banks was that they were often linked to a loan.* If a farmer had savings in the same bank he had borrowed from, the bank could demand that he use those savings to pay back the loan. Oftentimes farmers would rather defer the loan and simply save their cash in their homes, unbeknownst to the bank.

ITC believed that the system of trust engendered through the sanchalak would facilitate financial transactions. It could channel rural farmers into the mutual fund arena and earn a commission from banks on farmers' investments, using the technology introduced in the eChoupal. *Choupal* discussions would create data on likelihood to invest, and the results would be stored in a data warehouse for future campaigns.

Sustaining Success

Sivakumar wanted to know how to evaluate these new business opportunities for ITC. He knew that the eChoupal could not be everything to everyone, and he wanted to allocate ITC resources constructively. Once the sourcing of their core commodities was sufficiently strengthened, and cost-effective supply chains were running efficiently through the eChoupal, other business opportunities would develop. The poor and fragmented rural consumers were traditionally underserved, but the demand was growing with rising aspiration levels triggered by, for example, broader advertising. Deveshwar said "The pioneering eChoupal business model contributes to creating the market through improved farm incomes, whilst placing ITC in a unique position to reap benefits through its closeness to the potential consumer."

Sivakumar also needed to consider the sustainability of the existing eChoupal network and its success to date. Many nonprofit organizations had tried to introduce technology to rural India, but had not been able to sustain their initiatives. Computer equipment was expensive and had a finite lifespan. After a few years it would have to be replaced. Should (and could) ITC bear the cost of continually replacing IT equipment? ITC was convinced that as more farmers bought

* Total crop loans in 2001 came to Rs. 33,000 crore ($7 billion).

products through the upstream channel, many other marketers would follow. But how should ITC best jump-start its upstream commerce initiative to ensure its long-term success?

CASE STUDY 2 Akshaya Patra: Feeding India's Schoolchildren

David M. Upton, Christine Ellis, Sarah Lucas, and Amy Yamner

The big thing to worry about now is scalability. How to run such a huge operation on a consistent basis with worries about quality and consistency of food... We've gone from 60,000 to over 500,000 children. Now we want to get to one million. How do we do that?

Raj Kondur (**MBA '97**)
Akshaya Patra Board of Trustees, and CEO,
Nirvana Business Solutions

Bangalore, India: CC Das, Program Director of the Akshaya Patra Foundation, felt the heat and humidity from the kitchen. At 6:00 a.m., the kitchen was bustling. Workers from poor local neighborhoods transferred hot cooked rice from industrial-size boilers to individual delivery containers, while other employees stirred spices into large vats of simmering mixed vegetables and broth. Trucks waited outside, ready to load the proper quantities of food for their daily routes. With only an hour of cooking remaining, everything was going according to plan. CC Das breathed in the warm and fragrant aromas and was reminded that, while six years ago this operation was just an idea, the kitchen now fed 145,000 schoolchildren each day. He mulled over how Akshaya Patra could continue to construct new kitchens and grow in other regions.

Baran Village, India: Over 1000 miles away, in the rural Baran district of Rajasthan, Trilok Gautham, executive supervisor of the Baran program, had just arrived at his first stop of the day. He stood in the simple room that served as a kitchen for a village school. Every morning four brightly clad women prepared food for the schoolchildren, providing most of the children with the only meal they would have that day. The team alternated duties, sharing the chores of rolling dough into flat bread, tending the fire, and chopping vegetables for the *daal* (lentils).

Gautham had seven more schools to inspect that day, part of his routine to ensure cleanliness, hygiene, food safety, and quality for each of the 79 schools he covered. Less than a year ago, no food was served during the school day and children often went hungry. Today, 15,000 children in Baran would have a hot meal. On his dusty 40-minute bicycle ride to the nearest village, Gautham thought about the challenges of providing services to additional villages. There was no doubt in either of their minds that things were going well for Akshaya

Patra. But what were the implications of their current success? In one week the Board of Trustees would gather to discuss options for growth. No one questioned the need to expand services to feed more children. The question was, "How?"

INDIA OVERVIEW

Bordering the Arabian Sea and the Bay of Bengal, the Republic of India had a land mass of 3.2 million square kilometers, about one-third the size of the United States, and a population of approximately 1.1 billion, over three times as many people as the U.S.* For many years, India had been a colony of the British Empire. Nonviolent resistance to colonialism under the leadership of Mahatma Gandhi brought independence in 1947.[†] Since gaining independence, a succession of Indian governments worked to spur economic growth. Recent annual GDP growth of about 6.5% helped the country reduce by half the proportion of people living on less than a dollar a day. Nevertheless, 80% of India's population lived in rural areas, and poverty was concentrated largely in the regions that were often the most challenging to serve.[‡] According to a United Nations report, there remained many social needs to address in India, especially in the areas of health, primary education, and gender equality.[§] India faced multiple challenges in the education of its poor. The overall adult literacy rate was 61% in 2004. A gender disparity in literacy was prevalent, with male literacy rates at around 73% and female rates at 47%.[¶]

According to USAID, India had the world's largest concentration of desperately poor people. More than 300 million Indians lived in abject poverty, a number which exceeded the impoverished population in Africa and Latin America combined. Even with a range of other food assistance programs in place throughout the nation, India's efforts to feed the poor often did not reach the most vulnerable sections of the population.**

HISTORY OF MIDDAY MEAL PROGRAMS IN INDIA[††]

Midday meal programs (school lunch) emerged to address the multiple challenges of poverty, hunger, and access to education. Prior to receiving midday meals, many impoverished children performed poorly in school due to short attention spans associated with extreme hunger. Other children either did not enroll in school, or dropped out at a young age choosing to seek work during the school day to earn money to feed themselves and their families.

---

* http://www.cia.gov/cia/publications/factbook/geos/in.html

† CIA World Fact Book

‡ http://www.worldbank.org.in/WBSITE/EXTERNAL/COUNTRIES/SOUTHASIAEXT/INDIAEXTN/0,menuPK:295591~pagePK:141132~piPK:141121~theSitePK:295584,00.html

§ Mid-term appraisal of the Tenth Five Year Plan—Annexure 2.2.1 Table on "Progress in achieving the MDGs."

¶ http://www.uis.unesco.org/profiles/EN/EDU/countryProfile_en.aspx?code=3560

** http://www.wfp.org/country_brief/indexcountry.asp?country=356#, PDF document

†† The data in this section was derived from http://www.righttofoodindia.org/data/wsfmdm.pdf unless otherwise stated.

Although 50% of India's children were malnourished, the provision of midday meals was sporadic and in many places nonexistent. Responding to pressure from the Indian people, the Supreme Court of India passed an order on November 28, 2001, which mandated: "Cooked midday meal is to be provided in all the government and government-aided primary schools in all the states."

Inconsistent food quality, occasional food poisoning, poor hygiene, and operational concerns were among the complications to the provision of government-sponsored midday meals. The meals were prepared by teachers, who cooked the same meal every day: *ghoogri*, gruel made of boiled wheat. Children reported that that they grew tired of eating the same food daily, they did not like the taste, and it often made them feel sick. In 2004, a fire accidentally started by a teacher cooking the midday meal killed 90 children in Tamil Nadu, an event which underscored the safety issues inherent in meals prepared in makeshift kitchens based on school sites.*

By January 2004, nearly 50 million children received midday meals provided either by the government or by NGOs working in partnership with the government. However, given the scope of hunger in India and the difficulties faced by the government programs, the task of feeding schoolchildren was still a significant challenge.

AKSHAYA PATRA

In 2000, the Akshaya Patra Foundation (TAPF) was founded to address the dual challenges of hunger and education in India.† The organization provided nutrition-rich midday meals to extremely underprivileged children in India with the aim of increasing school enrollment, reducing drop-out rates, and improving academic performance. The Akshaya Patra program had a simple vision: "No child in India should be deprived of education because of hunger." Because of India's enormous population, this vision was difficult to realize.

Expansion of Operations

Akshaya Patra began feeding 1500 students in five schools in Bangalore. It was one of the first organizations in the region to provide freshly cooked, hot, nutritious, and balanced meals, and within six months of starting the program, it had requests from 3000 schools. Akshaya Patra soon scaled up services to feed 30,000 children. When the foundation's growth caught the attention of local government officials, the organization began receiving government financial support. By April 2003, it was feeding 43,000 children in Bangalore daily.

Akshaya Patra recognized the need for midday meal programs in other parts of the country and expanded the program to other areas. In August 2003, it opened a kitchen in Vrindavan, in northern India. In July 2004, in partnership with Mrs. Sudha Murty, the chairperson of Infosys Foundation, it began a

* http://washingtontimes.com/upi-breaking/20040719-111147-6417r.htm

† The founders desired a name for the organization that reflected their aspirations to provide unlimited food to underprivileged schoolchildren. They chose to use the Sanskrit term "Akshaya Patra," meaning "abundant and inexhaustible."

midday meal program in Hubli-Dharwad. By November 2004, Akshaya Patra had also commenced a pilot program in 25 schools feeding 5200 children in Jaipur, Rajasthan.

As the urban operations grew, Akshaya Patra recognized that in order to reach the majority of India's most undernourished children, it also needed to serve the rural districts. In August 2005, Akshaya Patra began services in the region of Baran, located in east Rajasthan, in response to the number of starvation deaths in the area. Because very few midday meal programs operated in rural districts, the Baran program was an experiment. Neither the government nor Akshaya Patra was able to serve these communities adequately on its own, but as partners, their mutual goal was achievable. With rapid growth in both urban and rural areas, TAPF had expanded to each day feed 567,622 children (equivalent to more than half the population of Boston) in 2000 schools in ten locations in India by March 2007.

Management and Funding

Despite growing in scale, TAPF's seven founders maintained the original Board of Trustees, consisting of three volunteers and four senior executives. The board provided strategic advice and worked on performance improvement. Although headquartered in Bangalore, it made operational decisions based on reports received from the managers at each location. Akshaya Patra was funded through a combination of government subsidies and private donations. The organization received

- 2.6 kilograms of rice or wheat per child for students in class 1 to 7 and excise duty exemption from the central government of India
- Rs 1.31 for students in class 1 to 7—state government of Karnataka
- Rs 1 for students in class 1 to 5—state government of Uttar Pradesh
- Rs 1.50 for students in class 1 to 5 and sales and road tax exemption—state government of Rajasthan
- Rs 1.65 for students in class 1 to 5 and sales tax exemption—state government of Orissa

In addition, the government also gave 100% income tax exemption for donations made to the Akshaya Patra program under section 35 AC/80GGA (bb) of the Indian tax code. The remainder of the funding came from corporate and individual donor contributions. By March 2007, there were over 16,000 private donors. While funding was coordinated out of the headquarters in Bangalore, the organization hoped to widen the footprint of its fundraising base. As it expanded to additional cities throughout India, Akshaya Patra's aim was that each location becomes self-sustaining. In addition, Akshaya Patra initiated fund-raising efforts in the United States, with a particular focus on Indian-American donors.

Maintaining Cost Efficiencies

When Akshaya Patra first began serving the rural areas, it cost 10 rupees per day to feed each child. Comparatively, the cost was 6 rupees with the centralized, urban model, which included raw materials, labor, distribution, and administrative overhead. Increased transportation costs associated with the lack of road infrastructure and the dispersed locations of schools made the decentralized model less cost-effective than the centralized model. Additionally, Akshaya Patra was unable to achieve economies of scale through vendor relationships and administration.

Site Selection

Schools were selected based on demonstrated need and physical location. With a centralized model of food preparation and delivery in the urban areas, schools that were close to each other allowed more efficient delivery of meals. It also improved transportation times when delivering food. In the rural areas, few schools were located nearby one another. Distance and location prohibited delivery from a central location, and thus were not primary factors in rural school selection.

Constant Learning and Improvement

Since its early development occurred largely through trial-and-error, a culture of constant improvement and learning was adopted throughout the organization and affected everything from kitchen design to delivery of service. Although the menu was standardized early on, much experimentation went into the recipe creation. Finding the appropriate spiciness to suit all tastes was one example of Akshaya Patra's use of trial-and-error.

At each school, a distribution supervisor was responsible for handling school complaints. Akshaya Patra received one to two complaints per week, often related to the time of delivery, insufficient quantity of food, the taste of the food, or the quality of the rice. The distribution supervisors determined if the problem lay with the school or with the organization and worked to find a solution.

Distribution supervisors provided feedback from schools and helped to implement necessary changes. At one school, for example, a teacher noticed that students were not eating the vegetables. Upon investigation, Akshaya Patra learned that the vegetables were too big for the children to chew comfortably and recalibrated the vegetable cutting machine to ensure smaller-sized pieces. The teacher subsequently reported increased vegetable consumption. Similarly, several schools complained that the curd tasted sour. Aware that they had taken great care to ensure that the time of transportation and vehicles would not allow spoilage to occur, Akshaya Patra approached the curd supplier and determined it was the quality of the curds that was leaving a bad taste. A new supplier remedied the problem.

Worker initiatives likewise led to improvements in kitchen design and operations. When workers complained that the Jaipur kitchen was hot, the organization found a solution. Supervisor Rajindar Sharma explained, "Because of the weather, everyone felt very hot. The whole kitchen became like an oven. At first, we panicked. What could we do? Then we decided to introduce coolers. It is better now." Similarly, a worker noted that if the vessels for the *chapatis* (similar to wheat tortillas) were lined, the *chapatis* would not get as dry by the time they reached the schools. His idea to line each vessel with paper succeeded in keeping the *chapatis* moist.

Dual-Pronged Distribution Strategy

Akshaya Patra first provided food to schoolchildren through a centralized kitchen in Bangalore, a bustling urban center. Using a hub-and-spoke model, they cooked mass quantities of food and distributed smaller amounts to individual schools in the surrounding slum and village areas.

As Akshaya Patra expanded services to the rural districts, a centralized model proved inefficient. Because of the dispersed geography of the villages in the rural districts, Akshaya Patra designed a decentralized model in which they built small kitchens to serve the local schoolchildren in each village. Through this two-pronged distribution strategy, Akshaya Patra was able to design services to fit the contrasting needs of the urban and rural regions.

Centralized Model: Operations in Bangalore

The Bangalore kitchen was designed by a team of expert engineers, and many modifications to the original design were made as the organization grew in scope and scale. The Bangalore kitchen was initially intended to feed 1500 students. By 2007, the kitchen was feeding 145,000 children daily. As Akshaya Patra increased the number of children fed, it increased the capacity of the kitchen. According to the operations manager, Bangalore was an "evolved kitchen" which had exceeded its expectations for growth. Changes to the kitchen were made on an as-needed basis. The organization repeatedly stressed the importance of process, design, and high quality of food. They determined that the organization must have replicable, hygienic kitchens that would be productive and process-oriented.

Supply Chain

The daily meals included rice, lentils, vegetables, spices, and curds (yogurt). In the urban areas, vegetables were procured from local markets through an ongoing relationship with third-party vendors. With nutritional balance always in mind, menus varied to incorporate whatever was plentiful at the markets, and thus less expensive. In the Bangalore kitchen, for example, curds were sourced from two different suppliers, with 3500–4000 tons of curd received and distributed daily. The Bangalore storage room was able to store up to three days worth of fresh food and substantially more dry goods such as rice and lentils.

Challenges with the Supply Chain

In many locations, Akshaya Patra received rice as a subsidy from the central government, which it reported was of poor quality. Rice from the Food Corporation of India (FCI) was sold by farmers to the FCI through a middleman. Since price was set by weight, the middlemen often added foreign objects to increase tonnage, including stones, nails, and metal. Akshaya Patra believed that within the supply chain, these objects were added to the product in order to maximize profit and called this practice a "very open secret." Akshaya Patra thoroughly cleaned the government rice by using a destoning machine to separate the rice from other nonfood products and often found that 20% of each bag of rice was unusable.* In the past, Akshaya Patra had avoided poor-quality rice by exchanging the government-issued rice for that which was available in the market and of higher quality, paying cash for the difference in price. However, new regulations prevented them from continuing this practice.

Around-the-Clock Operations

Operations in the Bangalore kitchen began at 10:00 p.m. Sunday evening and continued through Friday evening. The night shift, or the precooking shift, began with the sorting of all ingredients, from vegetables to spices, into the necessary quantities. The number of meals to be cooked was determined daily by the food requirements sheet, which noted any necessary changes in quantity the schools may require compared to the previous day. Though the workers were largely uneducated, the comprehensive standardization and training resulted in efficient and accurate operations.

Preparation and Cooking

Akshaya Patra cooked the traditional food of each region, being sensitive to the needs of the local palate. From the southern Indian Bangalore kitchen, students were served a customary hot meal of rice, *sambhar* (a south Indian native soup with lentils, vegetables and spices), sprouts, *ghee* (clarified butter), and curds.

At 2:30 a.m., steam generator boilers were turned on.† A system of hoses allowed the entire kitchen to be sanitized with hot water and steam; freshly cut vegetables and rice went into boilers; spices were mixed, and sauces were made. For the next six hours, the kitchen was alive with motion preparing 145,000 meals for the day. Every day, 95 batches of rice were cooked. Starting at 2:45 a.m., nine 100 kg-capacity and four 50 kg-capacity steam-powered rice boilers took 20 minutes to cook 100 kilograms of rice. The entire operation was designed to ensure minimal human contact with the rice, and thus minimal chance for contamination.

The *sambhar* contained a mix of vegetables, typically including carrots, potatoes, tomatoes, cabbage, beans, or eggplant. Variety ensured quality as well as economy, while also offering children different tastes each day. The *sambhar* took two hours to cook in 1200-liter containers, enough to feed 6000 students from each vessel. The entire cooking and packing process was completed by 7:15 a.m., and by 10:00 a.m. the kitchen was fully cleaned with preparations for the next day already underway.

* Akshaya Patra reported that people had come to the organization to collect the stones that had been removed from the rice. The foundation did not give these stones back as it worried that they would perpetuate the cycle.

† This actual start time could vary by as much as 30–45 minutes in either direction, depending on the quantity to be cooked as well as the type of rice.

Delivery

The packing supervisor determined the amount of food to go to each school based on figures calculated the evening before. Every morning starting at 2:45 a.m., 22 customized vehicles delivered prepared meals to 145,000 children in 486 schools.* Each vehicle had a driver, two loading/unloading workers, and a security guard. The security guard held the key to the food storage unit of the truck and ensured that the correct amount of food reached each school. He received a signature from the designated person at the school site and verified quantities needed for the next day. Meanwhile, the food was unloaded and brought to the appropriate area of the school where it was served to the students by teachers or other school personnel. The truck completed its delivery route, and on its way back to the Bangalore kitchen stopped at each school to collect the empty vessels.

Each security guard had a mobile phone, which connected him with the central kitchen and was useful when unexpected situations arose. Emergencies included a school that needed additional food, a vehicle breakdown, or a traffic delay. If there was a significant amount of food remaining after all of the school deliveries had been made, the security guard would contact the head office for permission to stop in the slum areas to distribute the excess.† Each day, the amount of time it took to move from one school to the next was tracked, and deviations were documented. TAPF also carefully monitored the fuel consumption of each vehicle.

Replicating the Model

Recognizing the need for midday meal programs in other regions, Akshaya Patra expanded services to northern India. The original Bangalore kitchen was located in southern India, where the diet was primarily rice-based. In contrast, the local diet in the northern region was predominantly wheat-based. Akshaya Patra devised a new menu, with the daily meal consisting typically of *rotis* or *chapatis*, *daal*, or curry and vegetable rice/sweet porridge. TAPF also made additional adaptations to the model depending on the needs of the region. In Jaipur for example, where large-scale unemployment among women was a concern, the kitchen chose to hire 90 women from local villages and only automate some of their operations.

---

* Akshaya Patra owned and maintained 17 vehicles and hired the additional 5. The cost of a vehicle was approximately 1.4 million rupees (45 rupees = US$1). Vehicles were designed to keep food warm and to be dust free. Akshaya Patra had recently moved to a model in which corporations sponsored vehicles.

† Reports were created to make sure that extra food was not prepared. It was not, however, always possible for schools to predict the number of absentees because of the possibility of widespread absences due to festivals, holidays, field trips, or other events. Akshaya Patra reported that this happened almost daily at one or two schools. To cope, they considered hiring an additional vehicle which could go out and meet the vehicle, take the extra food, and do the delivery to the slums while the original vehicle continued its route.

Incorporation of Automation

Akshaya Patra utilized automation and mechanization as much as possible. For example, after the vegetables for the *sambhar* were sorted and cleaned, they moved from the holding vessels to an automatic cutting machine. Imported from Germany, and equipped with a motor comparable to that of a BMW, the machine was able to cut 40 kilograms of potatoes in 60 seconds and could vary both the shape and the size depending on need.

According to Ganesh Thapa, assistant supervisor and one of the kitchen workers, "Initially we had more work because it was all manual cooking. Then automation came to the kitchen. Compared to the early days, it is a relief to have mechanized cooking." A new kitchen at Hubli went a step further. Opened in May 2006, it had the capacity to serve 200,000 children. In order to increase cost-efficiency and decrease the labor needed, this multistory kitchen was designed using basic concepts of gravity flow.* Vegetables were cleaned, sorted, and cut on the top floor, and dropped through holes in the floor into the cooking vessels below. From the cooking vessels, rice and vegetables moved through funnels into large containers. This system required less overall labor and ensured that health and safety standards were easier to maintain.

Decentralized Operations

Akshaya Patra began its efforts to feed rural schoolchildren in the state of Rajasthan, the largest state in India in terms of land area, but among the least densely populated. There was little infrastructure of any kind, including electricity and water, and villages in Baran were often not connected by any roads. Degraded forests covered nearly half of the land and numerous small rivers made transportation and communication very difficult. Akshaya Patra chose to work with the least developed and only remaining primitive tribe in Baran. The dispersed geography of Baran meant that Akshaya Patra needed to design a wholly different operations strategy: one based on small, decentralized kitchens. It helped the villages set up kitchens at or near the local school and provided basic infrastructure to start the kitchens. Construction of the stove, storage area, and washing area was done under the organization's guidance. By January 2007, the decentralized operations in Baran served 79 villages, feeding 15,000 children per day.

Human Resource Issues

Self-help groups of four to six village women were formed to be employed as cooks for the midday meals. Most of these women had no education or work experience, so the organization provided basic training in cooking, nutrition, and hygiene. Other training included maintaining accounts, inventory, and requisition slips, and often teaching the women to count. Many of the women had not learned the days of the week, so Akshaya Patra taught them to do different activities on different days, simultaneously providing meal variety. Many village

* Capacity utilization increased 40%.

cooks were able to tour the nearest Akshaya Patra kitchen to gain a better understanding of the work at hand.

Each village kitchen had one head cook who was responsible for purchasing vegetables, firewood, and supervising the daily operations; she earned 50 rupees for three hours per day, while the other women earned 1000 rupees per month. In contrast, many of the people in the village earned only 8 rupees per day and worked long hours. The head cook position was rotated on a monthly basis so that each woman had a chance to be responsible.

The central office in Baran was responsible for the bimonthly procurement of nonperishable food items, the distribution of key items, as well as the supervision of all village operations. A cluster supervisor was responsible for supervision of the rural kitchens' cooking, distribution, quality control, and hygiene, and oversaw eight to ten villages.

Overcoming Challenges

When Akshaya Patra first entered the district of Baran it faced many challenges, among them the task of educating the workers in hygiene. In all of its facilities, Akshaya Patra emphasized strong hygienic standards, but village hygiene standards were very different from those in the urban areas. Due to inadequate water supplies, villagers would bathe approximately every 6 to 7 days. Akshaya Patra told the cooks that they needed to bathe daily in order to maintain hygiene appropriate for food service. However, the women did not understand this request, thus they did not adhere to it. Akshaya Patra therefore constructed a communication that was more relevant to the village people. They pointed out that the first bite of the food should be blessed and offered to God. Thus, the women took their own initiative to wash in order to bless the food.

Akshaya Patra also faced difficulties in the transportation of goods. For example, when it rained, delivery trucks could get stuck in mud and ruts for up to three days. Because of the lack of electricity and refrigeration, spoilage of vegetables was a problem. Although the head cook went to the weekly *haat* (market) to procure vegetables, the organization needed to educate women to buy carrots, potatoes, and other vegetables with longer shelf lives.

Extension of Services

Akshaya Patra realized that they had built a large production and distribution network with uses beyond midday meals. The trustees chose to use this infrastructure to provide holistic services to underprivileged groups, including medical, educational, and adult services.

Distribution of Medical Services

Doctors found that more than 85% of children fed by Akshaya Patra suffered from worm infestations as a result of unhygienic living conditions. In addition, they were deficient in vitamins, particularly vitamin A. In 2002, Akshaya Patra began a medical intervention program, which administered deworming medicine and micronutrient capsules with folic acid, iron, and vitamin A. Medical assistance required additional expertise, so Akshaya Patra collaborated with the

Divakars Service Trust in Bangalore and Durlabji Hospital in Jaipur. They also proposed partnerships with dental college hospitals to include preventative dental care and worked with an eye-care institution to test children for eye-related ailments and provide free ophthalmologic care.

Inclusion of Infants and Expectant or Nursing Mothers

Akshaya Patra realized that children of all ages were in need of nutritious meals. They thus began to serve infants, as well as expectant and nursing mothers, in centers near the schools which participated in the program. In Bangalore, Akshaya Patra served over 3000 expectant and nursing mothers in partnership with the city municipal corporation. Similarly, in Jaipur, Akshaya Patra served 64 preschool centers, providing unlimited meals to approximately 100 people in each location.

Feeding Adult Laborers

In Jaipur, many low-skilled village workers migrated to the city for the week, sleeping on the street, and returning home on the weekends to provide wages for their families. They paid up to 30 rupees a day for a small meal. Earning only about 150 rupees per day, it was very difficult to save money and break the cycle of poverty. In 2005, Akshaya Patra began an extension service in Jaipur to provide meals to rickshaw drivers and other low-skilled laborers. For 5 rupees, the laborers received unlimited food at four set locations from 7:00–8:00 p.m. every evening. As of March 2006, Akshaya Patra served approximately 700 laborers per day and planned to increase to 2000.

Leveraging Corporations and Other NGOs

To extend its reach even further, Akshaya Patra worked to leverage relationships with other NGOs and potential corporate sponsors to train them in starting or expanding midday meal programs on their own. For example, Akshaya Patra worked with Havell's, a midsize electrical and industrial component manufacturer, to begin a school feeding program from a factory kitchen at Alwar, not very far from New Delhi. In 2006, Havell's fed approximately 10,000 children daily, with the ambitious goal of expanding to 30,000.

Measuring Success

Akshaya Patra measured success in the number of children fed. Additional performance metrics included an increase in school enrollment and attendance, and improvement of academic performance and student health.

Number of Children Fed

Six years after establishment, as of December 31, 2006, Akshaya Patra was feeding 522,000 underprivileged children in over 2000 schools.

Health and School Performance

An impact study done in one of the rural areas served by Akshaya Patra, which was conducted by the M. S. Ramaiah Medical College, revealed that the number of children below the optimal nutrition level was reduced from 60% to almost 0%.

Anemia was reduced from 40% to less than 5%. Skin infections decreased from 80% to almost 0%. In addition, children developed better resistance to diseases, and they showed significant improvements in height and weight. When Akshaya Patra first began the program in rural Baran, children gained 0.5 kilograms (1.1 lbs) of weight per month. In contrast, during a 10-day winter break in the villages, when the program was not in operation, the average child dropped 0.5 kilograms in weight. In a study the Akshaya Patra program conducted by the Department of Education, Government of Karnataka, 99.6% of students felt that they could pay better attention, and 93.8% of teachers reported overall academic improvement.

In Bangalore, the headmaster of a school that served 560 students reported that 25% of students were totally dependent on Akshaya Patra midday meals. Since Akshaya Patra began providing food, he claimed that attendance was more consistent, drop-outs and long absences decreased, concentration improved, as did height and weight, and students were more mischievous because they were more energetic.

Moving Forward

Akshaya Patra aimed to reach 1 million children daily by 2010. The chairman of the organization, however, felt confident that they would reach this goal by 2008. He commented, "Performance should always exceed promises." He identified three significant limitations to growth: perfecting and setting up operations, training a dedicated workforce, and funding.

Human capital posed particular challenges. In 2006, kitchen management in all eight locations was overseen by religious volunteers, which meant that less than 2% of Akshaya Patra's expenses were from operations, administration, and marketing, compared to more than 20% at comparable NGOs. However, questions remained as to whether the dependency on volunteers was a limitation to growth.

Operational Models for Growth

The Board of Trustees considered the possible courses of action. With a centralized model feeding urban children, and a decentralized model feeding rural children, the dual-pronged strategy presented a variety of challenges to overcome.

Centralized Distribution Model

The centralized kitchen model had the benefits of scale and the best promise for being able to feed the largest number of children for the lowest cost. Akshaya Patra achieved much success through the centralized model, and had recently invested in a new centralized state-of-the-art kitchen in Hubli. However, they still faced many challenges, including distribution issues, maintaining continual improvement, and creating a flexible standardized model while also allowing for local customization of labor needs and food preferences. Worker retention was low in some cities as workers trained in their kitchens were often recruited to work in high-end hotels where they would receive a better wage. In addition, as plans for construction of new kitchens developed, they needed a replicable model while also allowing room for improvement. Akshaya Patra also considered future capacity needs and wondered if they should build each kitchen for the current capacity allotted, or with room for increased capacity, on the assumption of additional funding in the future.

Decentralized Distribution Model

The rural areas were best targeted through the decentralized model, but Akshaya Patra questioned whether they could achieve the necessary scale to make a difference in these areas. Employee recruitment, particularly of trusted supervisors who would not fall prey to the corruption schemes that were so prevalent in India, was a further limiting factor. Furthermore, localized training of village women was extremely labor intensive. The trustees considered expanding to a training-based model, where Akshaya Patra would host 50–100 rural women in a central training location for a 15-day intensive instruction course in proper food preparation, hygiene, and accounting.* This option had the benefits of being able to reach more people, but did not solve the problems of quality assurance, corruption, and theft.

Preparing for the Trustee Meeting: What to do Next?

As the trustees prepared for their meeting, they contemplated how best to balance these challenges. With millions of children still hungry and out of school in India, the need for programs like Akshaya Patra was great. The trustees of the organization were committed to finding ways to serve more children every day, both urban and rural. The possibilities seemed endless, but funding was not, and they knew it was time to make some decisions.

ACKNOWLEDGMENT

The case studies were previously published by Harvard Business School Publishing as "The ITC eChoupal Initiative" (9-604-016) and "Akshaya Patra: Feeding India's Schoolchildren" (9-608-038). With permission of the publisher, they are reprinted here in their entirety, without illustrative exhibits.

* As part of this program, the government and other NGOs would also send women to Akshaya Patra to be trained.

7 Building Global Teams for Effective Industrial Research and Development

Juan M. de Bedout

CONTENTS

PREFACE

The last two decades have seen a flurry of activity by large multinational corporations to build global operations that take advantage of the immense opportunities afforded by the emerging markets in the developing world. Many of these companies have included research and development operations as part of their globalization strategy (Kuemmerle 1999). GE has been a pioneer in this regard. GE's Global Research organization has branched out over the last decade from its headquarters in Niskayuna, New York, adding three new sites around the world, including Bangalore, India; Shanghai, China; and Munich, Germany. Multiple motivations have prompted this global diversification, including access to top talent in technology spaces that is more readily available in some regions, as well as a desire to build technology teams with local awareness of important market trends. This chapter offers observations about how the local culture has influenced the maturation of the global sites in their role supporting the development of advanced technology for GE's businesses, and provides insight into how some of the cultural characteristics of these sites should be considered to accelerate the maturation. The chapter also offers thoughts on how the global sites will lead GE's efforts to drive local innovation for the emerging markets in which they are immersed, and how they will further be instrumental to the company's new drive for *reverse innovation*, where products designed and developed for emerging markets find new markets in the developed world.

INTRODUCTION

Most discussions about globalization over the last decade have centered around the establishment of global sales and manufacturing footprints by large corporations to access the emerging markets. Less discussed is the branching out of the research and development (R&D) operations for some of these companies, which can bring many advantages, including access to a broader pool of technical talent and the presence of deep technical resources in the heart of emerging markets, to name a few. How these new teams mature into productive and impactful branches for their parent companies is an interesting topic, and is of course dependent on many factors. This chapter explores the topic through the example of the global diversification of General Electric (GE)'s research and development operations, which are arguably the most diverse industrial labs in the world. The chapter highlights how the local cultures of GE's different R&D sites influenced how they have matured in their impact for the company.

GE Global Research was founded in Schenectady, NY, in 1900 by Charles Proteus Steinmetz, a luminary in electrical engineering and one of Thomas Edison's key partners in the early years of the General Electric Company. The mandate of the first industrial research lab of its kind in the United States has remained in place to this very day: to drive the development of advanced technologies that give GE's products a competitive advantage in the marketplace. The impact of this institution on GE has been profound, and includes pioneering work in a multitude of spaces, such as the creation of ductile tungsten filaments for light bulbs, the first medical X-ray machine, the first U.S. jet engine, the first semiconductor laser, magnetic resonance imaging, and the first digital X-ray machine. GE's Global Research operations today are home to around 2600 researchers with a broad coverage of science, engineering, physics, and mathematics disciplines.

Under GE's prior chairman, John F. Welch, the R&D labs took a strong first step to follow the company's push for globalization (see Figure 7.1). In 1999, the John F. Welch Technology Center was founded in Bangalore, India, and became home to local business teams for GE Energy, GE Healthcare, GE Transportation, GE Aviation and GE Consumer & Industrial, alongside a new team for GE Global Research. Under GE's current chairman, Jeffrey R. Immelt, the trend was accelerated, with the establishment in 2002 of the China Technology Center in Shanghai, which also became the home to several local GE business teams and a team for Global Research. In 2004, GE Global Research established its latest facility on the grounds of the Technical University of Munich, in Garching, Germany. In 2010, announcements were made for new R&D facilities to be built in Michigan and Brazil.

Global Research Center
Niskayuna, NY

John F. Welch Technology
Center
Bangalore, India

China Technology
Center
Shanghai, China

Global Research—
Europe
Munich, Germany

FIGURE 7.1 GE's four Global Research sites.

The push for the globalization of GE's R&D operations was prompted by several factors, including:

- The ability to access top technical talent around the world that either cannot or will not move far from their homes
- The ability to find skilled talent in disciplines that are difficult to recruit in the United States, such as power electronics for example
- The presence of deep technical resources in the heart of emerging markets

The first two factors cited above have been dominant, and the bulk of Global Research's efforts throughout the last decade have been directed towards developing next-generation technologies under guidance from the business engineering headquarters in the developed world; this could be called a traditional R&D role. As the global sites have matured, they have seen a growing role in leading R&D projects for the businesses, and have had a more profound influence on the technology strategy for these businesses as they look to understand the technology needs for their next-generation products. How local cultures have had an influence on the maturation of these global sites will be explored in the third section of this chapter.

While a smaller percentage of the overall effort, GE Global Research is increasingly focusing on the third factor cited above. In-country-for-country (ICFC) R&D programs sponsored by local business units in China and India are looking to leverage the talent of the engineering and R&D resources in these countries to develop locally grown product lines that cater to the needs of the local markets. Today, a portfolio of ICFC projects for GE Energy, GE Healthcare, GE Oil & Gas, and several other businesses are currently in progress. Early successes from this initiative have awakened the realization in the company that the opportunity is much broader than simply accessing emerging markets; several examples have surfaced where products designed specifically for developing markets are creating new markets in the developed world. Within GE, this new paradigm has been called *reverse innovation* (Immelt, Govindarajan, and Trimble 2009).

A great example of reverse innovation is GE Healthcare's portable ultrasound product (Figure 7.2), the development of which was prompted by the need of hospitals in remote and rural areas of China that could not afford the traditional imaging systems. Traditional ultrasound systems are large, hardware-intensive systems that are outside of the price range for these clinics. Leveraging local development teams, GE took the challenge to transfer much of the complexity to software, allowing the platform to be based on a laptop computer with an ultrasound probe. The new product traded off some performance and image quality for dramatically reduced price and size. In 2002, GE introduced its first compact ultrasound system for $30,000 dollars, followed by an enhanced version in 2007 at half the price. The new product has been very successful in China, where it is largely used for simple imaging procedures. Just as interesting perhaps is that the portable ultrasound product line has found new applications in the developed world in ambulances and emergency rooms,

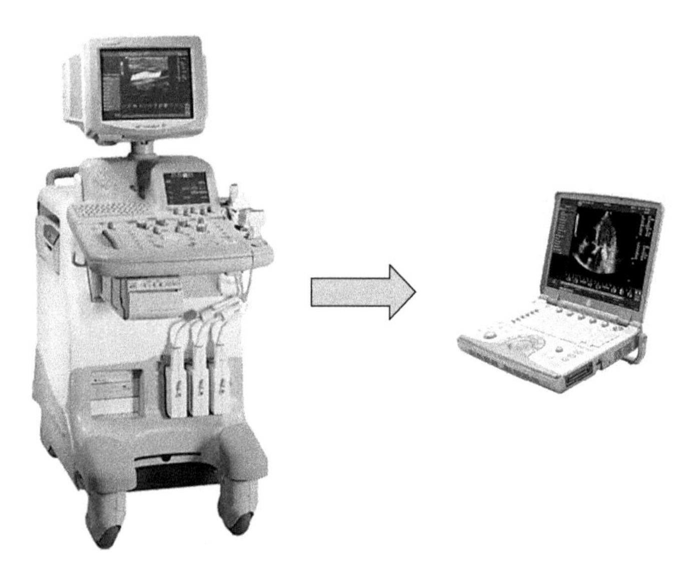

FIGURE 7.2 Traditional ultrasound, and GE's portable ultrasound product.

where its portability and low cost have created new uses that did not previously exist (Immelt, Govindarajan, and Trimble 2009).

As GE accelerates its ICFC and reverse innovation initiatives, the teams at the global sites will increasingly find themselves supporting the local business teams, as well as business teams in the developed world looking to bring their innovations into new markets. Early thoughts as to how the global sites may mature in this new role, and how they may use the local culture to achieve advantage in doing so, will be offered in the Global Research in ICFC and Reverse Innovation section of this chapter.

BACKGROUND: CULTURAL ATTRIBUTES OF COUNTRIES HOSTING GE'S GLOBAL SITES

To focus the discussions in these sections on how cultural issues have influenced the teams at the global sites, it will be useful to review the cultural attributes of the countries in which they are situated. Renowned organizational sociologist Geert Hofstede has developed a five-dimensional cultural model that includes the four countries where GE Global Research has sites, as well as for Brazil, which will host GE's newest R&D center. A thorough analysis of the meaning of these dimensions is left to the reader. The author's experience has found that two of these five dimensions are most significant in describing how the sites have evolved differently in their maturation within GE, namely power distance and individualism. The cultural model results for these two dimensions, addressing the five countries hosting Global Research facilities, are provided in Table 7.1. A brief synopsis of the meaning of these two dimensions can be provided as follows (Hofstede 2001):

- *Power distance*: A high score indicates acceptance of hierarchical power structures; a low score indicates preference for an equal distribution of power.
- *Individualism*: A high score indicates societies with loose relationships between individuals where people tend to watch out for their own personal interests; a low score indicates societies with strong relationships between individuals that collectively watch out for the group.

TABLE 7.1
Relevant Cultural Model Results

	US	India	China	Germany	Brazil
Power Distance	40	77	80	35	69
Individualism	91	48	20	67	38

Source: Hofstede, G. 2001. *Culture's Consequences: Comparing Values, Behaviors, Institutions and Organizations Across Nations*. 2nd ed. Thousand Oaks, CA: Sage. With permission.
Note: A higher score indicates a stronger tendency toward the labeled attribute.

MATURATION OF THE GLOBAL RESEARCH TEAMS IN SUPPORTING TRADITIONAL BUSINESS NEEDS

Big challenges arise when establishing a new team in a remote country. When building such a team, the leading objectives in many cases create conflicting situations. For example, one key objective may be to achieve a critical mass quickly that can take on meaningful responsibilities, have the visibility that can help recruit new team members, and justify the investment in infrastructure. However, building size quickly can be difficult because project managers prefer to leverage resources they trust, and on whom they know they can count for reliable execution. The experience over the last decade with building up the three global R&D sites found that reconciling this conflict usually resulted in lower-criticality work being assigned to the new team by the project managers, and a larger allocation of discretionary funds to help build knowledge of the businesses and relationships with their product line and engineering teams. As the teams in the new sites matured, the situation of course changed and the teams took on progressively more meaningful responsibilities, with all sites today operating in similarly impactful manners. The author has witnessed three stages of maturity that can describe the evolution of the global sites.

Stage 1: Building Credibility

The first stage constitutes building credibility with project managers and the businesses that are, essentially, their clients. With the establishment of each global site, strong leadership from the Global Research managers was needed to motivate the allocation of resources from the new teams to work on R&D projects. As mentioned in the previous section, during the early days of these new teams, work of a less critical nature was assigned to them from the project managers, who needed to balance the risk of an untested team insufficiently familiar with the technical details of the applications they were working on, with the directions from Global Research managers trying to ramp up work at the new site. In retrospect, this outcome was effective; the new teams working on these programs gradually built depth in the application spaces, learned GE's technology development processes and operational rhythm, and through execution earned the respect of their project managers and the GE businesses funding their programs.

To many in GE Global Research, it appeared that the speed with which the three global sites progressed through this first stage differed. In particular, many would argue that the global site in Munich took less time to develop credibility with project managers and the businesses relative to the teams in Bangalore and Shanghai. The difference is not associated with technical competence; in fact, outstanding researchers of comparable credentials are continuously hired at each of these sites. One obvious cause is that the Bangalore and Shanghai teams were simply established first, and that GE Global Research was essentially developing its integration processes on-the-fly, learning what it takes to bring these teams up to speed quickly. Munich, having been founded a few years later, was the beneficiary of the wealth of ramp-up knowledge built from the prior two sites, and should have been expected to mature more quickly.

However, cultural factors can also arguably be pointed to as one of the causes for the difference in rates of progression through this first stage. Referring to Table 7.1, consider the high score in power distance for the India and China sites, relative to the Germany and U.S. sites. In essence, the teams in China and India have a higher cultural predisposition for hierarchy than their counterparts in Germany and the United States. Several leaders in Shanghai and Bangalore feel that many individual contributors in their teams are less comfortable than their peers in New York and Munich in approaching their project managers and leaders, let alone their associates at the business, if they have project questions or are seeking guidance. Instead, they may look for an individual they know well who has a relationship with one of these managers, and seek their help in elevating their questions or concerns. The result is a lengthier resolution to questions within the team, and reduced meaningful exposure between the individual contributor and the project manager. In contrast, the team in Munich comes from a culture that is more predisposed to openness in engaging and even challenging project managers and business leaders.

Several leaders at Global Research also recall that the acceptance by the global teams of less challenging assignments than given to their counterparts in the United States during this first stage of development was different across the sites. Again, the predisposition to hierarchical power structures in Bangalore and Shanghai translated into acceptance of the project slate assigned to them by their leaders during these early stages. In contrast, the team in Munich was quick to expect and request projects of technical sophistication comparable to those assigned to their counterparts in the United States.

With time, the leadership teams at the sites in Shanghai and Bangalore have found means to get their teams to engage more freely and offer their opinions more voluntarily. Interestingly, one of the tools they have used takes advantage of the scores in the individualism dimension. Many of the lab managers at the Global Research sites in Bangalore and Shanghai work hard to form strong personal relationships with the members of their team, which in turn makes them more comfortable in engaging them with questions or ideas. For example, the leaders at both sites host off-site gatherings with the families of the team to build such a rapport. These approaches may be useful to help accelerate the progression through Stage 1 of the new Brazil site, which has power distance and individualism attributes similar to India and China.

Stage 2: Building Ownership

At the end of the first stage of development for the global teams, they have established credibility with the businesses and have a sustainable portfolio of projects that leverage their skills in a meaningful way. The next challenge for these teams becomes developing new project proposals to the businesses on their own that can be led effectively by a project manager at the global site. Achieving this is important on several levels. First, it means that the team is engaged in the strategic planning for the businesses, and is therefore aware not only of the strategic objectives of the business, but also the motivation for these objectives. In essence, this indicates maturity in the technical details of the applications they support. Second, providing a good slate of project proposals to business leaders provides exposure of the global site as a technically innovative resource, and enhances the probability that business leaders will think of that site, and the individuals proposing the projects, when considering future program needs. Third, the ownership of projects by the site serves to enhance the confidence and self-esteem of the team, who take pride in the important role they are playing for the company.

It would of course be interesting to trend the number of business program proposals per capita at each of the four GE Global Research sites over the last decade, and to trend as well the number of awarded business programs per capita at each site to provide quantitative insight into the speed with which the teams have matured through this phase. Unfortunately, the reconstruction of a database of this magnitude is not in the scope of the author's resources, and the treatment of the subject will have to remain qualitative.

As described in Stage 1, several leaders at Global Research would argue that the rate at which the global sites progressed through this second stage of development was different, with Munich exhibiting a slightly quicker pace. As with the prior stage, there is no appreciable difference in the technical competencies of the teams, and it is important to note that very rich and deeply impactful project ideas continue to emerge from all three global sites. As with Stage 1, the integration experience of the first two sites surely helped the Munich team. However, it is also plausible that a high power distance rating for the Shanghai and Bangalore teams meant that it was less natural for individual contributors to feel comfortable presenting their ideas to leaders in an open way, and hence the teams are less aggressive in pushing for their ideas to be noticed relative to their peers in the United States and Munich, who are culturally more predisposed to voicing their ideas with their leaders. A higher degree of respect for hierarchical power structures also plausibly introduces more constraints on the channels for communicating new ideas; for example, some employees may be less willing to present an idea in an open forum without running it by their current project manager or the Global Research leadership first, or perhaps not at all if they feel it inappropriate to do so. To some extent, this last item concerns the fundamental perception of how programs are created and how leadership roles are obtained. Several managers in Bangalore and Shanghai throughout the last decade noticed that many individual contributors have seen it as the privilege of their managers and leaders to identify the projects they are to work on, and hence have viewed them as responsible for identifying new opportunities for the teams.

Leaders at GE Global Research and from the businesses have used a variety of approaches to encourage the teams at the global sites to become more open in voicing their ideas, and to teach individual contributors how to more effectively engage in the strategic discussions with managers and business leaders, which provide venues to proposing new project ideas. Innovation challenges are held in a variety of forms, some being open to a broad class of ideas, and others narrowly focused on particular subjects. Some of the discretionary funding for the sites is also allocated to fund new ideas submitted by the employees to evaluation committees. Technical symposia and seminars are held periodically to provide researchers opportunities to share their work with their peers and leaders. Training courses are offered in a variety of development areas, including basic leadership, project management, and effective communication.

Another interesting approach involves a short-term relocation of a promising individual contributor to either the GE Global Research headquarters in New York, or to a business team that appears to be a good candidate for future engagement with the employee. These short-term assignments have been eye-opening for many of the individual contributors who have experienced them. In the assignments, they get to work with experts who not only mentor them technically, but also demonstrate how to build effective networks with the business teams, and how to engage in the operational rhythm more effectively. When these employees return to the global sites, they are not only better connected with business leaders, but also bring back the realization of how to get their ideas in front of the right audiences. The benefits multiply further, as the new behaviors of these emerging leaders serves as a role model for their peers. Longer-term exchanges that place key leaders from the global sites at the headquarters, and vice-versa, could serve to build even better global interactions through the development of networks and cultural awareness.

Stage 3: Providing Thought Leadership

Coming out of the second stage, the teams are recognized by the GE businesses as valuable and reliable technical resources that are knowledgeable about where those businesses are heading and why. The teams understand the strategic plans for product line evolution, and are successful in formulating technology development proposals that can help meet the objectives of those plans. Technical leaders have emerged from the teams who possess broad and deep technical knowledge and understanding of the applications they support, and are well-connected to leaders in the businesses. The final stage for the team consists in developing the ability not only to help fulfill the technical plan, but also to help shape the future technical strategy for the business. This influence is long-term in nature; it involves shaping technology plans that may take significant resources and time to develop to the point where they may be transferred to a product or service for the company. Substantial credibility must exist for teams proposing technology programs of this nature, as the return on the investment for this class of programs is far less certain than for programs that seek to develop product or service enhancements that would be expected to be commercialized within a few years.

At GE Global Research, roughly half of the program funding comes directly from GE businesses that are looking to leverage the R&D resources to advance the performance and functionality of their existing product lines. Roughly 20% of the program funding comes from external programs with government agencies and commercial partnerships. The final 30% comes from GE Corporate, to fund programs that are longer-term in nature and that would normally be deemed too risky to be funded directly by a business. Despite coming from GE Corporate, the chief executive officers (CEOs) of the different GE businesses have a strong influence on how this funding is allocated. The most significant programs funded by GE Corporate come in two forms: CEO programs, which are sponsored by the CEO of a business; and advanced technology (AT) programs, which are approved by GE's chief technology officer, with regular consultation from business leaders.

Again, it would be interesting to trend the number of CEO and AT proposals per capita at each of the four GE Global Research sites over the last decade, and to trend as well the number of awarded CEO and AT programs per capita at each site to provide quantitative insight into the speed with which the teams have matured through this phase. Again, the reconstruction of such a database is not in the scope of the author's resources, and the treatment will be qualitative.

It is factual that the bulk of the CEO and AT program proposals that have made it to the final round of evaluation have come from the team in Niskayuna, but proposals and subsequent awards are starting to increase at all three global sites. It would be incorrect to infer from this that the global sites have a smaller role in CEO and AT program execution; these programs are large multidisciplinary projects that draw resources from all of the global sites. However, it can be soundly argued that all three of the global sites are working to mature through this stage of development, and that they have yet to catch up with the Niskayuna site. At this point, cultural differences do not seem to be a strong factor in how the sites are maturing; instead, Niskayuna could be described as having the unfair advantage of an additional century in developing technical experts with substantial depth that have the career reputation to credibly lead this class of program. Furthermore, Niskayuna has the advantage of proximity between proponents of programs and the key influencers and decision makers that determine the outcome, which allows more interaction to take place during the shaping of the proposals. It's important to point out that the substance of the awarded programs at the global sites is strong, and it's refreshing to see technical leaders with solid reputations emerging there that can compete soundly with their peers from the United States.

It is interesting to note that unique mechanisms for developing innovative concepts at this stage have arisen in some sites, and cultural factors do play a role. A noteworthy example is the Bangalore site, where some leaders are leveraging the collectivist cultural tendencies of the team to drive involved discussions on new ideas. In particular, it has been noted that assigning a new idea topic to an individual contributor results in a subsequent flocking of the team to that individual to offer their ideas and suggestions. The ensuing collaborative discussions have consistently resulted in rich and well-thought program concepts. It is possible that this result could be replicable in Shanghai, where the culture also leans towards collectivist tendencies.

GLOBAL RESEARCH IN ICFC AND REVERSE INNOVATION

As mentioned in the Introduction, increasing attention is being devoted within GE to the development of products and services for the emerging markets, with help and leadership provided by the teams immersed in those markets. Multiple motivations drive this interest. The obvious one is that these emerging markets are growing at a faster rate than the markets of the developed world. Competing in these markets requires a good understanding of the consumers and the business environment of these markets, which in general are not as well understood by leaders from the developed world. The traditional philosophy of defeaturing Western products until they meet price points attractive to consumers in these markets is running up against disruptive innovation from local enterprises which, time and again, are able to conceive of products designed specifically to meet the needs of these consumers and are therefore more attractive to them (Govindarajan and Trimble 2005; Immelt, Govindarajan, and Trimble 2009). This is happening with automobiles (Tata), tractors (Mahindra), renewable energy (Suzlon), information technology (Wipro), and more.

The less obvious reason for the interest in developing products and services with teams from within those markets is self-preservation. In particular, products that are succeeding in the emerging markets are creating brand new markets in the developed world, and have the potential to disrupt conventional business practices (Immelt, Govindarajan, and Trimble 2009). Creating teams that can compete head on with the agile businesses emerging in the developing world may prove to be essential to protecting the long-term success of the company.

It is well understood within GE that changes need to occur within the company to provide more decision-making authority to the business teams in the emerging markets if they are to be successful in developing products for these markets. In a recent step to provide this decision-making authority, GE announced the establishment of a new Profit & Loss Center (P&L) in India, under the leadership of a senior vice president that reports to Vice Chairman John Krenicki. This new P&L has the mandate to increase GE's presence in India, and the ability to direct sales, marketing, engineering, and R&D resources within the country towards this objective. This contrasts with the traditional approach, where direction has been provided by the headquarters of GE's businesses, which have tended to focus on the technology needs for the established product lines.

It is unquestionable that the local R&D teams in China and India, and eventually Brazil, will be invaluable to GE as the company's local business units accelerate their ICFC efforts. These teams have strong opinions regarding how the customer needs within their countries differ from what is available from GE's established product lines. They also have relationships with local businesses, government officials, and academia that can be leveraged to help the ICFC agenda. Perhaps just as important, they know how to relate with and engage these external resources in an effective way, navigating interaction intricacies that are culturally obscure to foreigners. This does not however mean that these teams will operate in complete independence. They will be able to count on the wealth of expertise amassed by the broader company, and will have the ability to focus that experience and expertise to develop the best products and services for these markets, leveraging the best resources the company has to offer. This should provide a unique advantage to GE relative to the emerging local competitors.

It is interesting to hypothesize that a reversal of many of the global teaming interaction challenges may be awaiting the broader Global Research as ICFC programs become more prevalent. These may be most felt when the R&D teams in China and India leverage technical expertise from the broader Global Research and GE communities. Whereas in the past the global teams have had to adapt to better engage the GE culture, and to an extent the Niskayuna culture, ICFC and reverse innovation programs may find researchers and business leaders in the United States and Europe struggling to effectively engage the teams leading these efforts in China and India. In this case, the lower score in power distance of the Western teams may make it harder to engage business teams, partners, and customers in China and India where leaders may be culturally predisposed to expect a certain level of stature before seriously engaging them. Unfortunately, the author knows of no current examples to support or counter this hypothesis.

CONCLUSIONS

This chapter explored several features of how new global teams mature into productive and impactful branches for their parent companies, through the example of the global diversification of GE's R&D operations. Two different roles for the teams were considered. In the first role, the global teams work together to support the strategic technology plans for Western-headquartered multinational businesses, which emphasize traditional products for mature markets. In the second role, the global teams support local business units that are focused on developing products for their local and quickly growing markets.

Regarding the first role, this chapter described how the teams at GE's R&D sites in Bangalore, Shanghai, and Munich advanced through a series of three development stages, progressively building stronger technical maturity and impact for the company. Understanding that many elements play a role in this maturation, the influence of cultural factors in how quickly these teams developed was explored. It was argued that power distance may be the most influential of these cultural factors, affecting the rate at which these new teams built credibility with the company's established business units, and the rate at which they developed ownership for a portfolio of programs. In particular, the higher cultural predisposition for accepting hierarchical power structures in India and China can be argued to be a factor that decelerates useful interactions between individual contributors and their leaders at Global Research and the GE businesses. In contrast, the cultural predisposition of the team in Germany to relatively flatter power structures may result in more aggressive engagement of team members with relevant leaders, possibly accelerating the resolution of questions and concerns and achieving a timelier exposure of the team's new ideas to decision makers. Approaches used by Global Research leaders at all sites to enhance the open engagement of individual contributors were described. These approaches focus on making employees more comfortable in expressing their thoughts and ideas openly, and in providing opportunities for emerging leaders at the sites to immerse themselves in the business environment, allowing them to build relationship networks and gain first-hand exposure on how to engage in the company's operational rhythm. Today, all three global sites find themselves progressing through the third stage of development, where they are steadily increasing their roles in helping shape the technology strategy of the company.

Regarding the second role, it was described how the function of the local R&D teams will become increasingly important as the emphasis on ICFC and reverse innovation accelerates. The value of having deep technical expertise in the hearts of emerging markets, with a strong understanding of the fundamental needs of the clients in the region, will help GE better compete with products and services designed specifically to satisfy those needs. Thoughts were offered regarding how new global teaming challenges may arise from cultural differences between the local teams leading these programs and the researchers and business leaders elsewhere in the world that need to engage them, effectively reversing the burden of adaptation onto the Western teams.

As GE prepares to launch the development of its newest R&D centers in Brazil and Detroit, the lessons learned from the establishment of the prior three global centers will prove valuable in helping the new team ramp up in impact quickly. Similar power distance and individualism characteristics of the Brazil site to China and India indicate that care should be given to quickly helping the individual contributors in that team become comfortable in voicing their thoughts and opinions in an open manner with company leaders. For both sites, short-term assignments that immerse high-potential members of the team into relevant businesses as well as the Global Research headquarters can accelerate the development of meaningful networks and expose these employees to the GE culture and operational rhythm early. It will be interesting to compare the rate of maturation for the new center at Brazil with the rate for the new center at Detroit; this should help further clarify how influential culture truly is.

In closing, it's important to re-emphasize the values that have been created for GE through the globalization of its R&D operations. This effort has afforded the company access to top talent around the globe, in some cases in disciplines which are hard to fill with good researchers and scientists. This of course strengthens the ability to execute on the important research agenda of developing best-in-class technologies that can help differentiate the company's established product lines in the market. Augmenting this value is these employees' knowledge of local customs, consumer preferences, and market trends, and their relationships with local businesses, academia, and government. The company now has the opportunity to leverage these resources to lead the development of products and technologies specifically for their markets, with the unique advantage relative to local competitors of having access to the broader talents and resources of the parent corporation.

Many North American and European companies have branched out R&D, science, and engineering teams in emerging markets in a manner similar to GE, in some cases for similar reasons, and as such many of the observations discussed in this chapter should prove to be relevant and of value to them. A big twist on the topic are examples of companies born in the emerging markets that seek to globalize some of their engineering or R&D operations in the mature markets, such as Compton Greaves and Suzlon, in which case the motivations for globalization may be partially or entirely different from those that have prompted GE's actions. In undertaking this, these companies may have substantial experience with the hypothesis submitted in the previous section on the role of cultural factors in reversing the burden of adaptation and team integration onto the teams in the mature markets. It would be interesting and valuable to explore their experiences in ramping up their new teams, and their impressions on interaction challenges arising from the globalization effort, including those associated with cultural factors. Such work would be valuable in forecasting teaming and interaction issues that may arise in companies like GE embracing reverse innovation.

REFERENCES

Govindarajan, V., and C. Trimble. 2005. *10 Rules for Strategic Innovators: From Idea to Execution*. Boston, MA: Harvard Business Review School Press.

Hofstede, G. 2001. *Culture's Consequences: Comparing Values, Behaviors, Institutions and Organizations Across Nations*. 2nd ed. Thousand Oaks, CA: Sage.

Immelt, J. R., V. Govindarajan, and C. Trimble. 2009. How GE is disrupting itself. *Harv Bus Rev* 87(10):56–65.

Kuemmerle, W. 1999. The drivers of foreign direct investment into research and development: An empirical investigation. *J Int Bus Stud* 30(1):1–24.

8 Cultural Factors
Their Impact in Collaboration Support Systems and on Decision Quality

Shimon Y. Nof

CONTENTS

INTRODUCTION: COLLABORATIVE AUGMENTATION AND c-WORK EFFECTIVENESS

Decision making and action to implement the decision are fundamental to our civilization and industrial economy. Group decision making and team interaction further advance them by processes of collaboration. But how does collaboration impact the quality of decisions, and what is the role of cultural factors in the collaboration process? Is the impact always positive, and what are its dimensions? The objective of this chapter is to attempt to answer these questions in the context of c-work by reviewing the area of c-work, measuring the impacts of cultural factors, providing examples of impacts by cultural factors, and discussing implications, challenges, and emerging trends.

The purpose of effective, collaborative e-work, which we term c-work, is to augment humans' abilities at work and organizations' abilities to accomplish their missions. Much of this augmentation is by cyber-supported collaboration, enabling c-work. Developing such augmentation requires understanding the limits of humans' and organizations' abilities (i.e., determining what needs augmentation), and the anticipated limits of such augmentation. This chapter focuses on our attempt to understand such design limits with particular attention to cultural factors.

Augmenting human work by collaboration addresses bottlenecks, strengths, and weaknesses in human performance. Most of these bottlenecks, strengths, and weaknesses vary over ranges and over time, and are determined and influenced by individual differences, including physical, cognitive, and cultural. For instance, individual human limitations include working memory capacity, or the number of sensory stimuli that can be processed in a short period of time (e.g., visual [2–5 items] and tactile [3–5 items]; up to three times more for multimodality stimuli) (Horvitz et al. 2003; Miller 1956; Schmorrow et al. 2006). Also, inaccuracies exist in human statistical estimation of, for instance, sample means, event frequency, and risks (Bar-Hillel 1973; Clemen and Reilly 2001; Du Charme 1970; Tversky and Kahneman 1974). In statistical inference, limitations exist in aggregation of evidence, a tendency to seek confirming evidence and discount disconfirming evidence, and hindsight bias.

Weaknesses and advantages in the performance of human teams, including the impacts of cultural factors, are also well known. For instance, group decision-making limitations (Campion, Medsker, and Higgs 1993; Lehto and Nah 2006), community ergonomics initiatives (Taveira and Smith 2006), and cultures of participation (Fischer 2010) consider factors such as trade-offs, balancing, sharing, and partnerships that must be considered and taken into account. In game theory, competitive games imply that individuals seek self-centered solutions that may damage or even destroy other players, whereas in cooperative games players force group-optimal solutions on individuals. Group process bottlenecks include conflicts that are directly affected by cultural factors, such as ethics, social norms, biases, and perspectives (frames). Human information-processing bottlenecks for individual team members addressed in this context include working memory, perception, attention, and executive control functions (Proctor and Vu 2006). There are also ones associated with ability to calculate, to sort, merge, integrate, and engage in multitasking. Cyber-enabling collaborative environments for envisioning and discovery (collaboratories), memory aiding prompting systems, and community networks (e.g., Fischer 2010) are examples of efforts to overcome such limitations and enrich human cognitive and creative abilities.

Organizational and work group cultures have been known to influence the effectiveness and weaknesses in performance, for example, innovation and creativity, positive and negative impacts on quality and safety, and other error-related activities and services. Such cultures affect individual and team work by factors such as acceptance of authority, identification with community, and others classified as individualism-collectivism, power distance, and uncertainty avoidance (e.g., Hofstede 1991).

Augmented cognition strategies in human interactions and collaboration, which are increasingly supported by cyber-enabled and collaborative e-work, include task presentation, task sequencing, task delegation, task pacing, information filtering, mixed interaction, and decision support system strategies (by models, databases, and user interfaces). Other cyber-enabled collaboration strategies involving e-work and c-work include expert systems; knowledge-based systems; neural networks;

group support, negotiation support, and collaboration support systems (CSS); and enterprise systems for decision support, AI techniques, and intelligent agents. In terms of information and knowledge sharing, there are augmentation strategies of making existing knowledge more accessible, reliable, and available, including social media and social networking, leading to better opportunities for creativity and innovation (e.g., Shneiderman 2009).

Limits in collaborative e-work, as exemplified by supply and demand networks, are typically considered and rationalized relative to the three big factors of automation affordability:

1. *Flexibility to changes and disturbances*: Cultural factors include individual, team, and organizational cultural attitudes towards flexibility to plan, train, and respond to emergent and evolutionary changes and disturbances.
2. *Reliability and availability*: Similar cultural factors as in factor 1, and individual, organizational, and social attitudes towards error/conflict prevention and recovery.
3. *Quality and customer service*: Similar cultural factors as in factors 1 and 2, and attitudes of the individual, organization, and social groups towards customer-centric design.

Typical collaboration challenges mount in our increasingly interdependent global world. These include classical coordination over time of client-server and producer-consumer interactions; classical coordination over space, for example, buffers between robots/departments/enterprises; and over both time and space, to handle errors and other communications. Beyond effective augmentation, design objectives and benefits of collaboration must respond to the needs of sustainability and dependability with a growing world population. These needs include practically 24/7 service, availability, accessibility, and supply of goods, services, security, and information or knowledge, with minimal disturbances, conflicts, waiting, and distribution costs, social disruptions, and environmental damages. These needs also include personalization by variety and by comparison, including customization of product, process, service, and interactions; better quality of service (QoS) by following "best practices"; better coordination of suppliers; better techniques for collaboration, teamwork, community interactions; and other criteria and benefits.

Emerging production, service, and logistics involve networks of demand and supply. Smart teams enable them by managing rationalized workflow interactions, which require decentralized decisions and automation, and collaborative control with active workflow protocols. Organizations are also challenged by the growing interdependence, communication and cultural obstacles, safety and security threats, and design mismatches, which are well recognized by now but not well understood. Enabling methods and technologies to overcome these obstacles have been proposed and developed, and are now being tested against new constraints and demands.

Augmentation in emerging e-activities is generally enabled by "four wheels": (1) e-work; (2) integration, coordination, and collaboration; (3) distributed decision support; and (4) active middleware. The four wheels are as follows (Nof 2003; Velasquez and Nof 2009).

- *e-work and/or c-work*: e-work theory and models; agents; coordination protocols; workflow models and systems
- *Integration, coordination, and collaboration (ICC)*: ICC theory and models; human-computer interaction (HCI); computer-integrated manufacturing (CIM) and management; extended enterprises
- *Distributed decision support (DSS)*: Decision models; distributed control systems; collaborative problem solving
- *Active middleware*: Middleware technology; GRID and cloud computing; distributed knowledge systems; knowledge-based systems (KBS)

These four wheels and their 15 e-dimensions directly apply to, and are dependent on, automation and cyber functions. Not only are all of them critically dependent on human factors, but they also must take into account cultural factors, especially in the following three lifecycle phases:

1. *Design*: Human cultural factors and intelligence in conceiving, research, design, and specification of automation and cyber support
2. *Implementation*: Human cultural factors and intelligence in developing, building, maintaining, servicing, and updating the automation and cyber support
3. *Training and use*: Human cultural factors and intelligence in adopting, learning, applying, upgrading, and troubleshooting the automation and cyber support

The c-work elements are designed and integrated for collaboration support systems that aim at improving and enabling effective, high quality individual and team c-work. Examples are described next.

Emerging Principles of Collaborative Control Theory for c-Work

Examples of c-work include collaborative control and management of e-manufacturing, e-supply, e-service, e-production, and e-logistics. Collaborative control theory (CCT) has been developed to support the effective design of such systems (Nof 2007b), with research leading to six key principles, summarized in Table 8.1. Important features are the networks of collaborative teams and participating organizations or clusters, at many levels of implementation. Common to the CCT principles are augmentative improvements of systems' effectiveness by automatic collaboration support in the plan, structure, coordination, and execution under highly distributed interactions and disturbances. In these examples, the three phases influenced by cultural factors are evident. Table 8.1 includes the cultural rationale for the collaboration principles. This rationale is itself rooted in diverse cultural imperatives and perceptions.

MEASURING THE IMPACT OF CULTURAL FACTORS

It is known that cultural factors inherently influence the quality of decisions and actions in collaborative research, policy, design, planning, implementation, training, and use of products and services. It is not yet clear, however, how to model, measure, and evaluate these impacts. A general model is proposed in Figure 8.1. It incorporates three general components with a focus on collaborative decisions: the problem features, the knowledge available, and the process. Each of these components depends on cultural factors. For example, the culture of individual researchers and teams of researchers in collaborative knowledge exchange; the creativity, skills, and cultural biases of software engineers

TABLE 8.1
CCT Design Principles for Collaborative e-Work (c-Work) Effectiveness and Their Cultural Roots

c-Work Design Principle	Brief Definition	Cultural Rationale*
1. **CRP** Collaboration Requirement Planning	**Effective e-collaboration requires advanced planning and ongoing replanning.** • CRP-I: plan "Who does what, how, and when." • CRP-II: During execution, revise plan in real time, adapting to temporal and spatial changes and constraints.	"Think before you act"
2. **Parallelism & KISS** Parallelize and "Keep it Simple, Cyber System!"	**Optimally exploit the fact that work in cyber workspaces and human work-spaces can and must be allowed to advance in parallel.** • Effective e-Work systems cannot be constrained by linear (sequential) precedence of tasks—delegate! • "KISS": Keep It Simple, Cyber System! The cyber system may be highly complex as developed by team intelligence, but for human users it must be simple to use, learn, and relearn.	"Divide and conquer"
3. **CEDP** Conflict and Error Detection and Prognostics	**Minimize the cost of resolving conflicts among collaborating e-workers by automated EWSS (e-work support systems).** • Beyond reducing information and task overloads, design e-work to automatically prevent and overcome as many errors and conflicts as required for effectiveness.	"Learn from mistakes"
4. **FTT** Fault-Tolerance by Teaming	**Fault-tolerant collaboration can yield better results from a team of weak agents, than from a single, optimized and even flawless agent.** • Find how to achieve group advantages by smart collaborative automation, even with some faulty agents/channels.	"Team for synergy"
5. **JLR/AD** Join/Leave/Remain in a CNO, or Associate/ Dissociate to or from a Collaborative Network	**For individual organizations: Decide repeatedly when and why to JLR/AD a given CNO based on measured total participation gains and costs. For a CNO: Same, including increased coordination, relative to each member organization.** CNO = collaborative networked organization or cluster of autonomous entities, e.g., sensors, agents, robots, team-collaborators, suppliers, servers, enterprises	"Be selective"

(Continued)

Table 8.1 (*Continued*)
CCT Design Principles for Collaborative e-Work (c-Work) Effectiveness and Their Cultural Roots

c-Work Design Principle	Brief Definition	Cultural Rationale*
6. LOCC/BM Lines Of Command and Collaboration and Best Matching	**Design responsive/evolutionary mechanisms of interaction and organizational learning** for better ad-hoc and emergent decisions, effective improvisation, on-the-spot contact creation, and **Best matching protocols** to pair decision makers and executors, decision problems and tools, tutorial/repair advice and obstacles or concerns, performance measures and decision questions.	"Trust the backup"

Source: Nof, S. Y., *Management and Control of Production and Logistics*, eds. O. Bologa, I. Dumitrache, F. G. Filip, Vol. 4 (1), Elsevier IFAC Papers On-line, Availability, integrability, and dependability—what are the limits in production and logistics? (Plenary), 2007, http://www.ifac-papersonline.net/Detailed/39097.html. (accessed Feb. 2, 2011). With permission; Velasquez, J. D., and S. Y. Nof. Collaborative e-work, e-business, and e-service. In *Springer Handbook of Automation*, ed. S. Y. Nof, Heidelberg, Springer Publishers, Chapter 88, 1549–76, 2009. With permission.

* Variations of these cultural rationales as the logical basis for collaborative behaviors are rooted in many different cultures and are expressed in different yet logically similar parables and allegories.

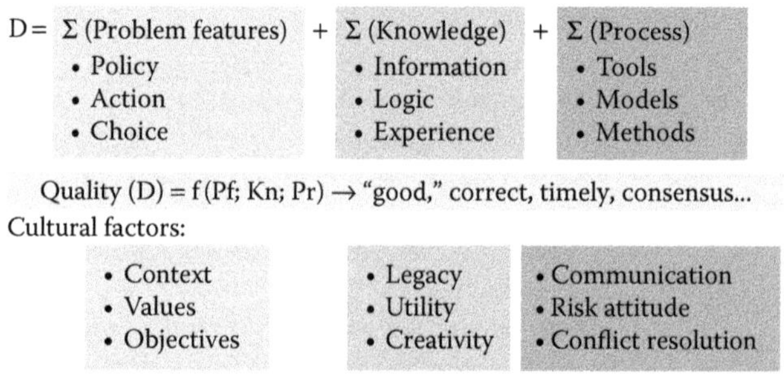

FIGURE 8.1 Decisions (D), decision quality, and cultural factors. f(Pf)—function of problem features; Kn—knowledge; Pr—process involving collaboration during decision making and action on the decision.

engaged in cyber-support tool building; and the testing models and methods applied by collaborative product inspectors, who bring their own cultural attitudes and legacies.

Traditional measures of system design, implementation, training, and use effectiveness are meaningful. There are, however, new dimensions and new features of collaboration, autonomy level, and work under highly distributed and decentralized interactions (Nof 2007a). These new features necessitate attention to additional measures and performance evaluation criteria. Certain measures have already emerged with the advent of CIM and networked enterprises: flexibility, agility, connectivity, integration ability, scalability, reachability, and more. Both the traditional and the new measures are highly influenced by cultural factors.

Two recent measures developed for c-work design and collaborative control are level of autonomy, and viability (e.g., Huang and Nof 2000a,b). The autonomy level implies the delegation of authority and resources, assignment of tasks, decentralization, and how active the agents, protocols, and other c-work system participants can be. All of these components of autonomy are clearly dependent on individuals' and groups' cultural factors. Viability in the context of collaborative e-work has been defined as a relative measure of the ratio between the cost of operating and sustaining agents, and the rewards from their services. Both the costs and the rewards are, again, highly dependent on individuals' and groups' cultural factors.

In terms of decision quality, theories of agent design and interaction have been developed using measures of autonomy and viability. Anussornnitisarn (2003) discovered that collaborative e-work performs significantly better when rationalized by viability-based protocols than by protocols that ignore viability. Examples of other emerging collaborative control measures for e-work design are based on: learning ability, collaboration ability, errors and conflicts severity, prevention ability, detectability, diagnosis ability, and recovery ability.

Some of the CCT-oriented measures include availability, dependability, and integrability (Nof 2007a). These measures, as discussed next, are inherently affected by the cultural factors. Traditionally, availability has been defined as an overall measure of *static* or *dynamic available capacity*, calculated based on the mean time to failure and mean time to repair (e.g., Cristian 1991; Hopp and Spearman 2000; Johnson and Malek 1988). *Instantaneous availability* has been defined as the probability that a system is performing properly at time t (equal to reliability), while *steady-state availability* is the probability that a system will be operational at any random point of time. A dynamic view of availability (e.g., Zakarian and Kusiak 1997) is based on the availability of given equipment units at a given time and at given cells comprising a facility or a system.

In production, logistics, service, and supply networks, it is often necessary to select suppliers based on a measure of their *selective availability* (e.g., Anane et al. 2002), which is determined by weighting several factors according to their relative importance, taking into account factors such as the precision tolerance level of supplied components, availability of substitute components to a given component, reliability of the considered supplier, and level of complexity (e.g., the number of steps required for obtaining the required components from a given supplier). The total value of selective availability is influenced by the relative contribution of each relevant factor. For instance, the lower the relative complexity level and the higher the availability of within-tolerance and substitute components, the higher the selective availability value.

Beyond these considerations, in the context of CCT, issues of autonomy, viability, and impact of collaboration, including effects of cultural factors, must be considered. The available capacity of collaborating participants to supply components and information, and to complete tasks, depends on how effectively they can overcome performance obstacles, minimize delays, and avoid errors and conflicts.

TABLE 8.2
Survey of CSSs and the Impact of Cultural Factors*

Collaboration Support System (CSS) Study Mode	Quality Objective	Cultural Factors Impacting Quality of c-Work			
		Designers	Decision makers	Operators	Customers
1. MERP (Mobile ERP) portal (Yoon et al. 2010) H-H-H	Better supply decisions and customer relations	Planning algorithms and software	Modes of sharing information, knowledge	(Decision makers are also the operators)	Supply delivery, satisfaction attitudes
2. Coordinating robots (UAVs) and sensors for surveillance (de Freitas et al. 2010) H-M-M	Faster alarm activation with swarm algorithms	Design strategy and software	Mode of responding to alarms	Strategy of activating the UAVs	Safety and security expectations
3. Massive group decision support interface (Zamfirescu and Filip 2010) H-H-H	Optimal solutions with cognitive swarming	Converge on cognitive swarming	Interaction modes with decision processes	(Decision makers are also the operators)	Quality of decisions differs with different tools
4. Resource sharing over testers network within a supply network (Ko and Nof 2010) H-M-H	Faster, timely, better tested assemblies	Task administration protocols	Accepting recommend-ations for resource sharing	Trusting the decisions made	Timely supply delivery, expected tested product reliability
5. Collaborative multirobot manufacturing (Nof 2009) H-M-M	More effective production process	Optimize collaboration modes plan	Follow optimal collaboration plan	(Decision makers are also the operators)	Timely and best quality cars and ships

*Impacts are surveyed on the c-Work participants: Designers (and programmers) of this CSS; decision makers applying the CSS; operators of this CSS; customers expecting better-quality outcomes gained by this CSS.

CI: Collaborative Intelligence, or Collective Intelligence

In c-work, as exemplified in Table 8.2 collaboration among multiple participants—humans, agents, robots, sensors, and other autonomous sources of knowledge, decisions, and actions—transforms traditional ways of creating knowledge and intelligence. The resulting collaboration-enabled behaviors are termed CI, or collaborative intelligence. CI has been defined as any system that attempts to tap the expertise of a group rather than an individual to make decisions (e.g., Fischer 2006; Sabherwal and Becerra-Fernandez 2011). Technologies that contribute to CI are all based on CSS, or collaboration support systems, including social media, collaborative publishing, common databases for sharing knowledge, and many others.

CI and collective intelligence, also known as distributed intelligence, collective knowledge, and collective wisdom, are different from other forms of thinking in that they improve as more people and knowledge sources contribute, exchange, possibly debate, and use it. While most knowledge systems involving work have emerged as cumulative layers of information and knowledge developed by generations of workers, thinkers, and experts, it is recognized that

$$\text{Work} \neq \text{e-work} \neq \text{c-work}$$

The work and its generated knowledge and intelligence according to traditional work methods and systems are different from the e-work, which is supported by electronic computers and communications. Such work and its associated and resulting knowledge are further transformed by c-work. However, several open questions remain:

- At what cost is the knowledge developed?
- With what resulting quality?
- When is this knowledge available for timely response and decisions?

Several studies addressing these questions are discussed in the next section.

CCT AND CSS EXAMPLES WITH IMPACTS AND CULTURAL FACTORS

Optimizing the process of knowledge and intelligence development, resulting quality, delivery and timeliness requires understanding how to design the collaborative control mechanisms. These mechanisms are based on CCT and CSS, which can enable and provide them. Several theoretical investigations and projects have addressed these questions.

Availability (A) as a CCT Measure of c-Work

All disturbances in collaborating processes can be considered to be errors and conflicts. Better performance resulting in better quality outcomes depends on the ability to avoid, prevent, and resolve errors and conflicts during collaboration. First, an error has to be detected by the system. Next, this system should have sufficient time and intelligence to resolve the error and complete prior tasks before the arrival of the next task that depends on their completion. In studies (Huang, Ceroni, and Nof 2000; Nof 2007a), it is assumed that given the probability of resolving an error or conflict as soon as it is detected, the system is available except during error or conflict detection and resolution. The availability can be derived using an equation where s is the probability of detecting an error or conflict in one inspection, and n is the number of repeated cycles of inspections, based on the probability to detect the first error during inspection and the available time to resolve this error. For a given s, availability increases as the number of inspections n increases. This result is intuitively true: when inspections are conducted more frequently, errors can be detected sooner and more time is available to resolve them, but cost increases. The trade-off between cost and availability can now be determined.

With effective c-work, the feasibility of ongoing automatic inspections and collaborative prevention, detection, and resolution of errors and conflicts is significantly more cost-effective. As a result, the probability that the ith conflict can be resolved in no more than n iterations is higher, and the value of n is lower, with a limit of $n = 1$, and possibly $n = 0$ with advanced prognostics aimed at preventing and eliminating most errors and conflicts. Similarly, the probability to resolve a conflict in the jth resolution attempt (iteration) is higher, with a limit of $j = 1$ and possibly $j = 0$. With measured availability, it is now possible to determine if the costs are justified by the benefits.

Dependability (D) as CCT Measure of c-Work

Dependability has traditionally been defined as the probability of a task/system to be successfully executed, producing justifiable intended results. With increasing security concerns and highly dispersed systems, it also implies the level of trust. Similar to availability, it is a performance requirement that depends on the severity and frequency of production and service interruptions. But its focus is not just on operational availability, but also on the quality level of decisions, services, and products.

If a system can complete its tasks with fewer service/quality errors or failures, in time, its dependability is higher (Anane et al. 2002; Avizienis et al. 2004; Johnson 2007; Johnson and Malek 1988; Laprie 1995). Dependability of a system can also be viewed by three characteristics: availability, reliability, and serviceability. A related term is maintainability, the probability of successfully completing a corrective maintenance action within a prescribed period of time. With attention to the quality of service results, for example, a computer system is *dependable* to the extent to which its operation is free of failures and its computations are free of errors.

Depending on the services expected from the system, different views of dependability can be defined, for example:

- Readiness for service, resulting in availability
- Continuity of service, thus increasing reliability
- Avoiding catastrophic environmental consequences, leading to safety and sustainability
- Eliminating improper alterations of information, thereby increasing integrity and timely maintainability

Increasingly, dependability is impacted by collaborations among agents that can provide ongoing recovery from failures, backup and support services, and anticipatory, optimized decisions. Dependability is also influenced by the autonomy of each agent, especially relative to rational response ability. Autonomous enterprise systems can collaborate intelligently with each other to adapt to varying conditions and customers' demand changes (Huang, Ceroni, and Nof 2000; Huang and Nof 2000a,b). Therefore, important emerging attributes of dependability are autonomy, viability, and agility based on multiagent behaviors, most of which are influenced by cultural factors: exchanging information and signals, negotiating, responding and adapting, and so on. The *degree of autonomy* can be defined as the number of successful strategies an agent in c-work can adopt, divided by the total number of possible strategies in this collaborative agent-based system. Since autonomy determines the agent's ability to handle decisions well, hence its dependability, agent *a*'s *static dependability* can be calculated as

$$D_a^s = \frac{\theta_a}{\Theta}$$

where θ_a is the number of successful strategies a collaborative agent *a* can adopt; and Θ is the total number of possible strategies in this collaborative agent-based system.

The *dynamic dependability* of agent *a* at time *t* can be defined relative to interdependence and mutual support of collaborating agents. It will be higher if the availabilities of the parties assigned to execute tasks are higher and the frequencies of information exchange between them are lower; higher exchange frequencies mean higher interdependence, resulting in lower dynamic dependability. Dependability can now be illustrated through the ability to resolve conflicts and recover from errors (Figure 8.2 a, b, c).

Agility, dependability, and availability are interrelated measures of c-work. In collaborating multiagent systems, for a given value of *D*, system agility increases with *A*. For a given *A*, to increase system agility, a system needs higher *D*. With the same *A*, higher *D* is associated with higher system agility. As system availability increases, however, system agility increases slower when increasing *D*. The influence of cultural factors on systems availability, agility, and dependability is clear but requires further understanding with two main considerations: (1) the influence on the direct human-based attitudes in decision making and action while detecting, preventing, and resolving problems, errors, and conflicts, and negotiating and creating solutions; (2) the influence of cultural factors on the ability of humans to accept and trust automation support and cyber-enabled detection, prevention, resolution, and negotiations as provided by CCT and CSS.

CCT and CSS Measure: Integrability (I)

Integration of information for decisions and of tasks execution in c-work depends on four main *types of collaboration* among participants, such as robots and agents (Nof 1994): *Mandatory*, when all parties must participate; *optional*, when only some of the parties must participate; *concurrent*, when collaboration occurs at the same

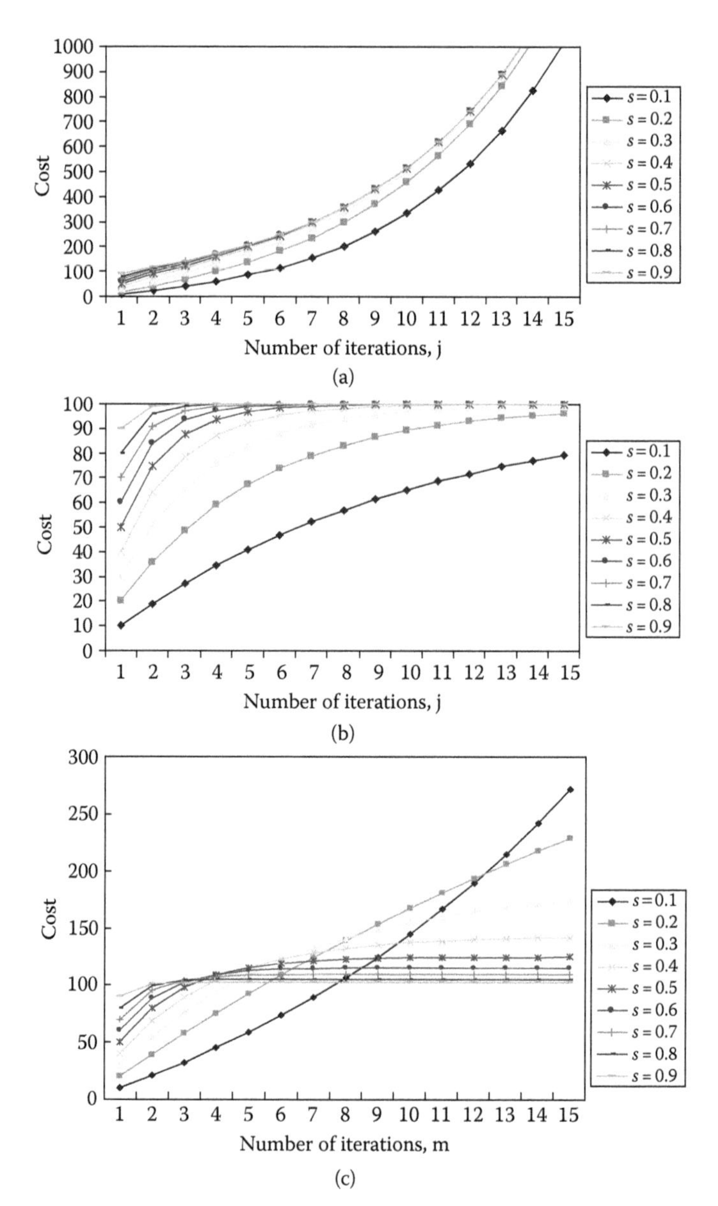

FIGURE 8.2 The cost of conflict resolution and error recovery with human participation for variable probability of resolution per iteration, *s*. Case (a): Cost increases exponentially (uncontrollable) when human participation is at or more than 20% of the time. Case (b): Cost reaches an upper bound (controllable), when CCT and CSS are applied to support errors and conflicts detection, prevention and resolution over 80% of the time. Case (c): Some costs are exponential, some are bounded, when human participation is just above and below 20% of the time. (Reprinted from *Management and Control of Production and Logistics*, Vol. 4(1), eds. O. Bologa, I. Dumitrache, F. G. Filip, Elsevier, IFAC Papers Online (http://www.ifac-papersonline.net/Detailed/39097. html), S. Y. Nof, Availability, integrability, and dependability—what are the limits in production and logistics? (Plenary), Copyright 2007, with permission from Elsevier.)

time, and can also be *competitive*; and *resource sharing*. Integration also depends on *interaction modes*, for example, by interface, group decision support system, or by cooperative information sharing or exchange. Examples of collaboration types by humans (H) and machines (M) for integration problems: Human planning of machine work (H-M) is optional collaboration with multiple processors; multirobot assembly (M-M) can be mandatory or optional, progressing in parallel; job-machine resource allocation is (H-M or M-M), concurrent and competitive; H-H collaborative decision making in supply network management can be mandatory, optional, concurrent, and competitive. Several illustrations of these c-work collaboration types and interaction modes are discussed next. In these cases, cultural factors have significant effect on the quality of the results.

Measures of integrability may be represented by the measure of collaboration type levels (Figures 8.3 and 8.4). While the theoretical limits of the effectiveness (or quality) represented by such measures are, by definition, between 0 and 1, in practice the limits are lower than 1. In the illustration, S_a increases with the number of agents/robots and reaches a maximum of about 0.7, but for more tasks with more agents, even lower levels are achieved. Similarly, for S_o levels above 0.7 are found only with few participants, but for any number of tasks.

Five recent collaboration support systems are surveyed as summarized in Table 8.2. Each of them enables c-work in various application areas, from service and logistics to production and manufacturing. The five CSSs are similar in that all

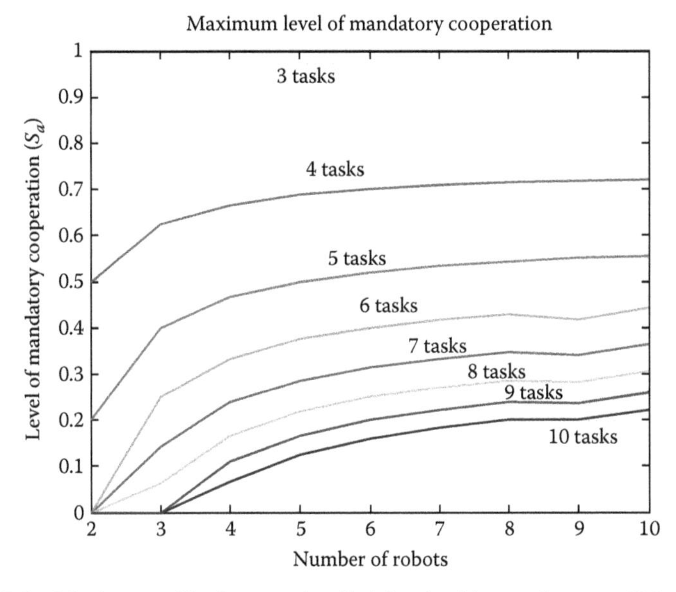

FIGURE 8.3 Maximum effectiveness (quality) level with mandatory collaboration (S_a) assuming each robot/agent can participate in at most 3–10 tasks and at least two robots/agents must perform the task simultaneously. (Reprinted from *Management and Control of Production and Logistics*, Vol. 4(1), eds. O. Bologa, I. Dumitrache, F. G. Filip, Elsevier, IFAC Papers Online (http://www.ifac-papersonline.net/Detailed/39097.html), S. Y. Nof, Availability, integrability, and dependability—what are the limits in production and logistics? (Plenary), Copyright 2007, with permission from Elsevier.)

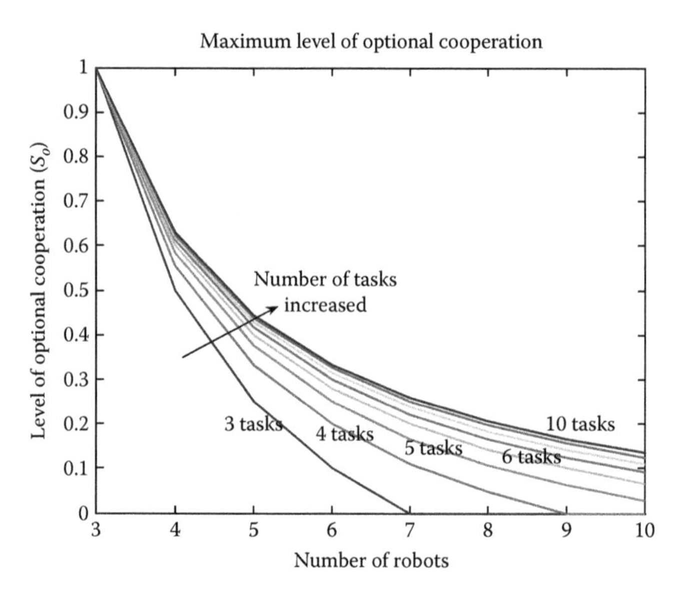

FIGURE 8.4 Maximum effectiveness (quality) level of *Optional collaboration* (S_o) assuming each task can be performed by at most three robots/agents and any one of several robots/agents can carry out the task, while only one is required; it is assumed that the effectiveness (quality) level increases when more than one robot/agent performs the tasks. (Reprinted from *Management and Control of Production and Logistics*, Vol. 4(1), eds. O. Bologa, I. Dumitrache, F. G. Filip, Elsevier, IFAC Papers Online (http://www.ifac-papersonline.net/Detailed/39097.html), S. Y. Nof, Availability, integrability, and dependability—what are the limits in production and logistics? (Plenary), Copyright 2007, with permission from Elsevier.)

of them incorporate cyber-enabled collaboration of c-work. Their particular collaboration type is indicated as follows:

- H-H-H when designers, decision makers and operators, and the beneficiaries, or customers of the given CSS are all human (Cases 1, 3)
- H-M-M when the designers and possibly some decision makers are human, some decision maker are machines, the operators are machines, there are always end-customers who are human (Cases 2, 5)
- H-M-H when the designers are human, the decision makers and some operators are machines, some operators and the customers are human (Case 4)

How do cultural factors impact the quality of c-work? As found in this survey, there are multiple effects by cultural factors, influencing all the human participants in the way they apply the CSS as decision makers and operators, and the way they are affected as customers (and end-customers, or consumers). In the case of designers and programmers, their cultural attitudes combine with their professional skills and expertise to impact the resulting quality of the designed and implemented CSS and c-work. Thus, they also influence the decision makers, operators, and customers of this CSS, who, in turn, add their own cultural attitudes and factors to influence the quality of results.

An illustration of designers' impact in Case 4 (Figure 8.5) shows that the designers influence quality by the way they select and design alternative collaboration protocols, called task administration protocols. Designers can apply in this particular system a coordination protocol (CP) which would be inferior to any task administration protocols (TAPs). But even if they select a relatively more complex TAP, they can select a better or worse logic (c-work performance under TAP 2 in this system is superior to performance under TAP 1).

The case of H-M-M c-work is illustrated in Figure 8.6 (Case 5). The designers can influence the number and type of robots/agents, their layout organization, and the exact program of how they perform c-work in ship building and in car manufacturing. The effectiveness (quality) of their design and its influence, through alternative methods of resource sharing and task collaboration, are illustrated in the graphs shown. In this case, the robotic machines are both decision-makers and operators (as in Case 2 the UAVs and sensors performing the intelligent surveillance); the customers are the human end-customers, but also include the machines and instruments which follow the CCT-based collaborative command and control decisions and operations.

Cultural factors as illustrated in the survey (Table 8.2; Figures 8.5 and 8.6) cover a wide array of influences, which can be measured by useful CCT-based measures of autonomy, viability, availability, agility, dependability, and integrability. They are also evident when focusing on information integration ability, *I* (e.g., Turban et al. 2006) from diverse sources of data and information, multiple databases, data-marts, data-warehouses, and knowledge bases. Such integrability includes the ability to integrate multimedia data from a number of diverse resources, across department, enterprise, and global boundaries, presumably increasing their usefulness; technology, information, and people integration, for example, adoption of electronic data interchange as a mechanism of digital integration. Two dimensions of information integrability in this case are connectivity, which is a technological quality measure, and *willingness to share* necessary information, which depends on cultural and behavioral attributes.

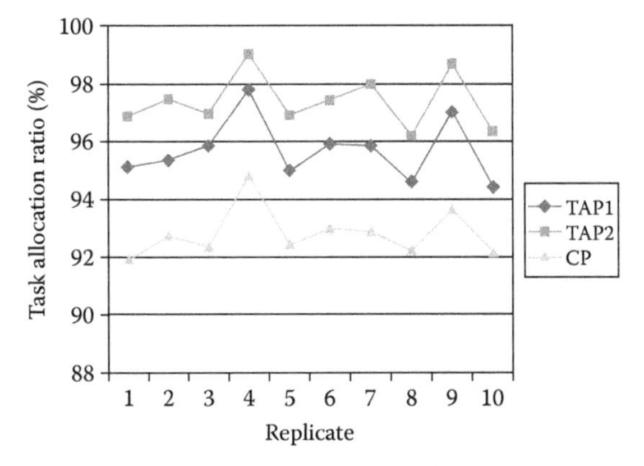

FIGURE 8.5 TestLAN collaborative performance quality, as measured by Task Allocation Ratio under three collaboration protocol types: one CP and two TAPs. (From Ko, H. S., and S. Y. Nof. 2010. *Int J Comput Commun Control* 5(1):91–105. With permission.)

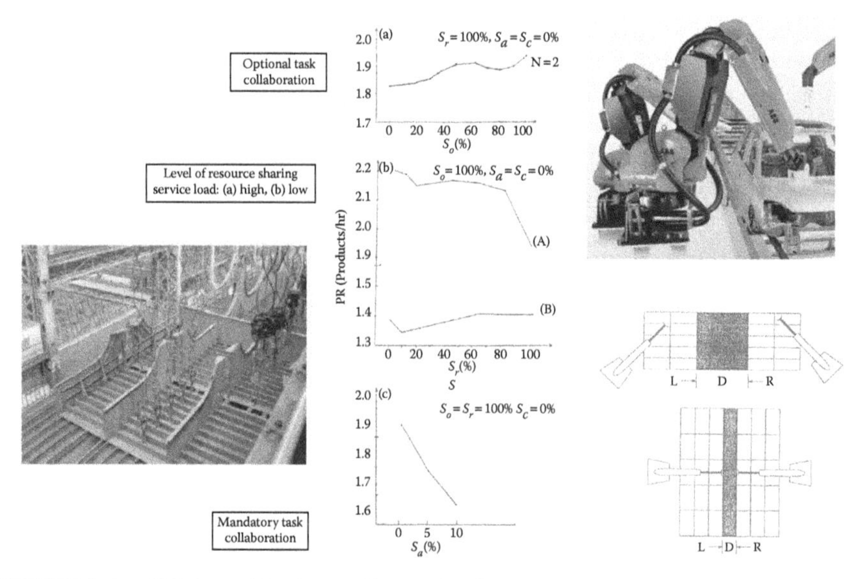

FIGURE 8.6 CSS in multi-robots collaboration for ship building and car manufacturing: Quality results depend on the design, plan, and execution of collaboration modes. (From Nof, S. Y. 2009. Automation: What it means to us around the world. In *Springer Handbook of Automation*. With permission.)

IMPLICATIONS, CHALLENGES, AND EMERGING TRENDS

Overall, cultural factors are expected to influence, and have been proven to enrich, design and decision-making activities. It is easier to dwell on hindrances:

- Products and services designed or supplied with disregard (and possibly disrespect) to the cultural factors, habits, and attitudes of their intended users
- People unwilling to share information
- Inadequate technology and information systems
- Misinterpretation of feedback and miscommunication among intended partners
- Lack of trust

For instance, the lack of compatibility may deter supply-chain integration initiatives and therefore confound companies involved in mergers and acquisitions. Our challenge is to recognize that cultural factors do influence the design quality and the outcome quality. Our further challenge is to better understand exactly what the implications of such influences are, and how they can help us benefit from the advantages of cultural factors, eliminating or limiting the potentially adverse effects. In this chapter, a number of measurements and evaluation criteria have been explained and illustrated for this purpose and for future research.

The emerging future of c-work enterprises depends on understanding how effective c-work can be designed, implemented, and applied. Collaborative control principles and techniques based on CCT provide useful insights to design guidelines and to further research in this area. Designing to achieve the benefits enabled by optimized e-activities will lead to needed levels of collaborative decisions, delivery, integrity, quality, performance, and QoS.

In this chapter, the nature of c-work, and the impact of cultural factors on the quality of design, decision making, operations, and use were discussed and illustrated. Useful measures that can explain the limits of expected quality, including viability, autonomy, agility, availability, dependability, and integrability were described. Clearly, c-work methods, systems and protocols need to be designed and selected carefully with specific attention to cultural factors. Measures related to cultural impacts on errors and conflicts, information assurance, and responsiveness are receiving increased attention because of the emerging challenges of social and environmental sustainability combined with economic sustainability.

In an effort to address these challenges, a new initiative by the Purdue School of Industrial Engineering began in 2009: *The Collaboratorium Initiative for Collaborative Intelligence* (Figure 8.7). Its goal: Enable and optimize human, system, and research collaboration while learning how to further improve them by the science of interactive collaboration. Our vision is a scientific exploration and integration resource to create/ share/ interact/ discover multidisciplinary knowledge, hands-on analysis, and research cooperation.

This initiative aims at studying and understanding the CI quality impacts. Several projects have begun over the Internet and over the platform of Purdue's HUBzero (McLennan and Kennell 2010) by adding CI functions and components. Cultural factors play a key role in these investigations. Some of the research questions that have been defined are how well can CSS for CI facilitate and deliver

- Significantly accelerated and better synthesis and integration of knowledge and discoveries
- Understanding the dynamics of interactive-collaborative c-work
- Timely delivery of critically needed discoveries and shared knowledge

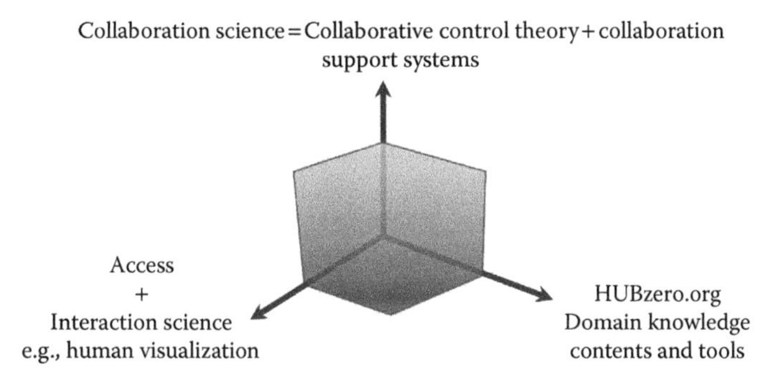

FIGURE 8.7 Purdue IE Collaboratorium Initiative for Collaborative Intelligence.

Several topics for consideration in future research include:

- Bio-inspired design and collaborative control principles
- Multilabs, research centers, and remote projects
- Networked factories and supply management with collaborative sensor- and RFID-systems
- Multienterprise work-flow innovation
- Multinational/multicultural/multidiscipline group decision making
- Multicultural grand challenge design by collaborative international teams

ACKNOWLEDGMENTS

Research reported in this chapter has been developed by the PRISM Center with support from NSF, Indiana 21st Century Fund for Science and Technology, and industry. Special thanks to my colleagues and students at the PRISM Lab and the PRISM Global Research Network, and in IFAC Committee for Manufacturing and Logistics Systems, who have collaborated with me to develop the c-work and collaborative control knowledge.

REFERENCES

Anane, R., M. Younas, C.-F. Tsai, and K.-M. Chao. 2002. Agent-based transactional framework for the supply chain. In *Proceedings of the 1st International Conference. on Machine Learning and Cybernetics*, Beijing, Vol. 4, 1956–61, IEEE.

Anussornnitisarn, P. 2003. Design of middleware protocols for the distributed ERP environment. Ph.D. dissertation, Purdue University, School of Industrial Engineering, West Lafayette, IN.

Avizienis, A., J.-C. Laprie, B. Randell, and C. Landweher. 2004. Basic concepts and taxonomy of dependable and secure computing. *IEEE Trans Dependable Secure Comput* 1(1):11–33.

Bar-Hillel, M. 1973. On the subjective probability of compound events. *Organ Behav Hum Perform* 9:396–406.

Campion, M. A., G. J. Medsker, and A. C. Higgs. 1993. Relations between work group characteristics and effectiveness. *Pers Psychol* 46:823–50.

Clemen, R. T., and T. Reilly. 2001. *Making Hard Decisions*. Pacific Grove, CA: Duxbury.

Cristian, F. 1991. Understanding fault-tolerant distributed systems. *Commun ACM* 34(2):56–78.

de Freitas, E. P., T. Heimfarth, R. S. Allgayer, F. R. Wagner, T. Larsson, C. E. Pereira, and A. Morado. 2010. Coordinating aerial robots and unattended ground sensors for intelligent surveillance systems. *Int J Comput Commun Control* 5(1):52–70.

Du Charme, W. 1970. Response bias explanation of conservative human inference. *J Exp Psychol* 85:6–74.

Fischer, G. 2006. Distributed intelligence: Extending the power of the unaided, individual human mind. In *Proceedings of ACM Advanced Visual Interfaces (AVI) Conference*, 7–14. Venice, Italy. New York: ACM.

Fischer, G. 2010. End-user development and meta-design: Foundations for cultures of participation. *J Organ End User Comput* 22(1):52–82.

Hofstede, G. 1991. *Cultures and Organizations: Software of the Mind*. New York: McGraw-Hill.

Hopp, W. J., and M. L. Spearman. 2000. *Factory Physics*. 2nd ed. New York: McGraw-Hill/Irwin.

Horvitz, A., C. M. Kadie, T. Park, and D. Hovel. 2003. Models of attention in computing and communications: From principles to applications. *Commun ACM* 46(3):52–9.
Huang, C. Y., J. A. Ceroni, and S. Y. Nof. 2000. Agility of networked enterprises—parallelism, error recovery and conflict resolution. *Comput Ind* 42:275–87.
Huang, C. Y., and S. Y. Nof. 2000a. Formation of autonomous agent networks for manufacturing systems. *Int J Prod Res* 38(3):607–24.
Huang, C. Y., and S. Y. Nof. 2000b. Autonomy and viability—measures for agent-based manufacturing systems. *Int J Prod Res* 38:4129–48.
Johnson, T. L. 2007. Improving automation software dependability: A role for formal methods? In *Control Engineering Practice*, Vol. 15, No. 11, November 2007, 1403–15 Special Issue on Manufacturing Plant Control: Challenges and Issues—INCOM 2004, 11th IFAC INCOM'04 Symposium on Information Control Problems in Manufacturing.
Johnson Jr., A. M., and M. Malek. 1988. Survey of software tools for evaluating reliability, availability, and serviceability. *ACM Comput Surv* 20(4):227–69.
Ko, H. S., and S. Y. Nof. 2010. Design of protocols for task administration in collaborative production systems. *Int J Comput Commun Control* 5(1):91–105.
Laprie, J. C. 1995. Dependability of computer systems: concepts, limits, improvement. In *Proceedings of the Sixth International Symposium on Software Reliability Engineering (ISSRE'95)*, 2–11. IEEE.
Lehto, M. R., and F. Nah. 2006. Decision-making models and decision support. In *Handbook of Human Factors and Ergonomics*, ed. G. Salvendy, 3rd ed., 191–242. Hoboken, NJ: John Wiley & Sons.
McLennan, M., and R. Kennell. 2010. HUBzero: A platform for dissemination and collaboration in computational science and engineering. *Comput Sci Eng* 12(2):48–52.
Miller, G. A. 1956. The magical number seven plus or minus two: Some limits on our capacity for processing information. *Psychol Rev* 63:81–97.
Nof, S. Y., ed. 1994. *Information and Collaboration Models of Integration*. NATO-ASI Series, Dordrecht, the Netherlands: Kluwer Academic Publishers.
Nof, S. Y. 2003. Design of effective e-work: Review of models, tools, and emerging challenges. *Prod Plann Control* 14:681–703.
Nof, S. Y. 2007a. Availability, integrability, and dependability—what are the limits in production and logistics? (Plenary). *Management and Control of Production and Logistics*, eds. O. Bologa, I. Dumitrache, F. G. Filip, Vol. 4(1), Elsevier, IFAC Papers Online (http://www.ifac-papersonline.net/Detailed/39097.html). Elsevier.
Nof, S. Y. 2007b. Collaborative control theory for e-work, e-production, and e-service. *Annu Rev Control* 31:281–92.
Nof, S. Y. 2009. Automation: What it means to us around the world. In *SpringerHandbook of Automation*, ed. S. Y. Nof, Chapter 3, 13–52, Heidelberg: Springer Publishers.
Proctor, R. W., and K.-P. L. Vu. 2006. The cognitive revolution at age 50: Has the promise of the human information-processing approach been fulfilled? *Int J Hum Comput Interact* 21:253–84.
Sabherwal, R., and I. Becerra-Fernandez. 2011. *Business Intelligence*. Hoboken, NJ: Wiley.
Schmorrow, D., K. M. Stanney, G. Wilson, and P. Young. 2006. Augmented cognition in human-system interaction. In *Handbook of Human Factors and Ergonomics*, ed. G. Salvendy, 3rd ed., 1364–83. Hoboken, NJ: John Wiley.
Shneiderman, B. 2009. *Designing the User Interface: Strategies for Effective Human-Computer Interaction*. 5th ed. Reading, MA: Addison-Wesley.
Taveira, A. D., and M. J. Smith. 2006. Social and organizational foundations of ergonomics. In *Handbook of Human Factors and Ergonomics*, ed. G. Salvendy, 3rd ed., 269–91. Hoboken, NJ: John Wiley.
Turban, E., J. E. Aronson, T. P. Liang, and R. Sharda. 2006. *Decision Support and Business Intelligence Systems*. 8th ed. Upper Saddle River, NJ: Prentice Hall.

Tversky, A., and D. Kahneman. 1974. Judgment under uncertainty: Heuristics and biases. *Science* 185:1124–31.

Velasquez, J. D., and S. Y. Nof. 2009. Collaborative e-work, e-business, and e-service. In *Springer Handbook of Automation*, ed. S. Y. Nof, Chapter 88, 1549–76. Heidelberg: Springer Publishers.

Yoon, S. W., M. Matsui, T. Yamada, and S. Y. Nof. 2011. Analysis of effectiveness and benefits of collaboration modes with information- and knowledge-sharing. *J Intell Manuf*, 22(1):101–12.

Zakarian, A., and A. Kusiak. 1997. Modeling manufacturing dependability. *IEEE Trans Rob Autom* 13(2):161–8.

Zamfirescu, C. B., and F. G. Filip. 2010. Swarming models for facilitating collaborative decisions. *Int J Comput Commun Control* 5(1):125–37.

9 Effects of Group Orientation and Communication Style on Making Decisions and Interacting with Robots

Pei-Luen Patrick Rau and Ye Li

CONTENTS

INTRODUCTION

Recent interest in cultural differences in decision making under a robot's influence has arisen from the internationally expanding use of service robots. According to a report of world robotics from the IFR Statistical Department, 63,000 service robots for professional use and 7.2 million units for personal and private use had been sold by the end of 2008. This number is expected to reach 112,000 and 18.8 million respectively by 2012 (IFR Statistical Department 2009).

Moving from industry to daily life, robots are designed to be more sociable and interactive, and increasingly engage in the decision-making process. For example, service robots in defense and rescue are able to collect and present sensor information about the environment and provide recommendations to end users (Stubbs, Wettergreen, and Hinds 2007), and robots for the health care of older adults could monitor the user's health condition and provide reminders and guidance (Pineau et al. 2003). In both cases, robots play the role of advisor and assistant in the decision-making process. As a result, analyzing people's perceptions of service robots is an interesting topic in the field of human-robot interaction. In addition, triggered by growing demand for service robots worldwide, interest in the effect of cultural differences in human-robot interaction has grown tremendously. Previous studies have shed some light on the diverse mental models and attitudes toward service robots across cultures (Bartneck et al. 2005; Kaplan 2004).

Studies in psychology, sociology, and communication science have provided valuable results concerning cultural differences in people's decision-making styles, the formation of trust, and the perception of others' opinions. As shown in previous studies, self-construal, communication styles, and group orientation play a role in predicting an individual's perception and reaction in decision making across cultures. Whether the findings in interpersonal communication are mirrored in human-robot interaction has not been fully verified.

In this chapter, we review the cross-cultural studies in the following aspects: self-construal, communication style, group orientation, and their impacts on decision making. The studies in robot-mediated decision making are then reviewed within relevant theoretical frames. Finally, we propose the direction of future work in the field of cross-cultural studies on robot-mediated decision making.

INDIVIDUAL LEVEL FACTOR: SELF-CONSTRUALS

Regarding cultural differences in decision making, scholars have made great progress in identifying characteristics of different cultures (Hall 1990; Hofstede 1984). Previous studies have contributed in linking Hofstede's cultural dimension of individualism versus collectivism with individuals' decision-making styles (Brew, Hesketh, and Taylor 2001; Gaenslen 1986; Mann et al. 1998). However, it is questionable to generalize the findings in cultural influence in decision making across national boundaries, since there is evidence to show that decision-making styles in closely related countries can be significantly different from each other (Chu et al. 2005).

Self-construal as an individual level was identified as one of the most influential factors in predicting individuals' perceptions, motivations, and behaviors in interpersonal interactions across cultures. In this section, we review the theoretical framework of self-construal and experimental findings in how self-construal predicts decision-making styles across cultures.

THEORETICAL FRAMEWORKS OF SELF-CONSTRUAL

The framework of self-construal has developed over many years. There are two widely discussed theoretical frameworks, one with two levels of self-construal and another with three levels. Markus and Kitayama (1991) proposed a framework of two levels of self-construal: independent self and interdependent self. Independent self-construal refers to the self as an autonomous, independent person, whereas interdependent self-construal refers to the more public components of the self that are more connected with others.

Through a thorough review of the psychological and anthropological studies, Markus and Kitayama (1991) found that Asian cultures emphasize interdependency between individuals, whereas American culture values independence more than interconnections among individuals. This theory has been supported and extended by further research (Fiske et al. 1998; Markus and Kitayama 1994; Markus, Kitayama, and VandenBos 1996; Oyserman, Coon, and Kemmelmeier 2002). For example, Oyserman, Coon, and Kemmelmeier (2002) conducted a cross-national meta-analysis of empirical literature evaluating the effects of individualism-collectivism on self-concept, well-being, cognition, and relationality. The results indicated that China stands out from other Asian countries as being both more collectivistic (e.g., feeling duty to the in-group) and less individualistic (e.g., valuing personal independence), whereas European Americans were both more individualistic and more collectivistic. However, this study found that self-construal could be merely moderately predicted by the cultural dimensions of individualism and collectivism.

In Markus and Kitayama's (1991) research in self-construal, cultural groups are not treated as completely homogeneous. Instead of generalizing their observation to the whole cultural group, they suggested in later studies that individual differences may exist within cultural groups (Fiske et al. 1998), but members belonging to a certain cultural group are highly likely to exhibit behavior and perceptions in compliance with cultural norms (Markus and Kitayama 1994).

However, other researchers found evidence that contradicted Markus and Kitayama's theoretical framework (Aron et al. 2004; Dinnel, Kleinknecht, and Tanaka-Matsumi 2002; Kim et al. 1996; Matsumoto 1999). For example, Kim et al. (1996) argued that when comparing interdependent self-construal of participants in the United States, Japan, Korea, and Hawaii, participants from the first three cultural groups did not differ on the scale of interdependent self-construal and participants from Hawaii had the highest scores on this scale, contrary to predictions based on Markus and Kitayama's theory. Through a thorough review of the literature directly testing cultural differences in self-construal, Matsumoto (1999) critically challenged the theory of two levels of self-construal. He concluded that this framework for explaining observed national differences in self-construal lacks validity.

Efforts have been made to modify the theoretical framework of self-construal. Cross, Bacon, and Morris (2000) proposed the concept of relational-interdependent self-construal, which describes the tendency to define self in terms of relationships with close others. Integrating relational self-construal with Markus and Kitayama's two levels of self-construal, Brewer and Gardner (1996) proposed a new theoretical frame of self-concept, which consists of three self-representations: the individual self, the relational self, and the collective self. They defined the individual self as the self that contains aspects differentiating the person from others, the relational self as the self that contains aspects shared with relational partners and defining a person in relationships, and the collective self as the self that contains aspects differentiating in-group members from out-group members. The three representations of the self were assumed to coexist in the same individual (Brewer and Gardner 1996; Sedikides and Brewer 2001).

The framework of three levels of self-construal was supported in a later study (Brewer and Chen 2007). In this study, researchers conducted a content analysis of the existing scales and clarified the conceptual confusion in defining collectivism. They suggested making a crucial theoretical distinction between relational collectivism and group collectivism, which further differentiates between the relational self-representation and the group self-representation.

Effect of Self-Construal

Previous study in cultural differences in decision making has indicated that people with different cultural backgrounds (e.g., individualism versus collectivism) hold distinct decision-making styles (Brew, Hesketh, and Taylor 2001; Gaenslen 1986; Mann et al. 1998). However, inhomogeneity in decision-making styles has been observed in similar cultures or even in the same culture (Chu et al. 2005). This finding has prompted researchers to seek the underlying causes of different decision-making styles and decision-making processes.

Self-construal has been recognized as one of the influential factors in predicting the decision-making style and interpersonal relationships. It has been found that a group composed of members with high independent self-construal preferred the competitive strategy to the cooperative one, whereas a group composed of members with low independent self-construal held an opposite opinion (Oetzel 1998). This finding was confirmed by Iyengar and Lepper (1999), indicating those with an interdependent nature showed an increased tendency to conform to others' decisions. In the study on the relational-interdependent self-construal by Cross, Bacon, and Morris (2000), participants with a high relational-interdependent self-construal tended to value others' needs and wishes when making decisions. These participants were more likely to think of previously unfamiliar partners as open and responsive and this consequently led to positive evaluations of the relationships. Similarly, the recent study of Morry and Kito (2009) indicated that participants with a high relational self-construal tended to present more relationship-supportive behaviors (e.g., trust and intimate disclosure) and have higher fulfillment of friendship functions (e.g., help, emotional security, and stimulating companionship), and consequently they evaluated the relationship as having a higher quality. The findings showed that relational self-construal could predict an individual's willingness to maintain or improve the quality of relationships with others.

We conclude that the different views of the self affect an individual's cognition and goals in social interactions, which consequently influence their behaviors and goals in decision making. People with a relational interdependent self-construal are more likely to make a decision that can maintain or promote social connectedness whereas those with an independent self-construal tend to make decisions to fulfill personal accomplishment.

EFFECT OF COMMUNICATION STYLES

In a face-to-face group discussion, verbal communication is an essential component. In this section, we review literature in the intercultural communication and psychology fields to identify cultural differences in individuals' communication styles and examine studies that indicate the influence of communication styles on decision making.

Cultural Differences in Communication Styles

Communication style refers to a "meta-message" reflecting the way individuals accept and interpret verbal messages (Gudykunst, Ting-Toomey, and Chua 1988, p. 100). One frequently used dimension of communication styles is indirect (implicit) versus direct (explicit) style, describing the extent to which speakers reveal their intentions through explicit messages. Someone who uses a direct style of communication will explicitly state his/her feelings, desires, and intentions, whereas someone who uses an indirect communication style will camouflage and conceal his/her true intentions when communicating verbally (Gudykunst, Ting-Toomey, and Chua 1988, p. 100). As found in cross-cultural communication and psychology research, Western people prefer to use a direct communication style whereas Eastern people favor an indirect communication style (Hall 1976; Kim et al. 1996; Sanchez-Burks et al. 2003).

The cultural dimension of high-low context proposed by Hall and Hall (1990, p. 6) provides a theoretical framework to explain the different preferences for communication styles across cultures. They defined context as the environment and circumstances around the transmitted message. People from low-context cultures (e.g., English, Nordic, and German) rely more on the explicit message and pay less attention to the surrounding information, whereas people from high-context cultures (e.g. Latin-European, Latin American, Arabic, and East Asian) pay more attention to the contextual information and rely less on direct information (Hall 1976).

In Hofstede's (1980) framework, the Eastern-Western differences in communication styles can be clarified along the cultural dimension of individualism versus collectivism. In a collectivistic culture, people value the interpersonal relationship and believe the implicit communication style better maintains interpersonal harmony (Ting-Toomey 1999). However, some researchers have suggested an individual's preference for implicit or explicit communication styles can be better predicted by the person's self-construal and values than by the cultural dimension of individualism-collectivism (Gudykunst et al. 2006).

Since cultural differences exist in communicative behaviors, we propose that communicating in a cultural normative way can enhance the understandability of the transmitted message. In the study of Gudykunst, Kim, and Cudykunst (2002), participants with a high-context cultural background (e.g., Japanese and Hong Kongers) indicated that they could understand their partners better when they adopted an implicit communication style. In addition, the scholars found that the culturally normative communication style enhanced people's responsiveness to the transmitted message and affected their perception of the relationship between the communicators

(Gudykunst, Kim, and Cudykunst 2002). Another study supported this finding by comparing the degree of misunderstanding of indirect messages across three cultures (U.S., Chinese, and Korean). The results indicated that compared with Eastern people, Americans made more errors in interpreting the indirect communication, particularly in a work environment (Sanchez-Burks et al. 2003).

In the process of designing sociable agents (e.g., social robots or interactive computer software), it is important to understand that each culture is linked to a unique set of communicative rules and patterns, since such agents are expected to engage and interact with users with different cultural backgrounds in a sociable manner. It has been found that designers tend to reflect their own communicative norms in the products, which may result in communication failure when the product is interacting with end users in another culture (O'Neill-Brown 1997).

Communication Styles Affect Decision Making

Previous studies suggested that communication styles had varying degrees of influence on decision makers. Ting-Toomey et al. (1991) carried out a study in Korea and the United States to analyze cultural differences in people's perception of requests from others. The results strongly indicated that U.S. participants found directness to be the most effective strategy for making requests, whereas Korean participants considered it the least effective strategy. This finding highlights the importance of adopting a culturally normative communication strategy for making requests, which may significantly influence the decision maker's acceptance of the request.

Gallois, Callan, and Palmer (2006) examined the influence of communication styles on hiring decisions. They compared the likeability and effectiveness of three communication styles (aggressive, assertive, and nonassertive), and concluded that the assertive applicants were more likely to be hired compared with other applicants; for aggressive and nonassertive candidates, the interviewers' sex-role beliefs affected their hiring decisions. The findings indicated that people were most responsive to a communication style that was similar to their own styles, and they were inclined to accept the requests transmitted in a familiar communication style.

According to the media equation proposed by Reeves and Nass (1998), people's reaction and response to media (e.g., television, computers) were similar to their reaction and response to other people in interpersonal interactions. We would expect that the effect of communication style on an individual's decision making could be generalized to communications between humans and artificial agents.

Robot's Influence on Decision Making Across Cultures

A social robot is a specific type of robot, designed to function as the user's partner in a social context rather than being used simply as a tool. Social robots have expanded capacity and improved autonomy, and therefore can function in entertainment, education, health care, or scientific work situations (Breazeal 2002).

In the scope of our study, we focus on social robots that play the role of information provider. In this circumstance, some crucial criteria for robot design are perceived trustworthiness, perceived reliability, and the active engagement capability.

TABLE 9.1
Identified Factors That Influence a User's Perception of Social Robots in HRI

Robot
- Appearance (Powers and Kiesler 2006; Syrdal et al. 2007)
- Autonomy (Syrdal et al. 2008)
- Communication style (Rau, Li, and Li 2009; Wang et al. 2010)
- Designed gender (Powers et al. 2005)
- Physical presence (Kidd and Breazeal 2004; Shinozawa et al. 2005)
- Social behavior (Kemper et al. 2009; Shinozawa et al. 2004)

Human
- Cultural background (Bartneck et al. 2005; Evers et al. 2008)
- Gender (Mutlu et al. 2006)
- Personality (Walters and Dautenhahn 2005)
- Previous experience with robots (Bartneck et al. 2007)

Task
- Matching between a robot's appearance and task (Goetz, Kiesler, and Powers 2003)
- Task structure (Mutlu et al. 2006)

Context
- Group relationship (Evers et al. 2008; Wang et al. 2009)
- Spatial relationship (Michalowski, Sabanovic, and Simmons 2006; Tasaki et al. 2005; Walters et al. 2006)

There have been many studies recently on the influence of social robots as information providers. Some factors that have been identified as influencing users' perceptions of social robots in human-robot interaction are listed in Table 9.1.

Among the vast influential factors, we are particularly interested in those that exert significant differences across cultures, since we expect such social robots will function in public to interact with users from different nations. From the viewpoint of a robot producer responding to the global market, this research is also very valuable. Since robotic products function in different countries, understanding the national cultural differences in users' perceptions of and reaction to these robotic products and integrating these cultural factors in the design process is crucial to develop culturally adaptable products.

When people interact with others from a different cultural background, they often subconsciously shift their value, behavior, and language corresponding to the cultural background of the counterparts (Briley, Morris, and Simonson 2005; Heine and Lehman 1997; Marian and Kaushanskaya 2004; Ross, Xun, and Wilson 2002). Mirroring the adaptive behaviors or language to robotic design, we expect future social robots will be able to detect the user's cultural background and adjust their behavior, language, and thinking mode correspondingly. To realize this desirable goal, we work on establishing a model that reflects the factors that influence users' perceptions of social robots across cultures. The findings in interpersonal communication and social science, as reviewed in the Self-Construals section, provide a solid foundation to establish our theoretical framework. We hypothesize that people's cultural backgrounds and culture-related behavior and cognition (e.g., communication styles, self-construal, and group membership) might exert similar effects on decision making when people interact with humans versus social robots.

However, directly applying findings in interpersonal interaction to interaction between a human and a social robot could cause problems since the relationship between a human and a robot is fundamentally different from that between human beings. For example, in human-robot interaction, the human typically plays the role of supervisor, operator, mechanic/programmer, teammate, or bystander (Yanco and Drury 2004), which is not comparable with interpersonal relationships.

Effect of Robots' Communication Style on Decision Making

Previous research found that people are more inclined to form trust with an unfamiliar partner based on culturally consistent signs (i.e., perceived ability and integrity for individualists, predictability, and benevolence for collectivists) without considering culturally inconsistent ones (Branzei et al. 2007). We are interested in verifying this cultural effect in the scope of human-robot interaction. We propose that when the initial trust is desired, for example, in a commercial promotion or rescue occasion, it would be beneficial and perhaps even necessary to design culturally consistent signs into the intelligent agents to improve their trustworthiness.

In 2007, we explored the effects of communication styles and cultural background on people's acceptance of a robot's recommendation in a laboratory experiment. The independent variables we chose were communication style (implicit versus explicit), participants' cultural backgrounds (Chinese versus German), and the robot's language (native language versus second language for Chinese and German participants). The research framework is shown in Figure 9.1.

We designed a price-estimation task with a robot assistant, in which the participants were presented with a series of product photos and asked to choose a price from two optional prices for each product. A Lego NXT robot was developed to suggest a price for the participants in the experiment (see Figure 9.2). In the implicit condition, the robot provided recommendations in an implicit and indirect way (e.g., saying "Why not think twice?" to express disagreement), whereas in the explicit condition, it expressed its recommendations in an explicit and direct way (e.g., saying "I do not agree with you" to express disagreement). The native languages for Chinese and German participants were Chinese and German respectively. The second language for all participants was English. We calculated the judgment changes in response to the robot's recommendations, and used a post-experiment questionnaire to measure likeability and trust in the robot as well as the perceived credibility of the robot.

We carried out the experiment with 16 Chinese and 16 German college students. The data showed that cultural differences existed in participants' perceptions and acceptance of the robot's recommendations. In contrast with German participants, the Chinese participants evaluated the robot as being more likeable, trustworthy, and credible; the Chinese participants were more inclined to accept the implicit recommendations from the robot (Rau, Li, and Li 2009).

We concluded that when the robot communicated in a style that was more familiar to the participants (i.e., implicit style for Chinese participants, explicit style for German

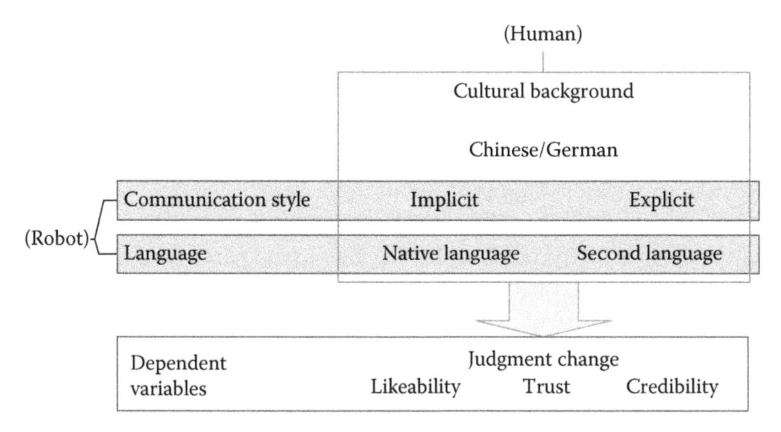

FIGURE 9.1 Research model of cultural background, communication styles, and language.

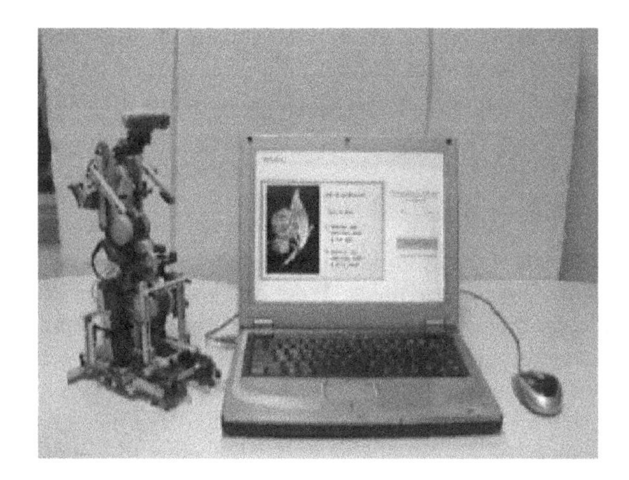

FIGURE 9.2 Lego NXT robot and the computer interface.

participants) they were more likely to give positive evaluations of the robot and to accept the robot's recommendations. This observation agrees with findings in interpersonal communication research, which suggested that people from high-context cultures (e.g., Chinese) are more sensitive to contextual cues and prefer to use implicit communication styles in social communication, whereas people from low-context cultures (e.g., German) prefer to communicate explicitly (Branzei et al. 2007; Hall 1976).

In addition, we found evidence that the Chinese participants had generally higher acceptance of the robot's recommendations than German participants did. This corresponded with the previous findings in psychology and communication science that indicated that the cultural dimension of individualism (e.g., German) versus collectivism (e.g., Chinese) predicted people's reactions to others' opinions when making decisions. According to Gaenslen (1986), individualistic decision makers are less influenced by interpersonal relations and are mainly reactive to the task environment when making decisions, whereas collectivistic decision makers highly value interpersonal relations and are primarily concerned with the behaviors of fellow participants. It was confirmed in another study that participants from Western countries were more confident about their decision-making abilities than those from Eastern Asian countries (Mann et al. 1998).

Although the contribution of this study was limited by the relatively small sample size and the special background of the German participants (i.e., they were studying in China), we believe the findings confirmed the cultural differences in people's perceptions of a social robot as a recommendation provider. The findings partially supported the generalization of the media equation on robots, that is to say, the way people treat and respond to media (such as the robots in our study) is similar to the way they treat and respond to other people in social interaction.

GROUP ORIENTATION: IN-GROUP VERSUS OUT-GROUP

In this section, we review literature from psychology and communication science to gain an overview of the impact of group orientation on decision making in a cross-cultural context. First, we present findings of in-group favoritism and review studies examining the impact of group orientation on the initial trust for strangers. Then we compare this impact in a cross-cultural context.

In-Group–Out-Group Bias

Differences in attitude towards an in-group member versus an out-group member have been recognized since the 1970s. Tajfel and his colleagues' series of experiments established the foundation for the research on the in-group bias. They analyzed participants' strategies of interest allocation for in-group and out-group members in a minimal group setting; the members were separated on a purely cognitive criterion, such as shape or pattern preferences. The findings indicated that in-group favoritism and intergroup discrimination were sufficiently triggered by the sense of belonging to two distinct groups. In some circumstances, participants even sacrificed their personal interest to increase the intergroup differences and favor the in-group (Tajfel 1974; Tajfel et al. 1971; Turner, Brown, and Tajfel 1979).

Later studies by other scholars confirmed Tajfel and colleague's findings and reported that membership in a group increases cooperation by increasing the positive evaluation of in-group members (Boldizar and Messick 1988; Brewer 1979; Chen, Brockner, and Chen 2002; Kramer and Goldman 1995; Yamagishi, Jin, and Miller 1998). For example, the behavior of the in-group members was considered to be more fair (Boldizar and Messick 1988), and the in-group members were perceived to be more trustworthy and cooperative than the out-group members (Brewer 1979).

It has been found that such perceived positive qualities could result in initial trust in strangers who are categorized as in-group members. Foddy, Platow, and Yamagishi (2009) reported that people placed higher trust in in-group members than out-group members. In their study, participants were asked to choose between an in-group (the same university or major) and an out-group (a different university or major) allocator in a money allocation task. Participants showed significant preference for the in-group allocator. However, this preference vanished when participants confirmed that the allocator did not know their group membership. They concluded from this experiment that the trust in in-group members was triggered by expectations of favorable treatment from fellow in-group members in comparison to expected treatment from out-group members, and such initial trust was not likely a result of perceived positive qualities of in-group members. This study stressed the importance of mutual awareness of shared group membership in triggering higher trust toward an in-group member than toward an out-group member.

Cultural Differences in Group Orientation

The formation of in-group bias is affected by an individual's cultural background. Triandis and colleagues (1988) studied people's individualism and collectivism constructs across cultures and linked individual differences in the constructs with the in-group perception and decision-making behaviors. They conducted factor analyses on the scale of the individualism and collectivism constructs and found that American individualism could be characterized as self-reliant, with competition, low concern for in-groups, distance from in-groups, and subordination of in-group goals to personal goals. They reported that U.S. participants tended to be less concerned with the in-groups' opinions and that they made decisions on their own rather than asking about the opinions of other group members. When comparing individuals' perception of the in-group membership among Japanese and Puerto Rican and American participants, they found no systematical differences in terms of individualism and collectivism constructs. They suggested that when predicting people's perception of in-groups, the constructs of individualism and collectivism should be considered critically, since it was closely related to the in-group type (e.g., family, coworkers, friends, etc.), the context, and the type of social behavior (e.g., perceiving similarity, submitting to others' views).

To gain insight into in-group favoritism across cultures, Yuki (2003) conducted a survey in Japan and the United States to analyze individuals' willingness to sacrifice for the in-group (in-group loyalty) and the strength of in-group identification (in-group identity). The results indicated that Japanese participants' in-group loyalty was predicted by their understanding of intragroup relational structures, differences among in-group members, and the sense of personal connectedness. In addition to the above items, American participants' in-group loyalty was also predicted by perceived in-group homogeneity and in-group status.

Based on the findings of cultural differences in group perception, Yuki et al. (2005) conducted a questionnaire survey and an online money allocation game to analyze the impact of group relationships on the initial trust across cultures. The results indicated that American participants were more highly identified with their in-groups than Japanese participants; for American participants, the in-group identification was significantly correlated with trust. They suggested that Americans had higher trust toward in-group members. On the other hand, they found that compared with American participants, Japanese participants more frequently assumed indirect connections with others, and their trust was more influenced by a potential relational linkage. Yuki et al.'s research highlighted the different ways of forming the group identity for Americans versus for Japanese and indicated the consequent effect on the formation of initial trust toward strangers.

Is an In-Group Robot More Trustworthy?

Although group orientation has a great impact on people's attitudes towards group members and consequently influences people's decision making across cultures, similar effects have not yet been fully verified in human-robot interaction. Evers et al. (2008) carried out an experiment to analyze the effects of cultural background (U.S. versus Chinese) and in-group strength (weak versus strong) on collaborative behaviors and intentions with a human assistant versus a robot assistant. Before the experiment, they collected situations reflecting weak and strong in-group strength from Chinese and U.S. participants, and identified a situation when the team member had a history with the team and had shared successes as the strong in-group condition for both nations. The result from the laboratory experiment critically challenged the idea of directly applying the findings in interpersonal interaction to human-robot interaction. They suggested that participants from the United States had stronger relational self-construal, resulting in overall higher trust of and compliance with the assistants, which agreed with the findings in interpersonal communications (Cross, Bacon, and Morris 2000; Iyengar and Lepper 1999). However, Chinese participants' positive attitude toward the in-group members (i.e., feeling comfortable) did not hold for the robot assistant but only for the human assistant. Furthermore, although carefully controlled, their in-group strength manipulation was effective with human assistants but not with robotic assistants (Evers et al. 2008). The different perception of in-group with a human partner versus a robot partner and the different positive attitudes towards the two types of agents indicated that participants might subconsciously develop group relationships with a robot in a different way to that with a human being. Although the results may have been biased by the machine-like appearance of the robot, they were useful in comparing the formation of relationship and the initial trust with social robots versus human beings.

To find further evidence to support or challenge Evers' findings, we carried out a cross-cultural experiment to examine the effect of cultural background (U.S. versus Chinese), a robot's communication style (implicit versus explicit), and its in-group strength (strong versus weak) on people's responsiveness to and trust in the robot (Wang et al. 2010). The research framework is shown in Figure 9.3.

Participants formed a team of two, and they were assisted by a mobile robot in performing a collaborative task. They needed to make a decision to establish an environmentally friendly and productive chicken cooperative. In the implicit situation, the robot gave recommendations in an implicit style (e.g., "More dense bedding could protect the eggs better"), whereas in the explicit situation, the robot spoke in an explicit way (e.g., "We should choose the most dense bedding to protect the eggs"). In the in-group situation, the robot and the two participants in the team were tagged with red tags; in the out-group situation, only the two participants were provided red tags. A Wizard of Oz method was used in the experiment, in which the robot was remotely controlled but appeared autonomous to the participants. The laboratory setting is shown in Figure 9.4.

A total of 80 Chinese teams and 80 U.S. teams participated in the experiment, recruited from a Chinese university and a U.S. university respectively. We found evidence to support that the participants built more trust in and were more influenced by a robot that communicated in the culturally normative style. Chinese participants showed higher trust in the robot when it communicated implicitly, whereas the U.S. participants decreased their trust in these conditions; Chinese participants aligned their decisions more frequently with the robot's recommendations when it communicated implicitly, and U.S. participants were more inclined to change their decisions when the robot communicated explicitly. These findings supported our previous observations of Chinese and German participants with a robotic recommender (Rau, Li, and Li 2009).

Similar to Evers' experiment, although carefully designed and controlled, the manipulation check on the robot's group orientation was not successful; in other

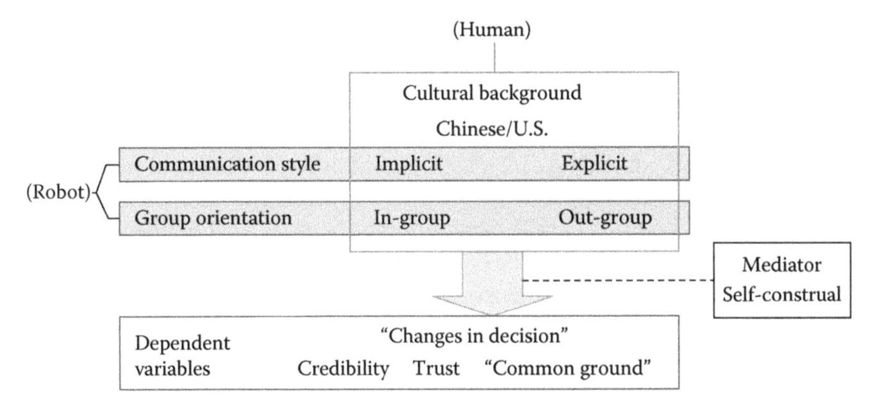

FIGURE 9.3 Research model of nationalities, group orientation, and communication styles.

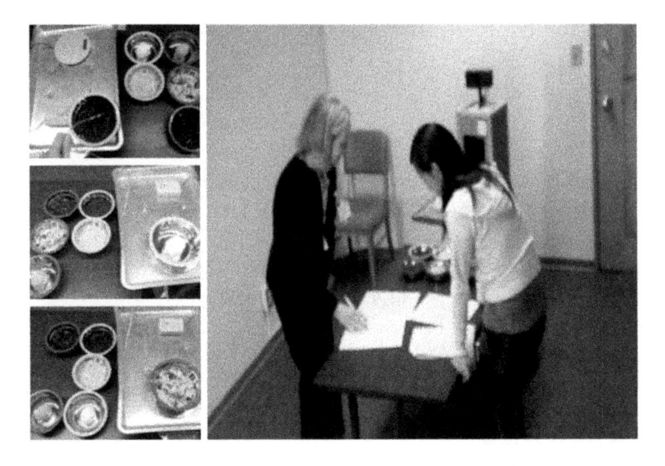

FIGURE 9.4 Setup for measuring temperature, weight and density (left, top to bottom), two U.S. participants working with the robot (right).

words, the participants did not see the robot as more or less of an in-group member corresponding to our manipulation. It is possible that our manipulation was insufficient to differentiate between strong and weak in-groups, but it is more likely that people's perception of the in-group might be fundamentally different with a robot versus with a human partner. If that were the case, the attitudes toward and trust in a social robot would not be effectively improved when it was designed to be an "in-group" member according to the interpersonal perception of in-group. Future work is desirable to find out people's real perception of in-group with a robot partner.

FUTURE WORK

As a rapidly progressing field, research about interaction between humans and social robots combines knowledge about artificial intelligence, psychology, communication science, and sociology, but is still in the initial stages. It calls for joint efforts of researchers from multiple disciplines to establish a theoretical framework to predict users' perceptions of and reaction to such artificial agents across cultures. When the robot plays the role of a decision supporter in an intercultural environment, the way people from different cultures perceive and react to them directly influences the efficiency of the decision-making process and the quality of decisions. Previous findings have found that people formed different mental models and perceptions of a robot partner versus a human partner, and cultural differences existed in this cognitive process. Future study should work towards a theoretical model with attributes of the interacting agents (i.e., the robot and the users) as well as the relationship between the agents, the task, and the context in which the task is carried out in order to predict people's perceptions and reactions in interactions with social robots. We suggest that caution must be taken when directly applying the findings in interpersonal communication and psychology to human-robot interaction, and the results from human-robot experiments should be critically explained. Comparative experiments with a human target versus a robot target are promising, especially when the effect of the relationship between the interacting agents is one of the independent factors, and such studies will provide persuasive evidence to support researchers' hypotheses.

REFERENCES

Aron, A., T. McLaughlin-Volpe, D. Mashek, G. Lewandowski, S. Wright, and E. Aron. 2004. Including others in the self. *Eur Rev Soc Psychol* 15(4):101–32.

Bartneck, C., T. Nomura, T. Kanda, T. Suzuki, and K. Kato. 2005. Cultural differences in attitudes towards robots. Paper presented at the AISB 2005 Symposium on Robot Companions: Hard Problems and Open Challenges in Robot-Human Interaction, University of Hertfordshire, Hatfield, UK.

Bartneck, C., T. Suzuki, T. Kanda, and T. Nomura. 2007. The influence of people's culture and prior experiences with Aibo on their attitude towards robots. *AI Soc* 21:217–30.

Boldizar, J., and D. Messick. 1988. Intergroup fairness biases: Is ours the fairer sex? *Soc Justice Res* 2(2):95–111.

Branzei, O., I. Vertinsky, I. Camp, and D. Ronald. 2007. Culture-contingent signs of trust in emergent relationships. *Organ Behav Hum Decis Process* 104:61–82.

Breazeal, C. L. 2002. *Designing Sociable Robots*. Cambridge, MA: MIT Press.

Brew, F., B. Hesketh, and A. Taylor. 2001. Individualist-collectivist differences in adolescent decision making and decision styles with Chinese and Anglos. *Int J Intercult Relat* 25:1–19.

Brewer, M. B. 1979. In-group bias in the minimal intergroup situation: A cognitive-motivational analysis. *Psychol Bull* 86:307–24.

Brewer, M. B., and Y.-R. Chen. 2007. Where (who) are collectives in collectivism? Toward conceptual clarification of individualism and collectivism. *Psychol Rev* 114:133–51.

Brewer, M. B., and W. Gardner. 1996. Who is this "we"? Levels of collective identity and self representations. *J Pers Soc Psychol* 71:83–93.

Briley, D. A., M. W. Morris, and I. Simonson. 2005. Cultural chameleons: Biculturals, conformity motives, and decision making. *J Consum Psychol* 15:351–62.

Chen, Y., J. Brockner, and X. Chen. 2002. Individual-collective primacy and ingroup favoritism: Enhancement and protection effects. *J Exp Soc Psychol* 38:482–91.

Chu, P., E. Spires, C. Farn, and T. Sueyoshi. 2005. Decision processes and use of decision aids: Comparing two closely related nations in east Asia. *J Cross Cult Psychol* 36:304–20.

Cross, S., P. Bacon, and M. Morris. 2000. The relational-interdependent self-construal and relationships. *J Pers Soc Psychol* 78:791–808.

Dinnel, D., R. Kleinknecht, and J. Tanaka-Matsumi. 2002. A cross-cultural comparison of social phobia symptoms. *J Psychopathol Behav Assess* 24:75–84.

Evers, V., H. Maldonado, T. Brodecki, and P. Hinds. 2008. Relational vs. group self-construal: Untangling the role of national culture in HRI. Paper presented at the 3rd ACM/IEEE International Conference of Human-Robot Interaction, Amsterdam, the Netherlands.

Fiske, A., S. Kitayama, H. Markus, and R. Nisbett. 1998. The cultural matrix of social psychology. In *The Handbook of Social Psychology*, ed. D. T. Gilbert, S. T. Fiske, and G. Lindzey, Vol. 2, 915–81. New York: McGraw-Hill.

Foddy, M., M. Platow, and T. Yamagishi. 2009. Group-based trust in strangers. *Psychol Sci* 20:419–22.

Gaenslen, F. 1986. Culture and decision making in China, Japan, Russia, and the United States. *World Polit* 39:78–103.

Gallois, C., V. Callan, and J. Palmer. 2006. The influence of applicant communication style and interviewer characteristics on hiring decisions. *J Appl Soc Psychol* 22:1041–60.

Goetz, J., S. Kiesler, and A. Powers. 2003. Matching robot appearance and behavior to tasks to improve human-robot cooperation. Paper presented at the IEEE International Workshop on Robot and Human Interactive Communication, October 31–November 2, 2003, MillBrae, California.

Gudykunst, W. B., Y. Y. Kim, and W. B. Cudykunst. 2002. *Communicating with Strangers: An Approach to Intercultural Communication*. 4th ed. New York: McGraw-Hill.

Gudykunst, W., Y. Matsumoto, S. Ting-Toomey, T. Nishida, K. Kim, and S. Heyman. 2006. The influence of cultural individualism-collectivism, self construals, and individual values on communication styles across cultures. *Hum Commun Res* 22:510–43.

Gudykunst, W., S. Ting-Toomey, and E. Chua. 1988. *Culture and Interpersonal Communication*. London: Sage Publications.

Hall, E. T. 1976. *Beyond Culture*. New York: Doubleday.

Hall, E. T. 1990. *Understanding Cultural Differences*. Boston, MA: Intercultural Press.

Hall, E. T., and M. R. Hall. 1990. *Understanding Cultural Differences: Germans, French and Americans*. Boston, MA: Intercultural Press.

Heine, S., and D. Lehman. 1997. The cultural construction of self-enhancement: An examination of group-serving biases. *J Pers Soc Psychol* 72:1268–83.

Hofstede, G. 1980. Motivation, leadership, and organization: Do American theories apply abroad? *Organ Dyn* 9:42–63.

Hofstede, G. 1984. *Culture's Consequences: International Differences in Work-Related Values*. London: Sage Publications.

IFR Statistical Department. 2009. Professional service robots are establishing themselves. In *World Robotics 2009—Service Robots*, ed. Gudrun Litzenberger. Frankfurt: IFR Statistical Department, available at http://www.worldrobotics.org/downloads/PR_Service_Robots_30_09_2009_EN.pdf.

Iyengar, S. S., and M. R. Lepper. 1999. Rethinking the value of choice: A cultural perspective on intrinsic motivation. *J Pers Soc Psychol* 76:349–66.

Kaplan, F. 2004. Who is afraid of the humanoid? Investigating cultural differences in the acceptance of robots. *Int J H R* 1(3):1–16.

Kemper, N., A. Amin, B. Wielinga, V. Evers, and H. Cramer. 2009. Give me a hug: The effects of touch and autonomy on peoples responses to embodied social agents. *Comput Animat Virtual Worlds* 20(2):437–45.

Kidd, C., and C. Breazeal. 2004. Effect of a robot on user perceptions. Paper presented at the IEEE/RSJ International Conference on Intelligent Robots and Systems, Sendai, Japan.

Kim, M., J. Hunter, A. Miyahara, A. Horvath, M. Bresnahan, and H. Yoon. 1996. Individual-vs. culture-level dimensions of individualism and collectivism: Effects on preferred conversational styles. *Commun Monogr* 63:29–49.

Kramer, R., and L. Goldman. 1995. Helping the group or helping yourself? Social motives and group identity in resource dilemmas. In *Social Dilemmas: Perspectives on Individuals and Groups*, ed. D. A. Schroeder, 49–67. New York: Praeger.

Mann, L., M. Radford, P. Burnett, S. Ford, M. Bond, K. Leung et al. 1998. Cross-cultural differences in self-reported decision-making style and confidence. *Int J Psychol* 33:325–35.

Marian, V., and M. Kaushanskaya. 2004. Self-construal and emotion in bicultural bilinguals. *J Mem Lang* 51:190–201.

Markus, H., and S. Kitayama. 1991. Culture and the self: Implications for cognition, emotion, and motivation. *Psychol Rev* 98:224–53.

Markus, H., and S. Kitayama. 1994. A collective fear of the collective: Implications for selves and theories of selves. *Pers Soc Psychol Bull* 20:568–8.

Markus, H., S. Kitayama, and G. VandenBos. 1996. The mutual interactions of culture and emotion. *Psychiatr Serv* 47:225–6.

Matsumoto, D. 1999. Culture and self: An empirical assessment of Markus and Kitayama's theory of independent and interdependent self-construals. *Asian J Soc Psychol* 2: 289–310.

Michalowski, M., S. Sabanovic, and R. Simmons. 2006. A spatial model of engagement for a social robot. Paper presented at the 9th IEEE International Workshop on Advanced Motion Control, October 31–November 2, 2003, San Francisco, CA.

Morry, M., and M. Kito. 2009. Relational-interdependent self-construal as a predictor of relationship quality: The mediating roles of one's own behaviors and perceptions of the fulfillment of friendship functions. *J Soc Psychol* 149:305–22.

Mutlu, B., S. Osman, J. Forlizzi, J. Hodgins, and S. Kiesler. 2006. Task structure and user attributes as elements of human-robot interaction design. Paper presented at the IEEE International Workshop on Robot and Human Interactive Communication, October 31–November 2, 2003, San Francisco, CA.

O'Neill-Brown, P. 1997. Setting the stage for the culturally adaptive agent. Paper presented at the AAAI Fall Symposium, Menlo Park, CA.

Oetzel, J. 1998. Culturally homogeneous and heterogeneous groups: Explaining communication processes through individualism-collectivism and self-construal. *Int J Intercult Relat* 22:135–61.

Oyserman, D., H. Coon, and M. Kemmelmeier. 2002. Rethinking individualism and collectivism: Evaluation of theoretical assumptions and meta-analyses. *Psychol Bull* 128:3–72.

Pineau, J., M. Montemerlo, M. Pollack, N. Roy, and S. Thrun. 2003. Towards robotic assistants in nursing homes: Challenges and results. *Rob Auton Syst* 42:271–81.

Powers, A., and S. Kiesler. 2006. The advisor robot: Tracing people's mental model from a robot's physical attributes. Paper presented at the 1st ACM SIGCHI/SIGART Conference on Human-Robot Interaction, Salt Lake City, Utah.

Powers, A., A. D. I. Kramer, S. Lim, J. Kuo, S.-L. Lee, and S. Kiesler. 2005. Eliciting information from people with a gendered humanoid robot. In *IEEE International Workshop on Robot and Human Interactive Communication,* August 13–15, 2005, Nashville, TN.

Rau, P., Y. Li, and D. Li. 2009. Effects of communication style and culture on ability to accept recommendations from robots. *Comput Hum Behav* 25:587–95.

Reeves, B., and C. Nass. 1998. *The Media Equation: How People Treat Computers, Television, and New Media Like Real People and Places.* Stanford, CA: CSLI Publications/ Cambridge University Press.

Ross, M., W. Xun, and A. Wilson. 2002. Language and the bicultural self. *Pers Soc Psychol Bull* 28:1040–50.

Sanchez-Burks, J., F. Lee, I. Choi, R. Nisbett, S. Zhao, and J. Koo. 2003. Conversing across cultures: East-West communication styles in work and nonwork contexts. *J Pers Soc Psychol* 85:363–72.

Sedikides, C., and M. B. Brewer. 2001. Individual self, relational self, and collective self: partners, opponents, or strangers? In *Individual Self, Relational Self, Collective Self,* ed. C. Sedikides and M. B. Brewer, 1–4. Philadelphia: Psychology Press.

Shinozawa, K., F. Naya, K. Kogure, and J. Yamato. 2004. Effect of robot's tracking users on human decision making. Paper presented at the IEEE/RSJ International Conference on Intelligent Robots and Systems, Sendai, Japan.

Shinozawa, K., F. Naya, J. Yamato, and K. Kogure. 2005. Differences in effect of robot and screen agent recommendations on human decision-making. *Int J Hum Comput Stud* 62:267–79.

Stubbs, K., D. Wettergreen, and P. Hinds. 2007. Autonomy and common ground in human-robot interaction: A field study. *IEEE Intell Syst* (March–April). 2(2):42–50.

Syrdal, D., K. Dautenhahn, S. Woods, M. Walters, and K. Koay. 2007. Looking good? Appearance preferences and robot personality inferences at zero acquaintance. Paper presented at the AAAI Spring Symposium 2007, Multidisciplinary Collaboration for Socially Assistive Robotics, Stanford, California.

Syrdal, D., M. Walters, K. Koay, and K. Dautenhahn. 2008. The role of autonomy and interaction type on spatial comfort in an HRI scenario. Paper presented at the 3rd ACM/IEEE Human-Robot Interaction Conference, Amsterdam, the Netherlands.

Tajfel, H. 1974. Social identity and intergroup behaviour. *Soc Sci Inf* 13(2):65–93.

Tajfel, H., M. Billig, R. Bundy, and C. Flament. 1971. Social categorization and intergroup behavior. *Eur J Soc Psychol* 1:149–78.

Tasaki, T., K. Komatani, T. Ogata, and H. Okuno. 2005. Spatially mapping of friendliness for human-robot interaction. Paper presented at the IEEE/RSJ International Conference on Intelligent Robots and Systems, Edmonton, Canada.

Ting-Toomey, S. 1999. *Communicating Across Cultures.* New York/London: Guilford Press.

Ting-Toomey, S., G. Gao, P. Trubisky, Z. Yang, H. Kim, S. Lin et al. 1991. Culture, face maintenance, and styles of handling interpersonal conflict: A study in five cultures. *Int J Confl Manage* 2:275–96.

Triandis, H., R. Bontempo, M. Villareal, M. Asai, and N. Lucca. 1988. Individualism and collectivism: Cross-cultural perspectives on self-ingroup relationships. *J Pers Soc Psychol* 54:323–38.

Turner, J. C., R. J. Brown, and H. Tajfel. 1979. Social comparison and group interest in in-group favouritism. *Eur J Soc Psychol* 9:187–204.

Walters, M. L., and K. Dautenhahn. 2005. The influence of subjects' personality traits on predicting comfortable human-robot approach distances. Paper presented at the Cognitive Science Society 2005 Workshop, Stresa, Italy.

Walters, M., K. Dautenhahn, S. Woods, K. Koay, R. Te Boekhorst, and D. Lee. 2006. Exploratory studies on social spaces between humans and a mechanical-looking robot. *Conn Sci* 18:429–39.

Wang, L., P. -L. P. Rau, V. Evers, B. Robinson, and P. Hinds. 2009. Responsiveness to robots: Effects of ingroup orientation and communication style on HRI in China. Paper presented at the 4th ACM/IEEE international conference on Human robot interaction, California.

Wang, L., P.-L. P. Rau, V. Evers, B. K. Robinson, and P. Hinds. 2010. When in Rome: The role of culture and context in adherence to robot recommendations. Paper presented at the 5th ACM/IEEE International Conference on Human-Robot Interaction, Osaka, Japan.

Yamagishi, T., N. Jin, and A. Miller. 1998. In-group bias and culture of collectivism. *Asian J Soc Psychol* 1:315–28.

Yanco, H. A., and J. Drury. 2004. Classifying human-robot interaction: An updated taxonomy. In *IEEE International Conference on Systems*. Man and Cybernetics, October 10–13, 2003, Alexandria, VA.

Yuki, M. 2003. Intergroup comparison versus intragroup relationships: A cross-cultural examination of social identity theory in North American and East Asian cultural contexts. *Soc Psychol Q* 66:166–83.

Yuki, M., W. Maddux, M. Brewer, and K. Takemura. 2005. Cross-cultural differences in relationship- and group-based trust. *Pers Soc Psychol Bull* 31:48–62.

Section IV

Decision Making in Healthcare

10 Deliberation and Medical Decision Making

Kathryn E. Flynn, Shamus Khan, Amy Klassen, and Erik Schneiderhan

CONTENTS

INTRODUCTION

Healthcare decisions are often difficult to make. The stakes are high; risks and outcomes are uncertain. To further complicate matters, the process of decision making in health care—including communication, deliberation, and decision making—is embedded within complex social contexts. Factors such as gender, race, culture, norms, and class are at play in how doctors and patients consider and make decisions about medical care. The literature on medical decision making has done well to focus on how these characteristics influence the process of making decisions. Yet, often missing from our understanding is the social context of communication itself. Fortunately, there is a large collection of literature on this topic in the social sciences. Unfortunately, there has been little dialogue between the social sciences literature on deliberation and the health sciences literature on medical decision making. In this chapter we begin this dialogue and generate two main insights.

First, we demonstrate the need for an expansion of the theoretical framework used to understand medical decision making. By grounding itself within a rational choice framework, the research and theory on medical decision making tends to adopt a thin conceptualization of communication. Specifically, this theoretical stance is rather asocial in its atomistic approach to action and thereby ignores the *social* dimensions of communication. As the deliberation literature suggests, communication is not just about the direct and easy transfer of information from one person to another (e.g., a health professional to a patient or vice versa). Instead, communication is dynamic and can either facilitate or inhibit the flow of information and increase or decrease the likelihood of information being understood. Research and theorizing on deliberation—with its emphasis on reason-giving and inclusivity—offers a broader, more textured theoretical approach than the rational choice framework and thereby holds the potential to reveal hidden complexities around healthcare communication.

Second, although the deliberation literature is theoretically rich, it is empirically weak. Put another way, much of the substance of deliberation theory is axiomatic. The theoretical concepts and premises of the deliberation literature would greatly benefit from an engagement with the rich empirical work on medical communication and decision making. These two bodies of work are a natural fit for one another. We demonstrate that fit and suggest that a further engagement of these two literatures could prove invaluable. On the one hand, the deliberation literature could better empirically evaluate its claims. On the other, the medical decision-making literature might develop a more robust understanding of the social process of communication, and thereby aid health outcomes through improved decisions.

This chapter is organized into three sections. First, we review the literature on medical decision making. We highlight what is currently known, and perhaps more importantly, what is not known. Second, we move to the literature on deliberation. We engage with its theoretical foundations and discuss its burgeoning body of empirical work. As with the previous section, we describe what the literature offers and discuss its lacunae. Third, we present and analyze the merits of engaging these two separate but complimentary literatures in a dialogue with an eye toward a single, reconstructed approach to how we deliberate about and decide on matters of health. We conclude this section with a brief discussion of the implications for future empirical and theoretical work in the medical and social sciences.

MEDICAL DECISION MAKING

For over two thousand years, dating back at least to Hippocrates, the relationship between doctor and patient was characterized by silent care (by doctors) and compliance (by patients) (Katz 1984; Zarin and Pauker 1984). Yet the case of *Salgo v. Stanford* (1957) opened these relationships to major changes by legalizing a patient's right to informed consent to medical treatment. Since then, greater patient involvement in health care has been a growing phenomenon, especially patient participation in making medical decisions. Today, medical decision making is often viewed as a negotiation between patients (and their families) and doctors (i.e., physicians, nurses, or other healthcare providers). Together, these participants are seen as responsible for promoting the health of the patient.

Medical Decision-Making Models

To make sense of this collective medical decision making, scholars have outlined three analytical stages to doctor-patient communication (Charles, Gafni, and Whelan 1999). The first stage is information exchange, which includes the type (medical, personal) and amount of information exchanged, as well as its flow and direction: either one-way from doctor to patient or two-way between doctor and patient. The second stage is deliberation, which is defined as the process of expressing and discussing treatment preferences. The third stage is decisional control, that is, who decides what treatment to implement. This framework assumes that patients and doctors have fixed preferences, and the purpose of communication is to gather information so as to make better decisions based on those preferences.

Within this framework, doctor-patient interaction can be marked by different communication ethics. In the traditional or *paternalistic model*, the doctor is the expert and the patient is in need of the doctor's expertise (Ballard-Reisch 1990). Doctors perform information management, assess options, and make treatment decisions for patients (presumably in patients' best interests) without consideration of patient preferences (Katz 1984). This model assumes that there are objective criteria to determine the best course of action and that the doctor is the only knowledgeable actor who understands these criteria. The role of the doctor is to choose the correct course of action; patients should assent to this course with gratitude (Emanuel and Emanuel 1992). Patients are envisioned as preferring not to make the final decision and not wanting to participate more than minimally in information exchange or deliberation. The role of the patient is to provide relevant, important information that the doctor uses to make the decision.

Conversely, in the informed or *consumer model*, doctors provide all relevant information to their patients, and patients alone assess their options and make the final decision. The doctor in this model serves as a technical expert to provide information and facilitate decisions made by a fully autonomous patient (Emanuel and Emanuel 1992). Patients are envisioned as wanting to make their own decisions and not wanting to deliberate with doctors (Charles, Gafni, and Whelan 1999). The doctor provides information, and the patient is responsible for making the decision.

In the *shared model*, the passivity of patients inherent in the paternalistic model is eliminated, and unlike the consumer model the responsibility for the final decision does not lie completely on patients (Charles, Whelan, and Gafni 1999). This shared model assumes that patients want to equally participate with their doctors at every stage in order to make the best possible treatment choice. The role of the patient and the doctor is the same: both provide information to one another and discuss it in order to come to a decision.

Since at least 1982, with the publication of a report by the U.S. President's Commission for the Study of Ethical Problems in Medicine and Biomedical and Behavioral Research, researchers and policy-makers have argued that the shared model is the ideal model. Shared decision making has been called the crux of patient-centered care (Weston 2001). It is thought to improve patient safety by reducing prescribing errors and increasing patient adherence to medications (Dowell, Williams, and Snadden 2007). And as patients are more likely to avoid discretionary and risky medical treatments when given the choice (Fraenkel and Peters 2009; Wennberg et al. 2007), it can also provide an opportunity to reduce costs. In light of such evidence, the Institute of Medicine (2004) has recommended promoting shared decision making in medical school curricula as a mechanism to improve care. This marks a considerable transformation, as for most of the history of medicine, medical practice has been based on a model that did not include the patient in decision-making tasks. However, the means by which these shared decision interactions are facilitated remains largely unknown. While shared decision making appears to change outcomes, it is not clear *how* and *why* this is so. As we shall show in our section on Empircal Foundations of Deliberation, this is also the case with the social science deliberation literature: we find that communication matters for making decisions, but we do not understand the mechanisms for explaining these outcomes.

Patient Preferences for Participating in Medical Decision Making

The interactive dynamic between patients and doctors is incredibly complex. There are the complexities of patient and doctor backgrounds and preferences, and importantly, how these mix as actors with different orientations must come together to exchange information and come to a decision. To support shared decision making, we must understand patients' and doctors' preferences for the distinct analytic stages of the decision-making process. The theoretical models deployed in this literature often assume a consistent preference for such information flows. Yet in practice there is wide variation in how actors feel about information exchange, deliberation, and control over the final decision.

Patients want to receive information from their providers almost universally (Deber, Kraetschmer, and Irvine 1996; Ende et al. 1989; Flynn, Smith, and Vanness 2006; Nease and Brooks 1995), and a majority of patients also want their doctors to have complete information about their medical histories (Flynn, Smith, and Vanness 2006). However, preferences over the amount and type of information, for deliberative interactions, and for control over final decisions differ substantially among patients (Flynn, Smith, and Vanness 2006; Robinson and Thomson 2001). Where one study suggests that patients do not want to be involved in problem-solving tasks but do prefer retaining control over decisions (Deber, Kraetschmer, and Irvine 1996), another study of hysterectomy decisions found that both the patients and the doctors felt that the doctor should play the role of information distributor, and the patient would take the lead in making the final decision, following the informed model (Richter, Greaney, and Saunders 2002). Previous work suggests that among older adults, patients who prefer discussing treatment options do not necessarily also prefer making the final decision about treatment and vice versa (Flynn, Smith, and Vanness 2006). However, for all these varied results, the concept of deliberation as distinct from decisional control is often ignored entirely, and measurement of these aspects of decision-making preferences has not been robust. Further, there has been little discussion resolving the potential conflict between promoting shared decision making as ideal and accommodating individual patient preferences for participation (or lack of participation) in decision making. As we will demonstrate in the next section, this lack of discussion marks a blind spot in the literature on health care and medical decision making. Deliberation researchers have recently demonstrated that, in fact, inclusion does matter in changing the decisions of individuals faced with choices.

Participant resources are an important part of shared decision making. For shared decision making to succeed, it is essential that patients have the necessary personal, health, social, and economic resources to participate as well as the willingness to participate (Institute of Medicine 2004). Numerous studies have examined personal characteristics that may be related to preferences for sharing in healthcare decisions. Younger age (Ende et al. 1989; Kaplan et al. 1995; Thompson, Pitts, and Schwankovsky 1993), female sex (Kaplan et al. 1995), higher education (Kaplan et al. 1995; Thompson, Pitts, and Schwankovsky 1993), and better health (Degner and Sloan 1992; Ende et al. 1989) have been associated with preferences for increased participation, but these factors still explain less than 20% of the variation in preferences between individuals (Benbassat, Pilpel, and Tidhar 1998; Ende et al. 1989).

Other social background characteristics also influence interactions. A study of acute intensive care units in France showed that family members who were of a foreign descent had a harder time understanding the acuteness of the medical situation (Azoulay et al. 2000). And scholars have shown a significant racial bias in communication between doctors and patients; African Americans and Latino Americans often receive poorer quality care relative to white patients because communication with these patients primarily focuses on the biomedical aspect of the problem to the detriment of including patient participation in either providing health information or in making health decisions (Ashton, et al. 2003). People who come from lower socioeconomic statuses tend to report lower levels of satisfaction with their healthcare experience and are more critical of their doctor's level of communication, perhaps because doctors have been shown to be less likely to actively engage lower-income patients and provide less feedback of patient progress compared to more well-to-do patients (Jensen et al. 2010). Language proficiency skills of the patient must also be taken into account. Limited health literacy not only impedes effective communication between doctors and patients, particularly in language discordant pairings (Spanish and English), but it also has been associated with lower levels of proactive and interactive communication (Sudore et al. 2009).

Flynn and Smith's (2007) framework for describing constellations of patient preferences among older adults found that, after adjustment for personal, health, social, and economic factors, the length of one's relationship with a particular provider was not associated with preferences for deliberation or decisional control, but highly relevant distinguishing factors included gender, cognitive ability, rural or farm origins, and multiple aspects of personality. While a study of undergraduate students concluded that personality was not significantly correlated with preferences for cognitive, decisional, or behavioral control in the healthcare context (Auerbach and Pegg 2002), the relatively healthy sample with little experience of the healthcare system or making treatment decisions may have biased the results. Flynn's work focused on preferences among a large sample of older adults, the vast majority of whom had common chronic conditions and regularly took prescription medicines. In that sample, increased conscientiousness (organization and self-discipline) and openness to experience (creativity and willingness to adjust to new things) corresponded to preferring the most active decision-making style compared with the least active. Decreased agreeableness (trust and cooperation) and neuroticism (nervousness and emotional instability) corresponded to preferring the most active decision-making style compared with the least active.

Provider Preferences for Involving Patients in Medical Decision Making

There is considerably less research available on provider preferences and characteristics related to sharing medical decisions with patients. A large survey of Norwegian physicians' attitudes in the doctor-patient relationship found that over one-third of physicians held ambivalent views about patient autonomy, physician paternalism, and deliberation (Falkum and Forde 2001). Deliberationists (high on deliberation, low on autonomy and paternalism) made up 12% of the sample. Classical paternalists (high on paternalism and low on deliberation and autonomy) made up 15%. Modern paternalists (high on paternalism and deliberation and low on autonomy) made up 18%. Autonomists (high on autonomy and low on paternalism and deliberation) made up 19% of the sample. This group was the only one easily distinguished from the others based on physician characteristics: autonomists were more likely to be female than male, middle-aged (compared to younger than 30 or older than 50), and psychiatrists compared to other specialties. Other research has found that female doctors may be less likely to be involved in patient conflicts (Weingarten et al. 2010). Women may have a more interactive dynamic that patients like, that is, they are more likely to promote preventive strategies and psychological counseling with their patients, relative to male physicians, who focus more on the patient's medical history and technicalities of physical examinations (Bertakis 2009). However, the gender of the patient also plays a role in the perceived satisfaction of the healthcare interaction. Bertakis (2009) found that gender concordance between the doctor and the patient predicted greater patient-centered communication. This limited understanding of how social background impacts decision interactions is shared by the social science literature on deliberation, and it is a clear place wherein more research might build both of these literatures.

While relatively little is known about provider preferences, when doctors enact a decision-making process of any kind, patients tend to prefer the care they receive. Scholars have found that patients are more satisfied with their decision-making interactions when doctors are more positive toward sharing the decision-making responsibility (Carlsen and Aakvik 2006). This work suggests that physicians who are more supportive of shared decision making are more likely to be more acutely in tune with their patients' needs, thus creating a more satisfying outcome for the patient. Yet this is not a one-way street, with physicians simply deciding how a patient interaction will go. How physicians interact with patients likely differs depending on the level of participation that patients have in the decision-making process. In one study, doctors tended to use more patient-centered methods in their interactions for highly participatory patients (Cegala and Post 2009). This suggests that the amount of participation patients have in their healthcare interaction impacts how doctors structure their interactions. It therefore becomes very important to think of this process as a dynamic interaction. In short, taking interaction seriously might mean thinking about more than just information and preferences, and instead lead one to consider the social context of the interaction itself.

Decision Aids

The movement toward a shared decision-making model and greater inclusion of patients in medical decision making has stimulated the development of formal tools to aid medical decision making. These *decision aids* aim to help patients participate in their health care and to prepare them to make informed treatment decisions by providing information about known options and their likely outcomes (O'Connor et al. 2003; Nelson et al. 2007). Hundreds of decision aids have been created across a wide range of medical conditions and situations, and they typically come in the form of information booklets or cards, interactive computer programs, workbooks, or DVDs.

A systematic review of trials testing the effectiveness of decision aids suggests that decision aids have a positive impact on many healthcare outcomes (O'Connor et al. 2009). When decision aids are incorporated into the counseling aspect of the medical encounter, they produce higher knowledge scores, increase the percentage of patients who express realistic understanding of the risk and harms of each choice, lower decisional conflict, reduce the number of patients unable to make a decision, and improve the level of consensus between doctors and patients. What these decision aids do well is to provide patients with relevant information about their health conditions and potential treatments. In the best scenario this includes information about treatment options and the related risks and benefits of each option. Yet often this information is unavailable or incomplete. A major challenge for decision aids is that for many of the healthcare decisions these aids are developed to address, there is no clear-cut right or wrong choice.

A consistent but contested theme in the decision aid literature is the desire to reduce decisional conflict. Decisional conflict encompasses the potential anxiety and distress that can come from an inability to make an important healthcare choice (O'Connor 1995). However, some researchers question whether clarifying patients' values and reducing decisional conflict are actually appropriate outcomes (Nelson et al. 2007). Nelson and colleagues note that the effectiveness of decision aids to clarify values is based upon two key assumptions. First, it is assumed that people need to have their values clarified and that formal interventions do a better job making values explicit than intuition. Second, values clarification interventions assume that people already have a well-defined set of values that simply need to be illuminated. Nelson and colleagues argue that the negative connotations of decisional conflict may be more harmful in times of uncertainty to patients than the conflict itself. Patients may come to see the decision-making process as an irrevocable, one-time event rather than being a work in progress. This desire to make a decision may actually push patients into making hasty choices that may jeopardize the goal of making well-informed quality healthcare decisions.

Other researchers have supported this view, challenging the assumption that argumentation and confrontation within the decision-making situation is detrimental. Rubinelli and Schulz (2006) contend that there is an important place for argumentation in doctor-patient relationships. They posit that providing clear arguments about healthcare options helps patients get a clearer understanding about their particular health issue, stimulates action on the part of the patient, and helps to even out the power dynamic between doctors and patients by promoting a shared decision-making relationship (Rubinelli and Schulz 2006). These findings suggest that in order for satisfactory healthcare decisions to be reached, both parties need the opportunity to convey their preferences, worries, and understandings about the health concern. In this case, deliberative processes may prove to be more beneficial, even if they increase decisional conflict, in bringing about more satisfactory healthcare decisions.

Thus, it may come as no surprise that the systemic review of decision aid trials found that while decision aids increase patient knowledge and patient participation in decision making, they do not, as a rule, improve patient satisfaction with decisions, anxiety, or health outcomes (O'Connor, Llewellyn-Thomas, and Flood 2004). This leads us to a core question, namely, what constitutes the best or most effective decision? Some researchers contend that the best decision is one that promotes the treatment option that is most in line with the latest medical practices and promotes patient adherence (Arbuthnott and Sharpe 2009), whereas others argue that the best decisions correspond most closely with a patient's values and preferences, are well-informed, low in decisional conflict, and promote patient confidence to take an active role in their health care (O'Connor, Llewellyn-Thomas, and Flood 2004). Patient satisfaction with their treatment process is also considered a key determinant of the quality of the medical decision (Ballard-Reisch 1990; Bertakis 2009; Jensen et al. 2010). With this lack of consistency in what outcomes are used to judge decision quality, it is difficult to systematically evaluate decision aids. However, we do know that decision aids are used to increase patient participation (Légaré, Stacey, and Forest 2007), and implicitly, more communication is equated with better communication. To the extent that research has looked *inside* the decision-making process, it is focused on preference-oriented and value-appropriate choices, as well as potential information asymmetries in the communicative process (Charles, Gafni, and Whelan 1999; Nelson et al. 2007; O'Connor, Llewellyn-Thomas, and Flood 2004; O'Connor et al. 2007; Teutsch and Berger 2005).

Strengths and Limitations of the Medical Literature

As should be clear from our review of this literature, there is extensive empirical material on medical decision making and health. This empirical foundation is the great strength of this literature. Importantly, researchers have noted positive health impacts to effective communication; the better the communication among participants, the better the health outcomes (Azoulay et al. 2000; Leonard, Graham, and Bonacum 2004; Ong et al. 1995; Perloff et al. 2006; Stewart 1995). Yet missing from the literature is a discussion of what effective communication looks like. Stewart (1995) calls for this work, noting that studies of the interactive context of communication are essential for both producing and understanding better health outcomes.

As part of this work, we point to the importance of how decision making is understood. Within this literature, deliberation is often little more than information exchange, but we might better understand how different kinds of communication influence health decisions and outcomes. At its core, this literature is interested in information and preferences, asking, what are patient preferences and how can information be properly conveyed to and by patients in order for those to be a key component of health decision making?

These are important questions. But they miss something about the social process of communication. First, we would argue that preferences are often emergent from social interactions and not fixed before them. Therefore, understanding interactions is central to understanding preferences. Second, while more-informed decisions are often better decisions, in medical decision making, perfect information is impossible. The "more is better" approach to information may lead some patients feeling overwhelmed, and still might not provide the *kind* of information that patients need.

From this we propose two ideas: (1) the giving of reasons within a social interaction helps coalesce preferences and (2) interactions wherein all those influenced by a health decision are included in the discussion will facilitate in identifying the relevant concerns around which information is required. In short, we are suggesting a move away from a sole reliance on a rational choice framework (where preferences are stable and the drive is for perfect information) toward a deliberative one (where preferences are constituted through reason-giving and relevant information is identified through inclusion). In this move, we would emphasize the social nature of deliberation—not simply how it might influence an individual person's decision, but instead how understanding the dynamic context of interaction is central to both producing and making sense of health outcomes.

DELIBERATION

The deliberation literature has developed parallel to and separate from the medical decision-making literature. Scholars in this field often conceptualize deliberation as an inherently political concept. "Deliberation" serves as a model for the ideal way to make democracy work through the creation of engaged public citizens, good decisions, and outcomes that are just (Habermas 1984; Rawls 1971). Such deliberation is defined by two key elements: (1) interactions are inclusive of all members of a society who are impacted by a decision, and (2) participants provide reasons for their positions (Bachtiger et al. 2010; Bohman 1996; Gutmann and Thompson 1996, 2002, 2004; Schneiderhan and Khan 2008).

Theoretical Foundations

Rawls (1971) and Habermas (1984, 1990) built the modern theoretical foundations of the deliberation literature. Rather than think of deliberation as a kind of individual contemplation that centers on individual preferences and information, both Rawls and Habermas think of deliberation as an interactive process between subjects. Rawls generates a theory of justice by imagining a group of people placed behind a "veil of ignorance"—wherein participants are ignorant of the conditions of their lives. Rawls (1971) believes that just decisions are reached when actors deliberate behind the veil. Such deliberation requires the two conditions outlined above: equal participation and the provision of reasons (Rawls 1971, pp. 231–234, 138–139). Likewise, Habermas's (1990, p. 66) view of norm and value generation envisions deliberation as a social process that meets the same two conditions: first, that actors offer up reasons in attempts to provide "justification of norms," and second, that all those affected are "participants in a practical discourse." The basic position staked out by deliberation theorists building on the work of Rawls and Habermas is that when humans engage in particular kinds of talk with one another, special insights and capacities are enabled that allows them to make better decisions than if they simply considered scenarios on their own or talked with others in ways that were neither inclusive or justificatory. There is a specialness to certain kinds of interaction that is productive. And, as such, interaction transcends information-giving/receiving.

There are three insights from the theoretical foundations of deliberation that might aid the medical decision-making literature. First, how people communicate influences the kinds of decisions they reach. Rather than think of communication as a binary "yes/no" or a process that has "more/less" talk, the deliberation literature suggests that some kinds of communication are more fruitful than others. "Just talk" is not enough. Instead, talk must happen within a fairly constrained context. That context provides the other two insights: talk must be inclusive of those impacted by a decision, and talk must be oriented around modes of justification. Utterances that just convey information are not deliberation. Instead, position-taking is important, and such positions must be justified by reason-giving.

Ironically, although both Rawls and Habermas emphasize the importance of inclusion to robust deliberation, they present arguments wherein members of political discourse look strikingly homogeneous. For Rawls, the veil of ignorance means that subjects have no class, no race, no gender, only their voice and their reason. For Habermas, the "public sphere" of deliberation was initially a highly exclusive space of white bourgeois men. As Fraser (1992) points out, expanding this public space to be inclusive of diverse members is more than a simple process of addition. The ways in which the interactions themselves work may need to be transformed as the space of political talk and decision making is expanded to be inclusive of participants from different social backgrounds (Young 1999). Following Mouffe (1999), perhaps the Habermasian model is simply unable to acknowledge the potential antagonism entailed by a pluralism of values. In other words, the foundations of the deliberative decision-making literature are predicated on a set of rather unrealistic and unlikely conditions, wherein people are either ignorant of who they are, or participants model their discussions on a discourse that was developed within a small, bourgeois, male world. As a result, deliberation has the potential "to exacerbate rather than diminish power differentials" (Hibbing and Theiss-Morese 2002, p. 191). The point here is that we require a reconstructed deliberation theory asks us to consider the question of *who* is included, *how* this influences the communicative process, and *what* is at stake in that process. The emphasis within the medical decision-making literature on some of these questions helps us see how it might aid in this reconstruction.

Habermas' and Rawls' theories of deliberation, however, have not limited work solely to those in this philosophical tradition. Given that Rawls is critical of utilitarianism and Habermas of rational action theory, one would not expect rational choice scholars to embrace deliberation. Yet those who deploy rational choice theory have wholeheartedly engaged with the concept (Austen-Smith 1992; Johnson 1991, 1993; Knight and Johnson 1994; Shapiro 1999). Unfortunately, their model is insufficient to reconstruct deliberation theory, insofar as it relies upon a heavily atomized view of social action. Nonetheless, we can extrapolate from this rational choice work to provide an account of the individual mechanisms at work in the deliberative process.

The standard, orthodox rational choice approach assumes that "all human behavior can be viewed as involving participants who maximize their utility from a stable set of preferences and accumulate an optimal amount of information and other inputs in a variety of markets" (Becker 1976, p. 14). There are two general ways in which this approach might explain changes in decisions due to deliberation. First, a position change may result from the acquisition of additional information about the situation concerning preferences or potential consequences (Austen-Smith 1990, p. 125, 1992). New information may change what an actor does insofar as it presents the actor with a new opportunity structure. Illustrating this point, Przeworski (1991) uses an example of actors' preferences over ice cream flavors. "Now, suppose that the chocolate fan is told that this flavor leaves indelible spots on her dress. Having received this information, she alters her preference, relegating chocolate to second place ... This is deliberation" (p. 17). The actor had a preference for keeping her dress clean over the taste or flavor of ice cream, and having acquired information about the potential risk of soiling her dress, she adjusted what she wanted within a given situation; she did not form new preferences. Yet as we have noted, this is a very limited view of deliberation—one that sees social interactions as little more than moments to facilitate the exchange of information. *How* this happens, in Przeworski's rational choice approach, is under-theorized.

A second way that deliberation might influence decision making is that actors may realize how outcomes affect other people. Through the communicative process, actors may become cognizant of effects on others and act accordingly. For example, in their quasi-experiments on deliberative polling, Luskin, Fishkin, and Jowell (2002, p. 484) suggest "some participants [in deliberation] may come to see the alternatives through different eyes." Participants, in becoming aware of the effects of decisions on others, may change their positions. This is only likely if actors have a preference over how a decision affects others. In both of these models deliberation is a way to increase information gains through reason-giving and inclusion and therefore affects individual decision making. These conceptualizations of deliberation improving information tie in nicely with the literature on medical decision making, wherein the social communication between a patient and doctor is meant to improve information flows. The decider has fixed preferences but imperfect information; deliberation improves information and thereby improves decisions.

However, this leaves a rather thin understanding of the importance of social interactions to decisions and it tends to underemphasize the importance of social background characteristics to interactions. Decisions are still highly individualized, determined by fixed preferences and information, with interaction merely providing the latter. Missing from this model is how interacting with others might constitute preferences, or how hearing how others justify their positions might influence one to change one's preferences. We acknowledge that reasons and inclusion *could* simply be folded into (or stretched to fit) the rational choice framework; a participant in a deliberation provided with a reason (i.e., information) from a variety of actors (inclusion increasing information) may come to some sort of realization about the situation and/or possible effects of future action. For example, a patient considering treatment options for a medical condition might be given information from several doctors, family members, and even individuals participating in online discussions.

However, David Austen-Smith's work (1990, 1992) questions whether deliberation could be solely about information getting, insofar as the information acquired would be volunteered in other interactive arenas. The results of his modeling on legislative debating show that "no information can be elicited in debate that would not otherwise be volunteered through the making of proposals and amendments" (1990, p. 126). This suggests that there is more to debate than just information exchange and leads us to wonder how strong of an information effect deliberation could have over other forms of interaction. In subsequent work, Austen-Smith (1992) expands this notion, arguing that talk is strategic; actors seek to influence decisions by changing what actors see as potential outcomes of action. Overall, we find the deliberation literature to be less than clear on the equivalence of or difference between inclusive reasoning and simple provision of information in a rational choice framework. This lack of clarity marks a blind spot in the deliberation literature and is worth addressing with empirical work, particularly that of researchers of medical decision making.

Johnson has done a considerable amount of work that attempts to bridge the gap between rational choice and deliberation theory. He has made programmatic calls for communication between game theory and critical theory (1991), pointing to places where theorists on both sides could and should dialogue with one another (1993). Together, Johnson and Knight have combined social choice theory with models of deliberative democracy to try and get beyond what they see as the problems of aggregation in deliberative theory (1994). Knight and Johnson (1994) propose a pragmatic form of deliberation that mediates and transforms conflict rather than transforming preferences. In short, by combining rational choice and deliberation theory, this collected work asks that we make sense of the social context of decisions and the social potential to human interaction in making better decisions. This indicates that there is not that great of a distance between rational choice and deliberation theory. This is heartening, for in suggesting that the medical decision-making literature take the deliberation literature more seriously, we are not asking scholars to search for a radically new and different theoretical orientation when making sense of the findings. Instead, small adjustments, some of which we outline below, are all that is required to marry these theoretical and empirical approaches. These adjustments require taking the social context of communication more seriously and making sense of how and why human interaction can be productive for decision making.

Empirical Foundations of Deliberation

Unlike the medical decision-making literature, the deliberation literature lacks a strong empirical base. The vast majority of this literature is either theoretical, or it assumes the effects of deliberation without demonstrating them. As Polletta and Lee

(2006) point out, most social scientific work on deliberation is either historical (Fraser 1992; Ryan 1992; Schudson 1992, 1997) or ethnographic (Eliasoph 1998; Eliasoph and Lichterman 2003; Hart 2001; Lichterman 1996). In these works, "deliberation" is often so densely integrated with other variables of interest that we cannot be sure (1) if it explains the observed outcomes, or (2) if the continually posited mechanisms of reasons and inclusion truly matter. As such, the deliberative decision-making process remains a "black box," whose internal mechanisms are poorly understood.

This is not to say that there is no empirical work on deliberation. Interactions marked by deliberation have been shown to help build successful democracies and instill in citizens a sense of their shared political project (Baiocchi 2001, 2003; Fishkin 1995; Fishkin and Luskin 2005; Heller and Isaac 2003; Polletta and Lee 2006; Schneiderhan and Khan 2008). How citizens interact within their democracy matters; deliberation has been shown to be a kind of interaction that changes how publics think of politics and make decisions. In short, deliberation helps make democracies work. Yet while there is general consensus among policymakers and scholars that deliberation is important to political outcomes, there is little empirical evidence explaining *how* and *why* deliberation matters.

Further, given the deliberation literature's theoretical emphasis on justice and inclusiveness, it is surprising that there is scant empirical work examining whether or not race and ethnicity matter in the deliberative process (or for that matter, any host of background characteristics). What evidence exists suggests that variation in deliberative processes exists across ethnic groups, with Jacobs, Cook and Carpini (2009, p. 58) finding that "ethnicity and gender continue to depress discursive participation." Deliberation may in fact "perpetuate marginalization" (Walsh 2007, p. 25). *Why* this is so is less apparent. Perhaps, as some argue, minority groups lack interest, skill, expertise, resources, knowledge, information, or even representation in deliberative processes, putting them at a relative disadvantage (Dryzek 2000; Fraser 1997; Griffin and Newman 2008; McGann 2006; Sanders 1997; Walsh 2007; Williams 2000). The explanatory mechanisms, however, have not been fully identified, specified, and analyzed.

Similarly, in terms of the deliberation in the context of health care, there is much left to explore. There is a body of work looking at deliberation around healthcare policy and emphasizing the importance of citizen inclusion and involvement in setting priorities (Abelson et al. 2003, 2004; Dolan, Cookson, and Ferguson 1999; Gutmann and Thompson 2002; Litva et al. 2002; Maxwell, Rosell, and Forest 2003). At the level of the doctor-patient interaction, however, there is almost no work looking at whether or not deliberation, as opposed to just plain talk, matters. One exception is a recent study of deliberation around pediatric primary care that shows the importance of inclusiveness in the process: "Using a reliable and valid technique, deliberation was demonstrated to occupy a substantial portion of the visit and include multiple proposed plans, yet passive involvement of parents and children predominated. Results support the need to develop interventions to improve parent and child participation in deliberation" (Cox, Smith, and Brown 2007, p. 74). Another exception is work by Caron, Ducharme, and Griffith (2006), who argue that "contextual factors" in the decision-making process matter and point to the importance of an "evolving decision-making process" that is "highly relational" and "contextually specific" (p. 202).

Strengths and Limitations

The social science literature on deliberation provides us with a more dynamic framework than rational choice for understanding social interactions. In the rational choice framework what matters is information, and questions of communication are basically over how talk might aid in the acquisition of information. "Choice" is the work of internalized, black-boxed preferences. In the deliberation framework, particularly that developed by Rawls and Habermas, it is not simply the aggregated individual preferences that matter. Instead, the *social context of interaction* is central to explaining good outcomes.

The intuition here is a simple one: there is something productive about social encounters between people. And this goes beyond others serving as output mechanisms for the requisite information. Instead, there is something that happens *between* people when they encounter one another that is socially productive in generating desirable outcomes. This insight is a powerful one when trying to make sense of how communication influences decision making and, thereby, outcomes.

Further, the context and content of the social process of interaction matters, and not just any kind of communication will do. Particular kinds of interactions are more likely to influence actors' decisions than others (Schneiderhan and Khan 2008). For readers who have sat through meetings consisting of lots of "just talk" this should come as no surprise. Not all social encounters are productive. Deliberation provides us with suggestions as to what might make specific encounters better at reaching desirable outcomes: providing reasons and being inclusive of those affected by decisions.

Though there is some limited evidence that reasons and inclusion matter (Schneiderhan and Khan 2008), considerably more empirical work is required for the deliberation concept to develop and for scholars to better understand how communicative processes impact social outcomes. This is the great weakness of the deliberation literature—empirically it does not have a firm foundation, and some of its basic principles are more axiomatic than established.

Finally, while the deliberation literature takes seriously the social context of interactions, it has done less well in thinking through the social background characteristics of people within interactions. No doubt the racial, gender, class, cultural, and national composition (to name a few) of a group influences group dynamics and thereby, outcomes. But *how* this happens is less than clear, and *what* the consequences are of diverse versus homogenous group interactions is unknown. In short, the deliberation literature provides us with a useful and provocative guide to understanding the importance of social interaction to communication and decision making, but it requires a deeper empirical basis to refine these insights and test their usefulness.

SOCIAL DIMENSIONS OF DELIBERATION AND MEDICAL DECISION MAKING

There has been little dialogue between the deliberation literature and the literature on medical decision making. Yet we hope to have shown that both have something to offer the other. As we conclude, we have four suggestions for the development of both the medical decision-making and deliberation literatures. They can be achieved by combining the theoretical and empirical insights of these two different fields. For both literatures we suggest

1. Moving beyond the atomistic model of rational choice theory and developing a more social model that takes interaction between those involved in decisions seriously
2. Recognizing the importance of different kinds of communication, and evaluating how they matter in real-world contexts
3. Disaggregating explanatory mechanisms to see why they matter; making sense of the *how* and *why* of different forms of communication
4. Taking seriously and problematizing the social background of participants in interaction

The first two suggestions are what the deliberation literature might bring to medical decision-making scholarship. The last two suggestions are where medical decision-making scholarship might help those working within the deliberative tradition. But we suspect that the strongest insights will be generated when scholars from these different areas work together.

Taking Social Interaction Seriously

Underpinning decision-making interactions are deep, dynamic social processes influenced by culture, history, and institutional contexts. These processes are best understood through an in-depth examination of what actually happens on the ground. At the level of theory, we need to move beyond a sole reliance on a rational choice model of decision making. As we have indicated, communication is more than just information transfers. We might think of social encounters more humanistically, as things that do not just aid us, but help make us in particular ways. This is one of the main benefits that the deliberation literature can bring to the medical decision-making literature.

Expanding How We Understand Communication

Our second point is an extension of our first. Communication is not simply a question of less or more, nor does it simply happen or not. Instead, there are modes of communication that facilitate information transfer and understanding as well as preference constitution and adjustment. Not all kinds of communication are the same. Scholars have noted that interactions marked by reason-giving and inclusion are notably different than those wherein people "just talk." This insight is of potential immediate benefit to the health literature. If participants in discussion do not simply state their position, but explain why they hold it, interactions will be of a different character. We would predict that such interactions would be beneficial in generating desirable outcomes, whether that is judged as patient/doctor satisfaction or health outcomes.

Disaggregating Communication: An Empirical Warrant

We need to understand how different forms of communication work and why. If different kinds of communication lead to different kinds of decisions and outcomes, we must do the work of social scientists and understand what about the communicative process explains these differences. There is certainly no perfect mode of communication. But understanding what elements help generate "best practices" in different empirical contexts can be enormously beneficial in aiding with the health outcomes of patients.

Further, at the level of the empirical, we need to look more closely at the interactions between stakeholders in the healthcare decision-making process, including nurses, doctors, patients, and family members. One particularly promising place to begin is with the discourse quality index (DQI) developed by Steenbergen et al. (2003). Grounded in Habermasian theory, the DQI offers seven coding categories for analyzing speech and evaluating its overall quality. Such work that disaggregates elements of communication to understand how and why they influence decisions and outcomes is central to any project that seeks an empirical understanding of communication. We feel this is one of the most fruitful arenas for research that might combine the health and deliberation literatures. This is in part because the DQI might be productively combined with the Roter Interaction Analysis System (RIAS) (see Roter et al. 1997, 1998, 2000) and the OPTION scale (Elwyn et al. 2005). While DQI provides a scale to better understand the quality of discourse in a social context, RIAS and OPTION help provide some of the key characteristics to the exchange of medical information in particular. Bringing together these metrics might be the first step to understanding quality deliberation in the healthcare context.

Taking Social Background Seriously

Finally, we need to do more to understand the role of personality and other personal characteristics in preference formation with the goal of helping providers encourage patient participation in healthcare visits and tailor decision support. We need to particularly focus on race, ethnicity, gender, and class, not just as variables, but as socially constructed, dynamic, relational identities/subjectivities that are at play and need to be understood both qualitatively and quantitatively. If we think of communicative interactions as dynamic processes, then we must understand how social background and personal characteristics influence that process. Taking these steps, we will gain a greater understanding of two enormously important parts of our communities: how we communicate with one another and how we can use that knowledge to improve health outcomes and satisfaction.

REFERENCES

Abelson, J., J. Eyles, C. B. McLeod, P. Collins, C. McMullan, and P.-G. Forest. 2003. Does deliberation make a difference? Results from a citizens panel study of health goals priority setting. *Health Policy* 66(1):213–25.

Abelson, J., P.-G. Forest, J. Eyles, A. Casebeer, G. Mackean, and T. E. P. C. P Team. 2004. Will it make a difference if I show up and share? A citizens' perspective on improving public involvement processes for health system decision-making. *J Health Serv Res Policy* 9(4):205–12.

Arbuthnott, A., and D. Sharpe. 2009. The effect of physician-patient collaboration on patient adherence in non-psychiatric medicine. *Patient Educ Couns* 77:60–7.

Ashton, C. M., P. Haidet, D. A. Paterniti, T. C. Collins, H. S. Gordon, K. O'Malley, L. A. Petersen et al. 2003. Racial and ethnic disparities in the use of health services: Bias, preferences, or poor communication. *J Gen Intern Med* 18:146–52.

Auerbach, S. M., and P. O. Pegg. 2002. Appraisal of desire for control over healthcare: Structure, stability and relation to health locus of control and to the "big five" personality traits. *J Health Psychology* 7:393–408.

Austen-Smith, D. 1990. Information transmission in debate. *Am J Pol Sci* 34:124–52.

Austen-Smith, D. 1992. Strategic models of talk in political decision-making. *Int Polit Sci Rev* 13:45–58.

Azoulay, E., S. Chevret, G. Leleu, F. Pochard, M. Barboteu, C. Adrie, P. Canoui, J. R. le Gall, and B. Schlemmer. 2000. Half the families of intensive care unit patients experience inadequate communication with physicians. *Crit Care Med* 28:3044–9.

Bachtiger, A., S. Niemeyer, M. Neblo, M. Steenbergen, and J. Steiner. 2010. Disentangling diversity in deliberative democracy: Competing theories, their blind spots and complementarities. *J Polit Philos* 18:32–63.

Baiocchi, G. 2001. Participation, activism, and politics: The Porto Alegre experiment. *Polit Soc* 29:43–72.

Baiocchi, G. 2003. Emergent public spheres: Talking politics in participatory governance. *Am Soc Rev* 68:52–74.

Ballard-Reisch, D. S. 1990. A model of participative decision making for physician-patient interaction. *Health Commun* 2:91–104.

Becker, G. 1976. *The Economic Approach to Human Behavior*. Chicago: University of Chicago Press.

Benbassat, J., D. Pilpel, and M. Tidhar. 1998. Patients' preferences for participation in clinical decision-making: A review of published surveys. *Ann Behav Med* 24(2):81–8.

Bertakis, K. D. 2009. The influence of gender on the doctor-patient interaction. *Patient Educ Couns* 76:356–60.

Bohman, J. 1996. *Public Deliberation: Pluralism, Complexity, and Democracy*. Cambridge, MA: The MIT Press.

Carlsen, B., and A. Aakvik. 2006. Patient involvement in clinical decision making: The effect of GP attitude on patient satisfaction. *Health Expect* 9:148–57.

Caron, C. D., F. Ducharme, and J. Griffith. 2006. Deciding on institutionalization for a relative with dementia: The most difficult decision for caregivers. *Can J Aging* 25:193–205.

Cegala, D. J., and D. M. Post. 2009. The impact of patients' participation on physicians' patient-centered communication. *Patient Educ Couns* 77:202–8.

Charles, C., A. Gafni, and T. Whelan. 1999. Decision-making in the physician-patient encounter: Revisiting the shared treatment decision-making model. *Soc Sci Med* 49:651–61.

Charles, C., T. Whelan, and A. Gafni. 1999. What do we mean by partnership in making decisions about treatment? *Br Med J* 319:780–2.

Cox, E. D., M. A. Smith, and R. L. Brown. 2007. Evaluating deliberation in pediatric primary care. *Pediatrics* 120(1):e68–77.

Deber, R. B., N. Kraetschmer, and J. Irvine. 1996. What role do patients wish to play in treatment decision-making? *Arch Intern Med* 156:1414–20.

Degner, L. F., and J. A. Sloan. 1992. Decision-making during serious illness: What role do patients really want to play? *J Clin Epidemiol* 45:941–50.

Dolan, P., R. Cookson, and B. Ferguson. 1999. Effect of discussion and deliberation on the public's views of priority setting in health care: Focus group study. *Br Med J* 318:916–9.

Dowell, J., B. Williams, and D. Snadden. 2007. *Patient-Centered Prescribing: Seeking Concordance in Practice*. Oxford, UK: Radcliffe.

Dryzek, J. S. 2000. *Deliberative Democracy and Beyond: Liberals, Critics, Contestations*. Oxford, UK: Oxford University Press.

Eliasoph, N. 1998. *Avoiding Politics: How Americans Produce Apathy in Everyday Life*. New York: Cambridge University Press.

Eliasoph, N., and P. Lichterman. 2003. Culture in interaction. *Am J Soc* 108:735–94.

Elwyn, G., H. Hutchings, A. Edwards, F. Rapport, M. Wensing, W. Y. Cheung, and R. Grol. 2005. The OPTION scale: Measuring the extent that clinicians involve patients in decision-making tasks. *Health Expect* 8:34–42.

Emanuel, E. J., and L. L. Emanuel. 1992. Four models of the physician-patient relationship. *J Am Med Assoc* 267(16):2221–6.

Ende, J., L. Kazis, A. Ash, and M. A. Moskowitz. 1989. Measuring patients' desire for autonomy: decision-making and information-seeking preferences among medical patients. *J Gen Intern Med* 4(1):23–30.

Falkum, E., and R. Forde. 2001. Paternalism, patient autonomy, and moral deliberation in the physician-patient relationship. Attitudes among Norwegian physicians. *Soc Sci Med* 52:239–48.

Fishkin, J. S. 1995. *The Voice of the People: Public Opinion and Democracy*. New Haven, CT: Yale University Press.

Fishkin, J. S., and R. C. Luskin. 2005. Experimenting with a democratic ideal: Deliberative polling and public opinion. *Acta Politica* 40:284–98.

Flynn, K. E., and M. A. Smith. 2007. Personality and health care decision-making style. *J Gerontol B Psychol Sci Soc Sci* 62(5):261–7.

Flynn, K. E., M. A. Smith, and D. Vanness. 2006. A typology of preferences for participation in healthcare decision-making. *Soc Sci Med* 63:1158–69.

Fraenkel, L., and E. Peters. 2009. Patient responsibility for medical decision-making and risky treatment options. *Arthritis Rheum* 61:1674–6.

Fraser, N. 1992. Rethinking the public sphere: A contribution to the critique of actually existing democracy. In *Habermas and the Public Sphere*, ed. C. Calhoun, 109–42. Cambridge, MA: MIT Press.

Fraser, N. 1997. Communication, transformation, and consciousness-raising. In *Hannah Arendt and the Meaning of Politics*, ed. C. Calhoun and J. McGowan, 166–78. Minneapolis: University of Minnesota Press.

Griffin, J. D., and B. Newman. 2008. *Evaluating Political Equality in America*. Chicago: The University of Chicago Press.

Gutmann, A., and D. Thompson. 1996. *Democracy and Disagreement*. Cambridge, MA: Belknap.

Gutmann, A., and D. Thompson. 2002. Deliberative democracy beyond process. *J Polit Philos* 10(2):153–74.

Gutmann, A., and D. Thompson. 2004. *Why Deliberative Democracy?* Princeton, NJ: Princeton University Press.

Habermas, J. 1984. *The Theory of Communicative Action*. Boston: Beacon Press.

Habermas, J. 1990. Discourse ethics: Notes on a program of philosophical justification. In *Moral Consciousness and Communicative Action*, ed. J. Habermas, C. Lenhardt, and S. W. Nicholsen (transl.), 43–115. Cambridge, MA: MIT Press.

Hart, S. 2001. *Cultural Dilemmas of Progressive Politics: Styles of Engagement among Grassroots Activists*. Chicago: University of Chicago Press.

Heller, P., and T. M. T. Isaac. 2003. Democracy and development: Decentralized planning in Kerala. In *Deepening Democracy*, ed. A. Fung and E. O. Wright, 77–110. London: Verso.

Hibbing, J. R., and E. Theiss-Morese. 2002. *Stealth Democracy: Americans' Beliefs about How Government Should Work*. Cambridge, UK: Cambridge University Press.

Institute of Medicine. 2004. *Improving Medical Education: Enhancing the Behavioral and Social Science Content of Medical School Curricula.* Washington, DC: National Academies Press.

Jacobs, L. R., F. L. Cook, and M. X. D. Carpini. 2009. *Talking Together: Public Deliberation and Political Participation in America.* Chicago: The University of Chicago Press.

Jensen, J. D., A. J. King, L. M. Guntzviller, and L. A. Davis. 2010. Patient-provider communication and low-income adults: Age, race, literacy, and optimism predict communication satisfaction. *Patient Educ Couns* 79:30–5.

Johnson, J. 1991. Habermas on strategic and communicative action. *Polit Theory* 19:181–201.

Johnson, J. 1993. Is talk really cheap? Prompting conversation between critical theory and rational choice. *Am Polit Sci Rev* 81:74–86.

Kaplan, S. H., B. Gandek, S. Greenfield, W. Rogers, and J. E. Ware. 1995. Patient and visit characteristics related to physicians' participatory decision-making style. Results from the Medical Outcomes Study. *Med Care* 33:1176–87.

Katz, J. 1984. *The Silent World of Doctor and Patient.* New York: Free Press.

Knight, J., and J. Johnson. 1994. Aggregation and deliberation: On the possibility of democratic legitimacy. *Polit Theory* 22:277–96.

Légaré, F., G. Stacey, and P.-G. Forest. 2007. Shared decision-making in Canada: Update, challenges and where next! *J Qual Health Care* 101:213–21.

Leonard, M., S. Graham, and D. Bonacum. 2004. Human factor: The critical importance of effective teamwork and communication in providing safe care. *Qual Safe Health Care* 13:85–90.

Lichterman, P. 1996. *The Search for Political Community: American Activists Reinventing Commitment.* New York: Cambridge University Press.

Litva, A., J. Coast, J. Donovan, J. Eyles, M. Shepherd, J. Tacchi et al. 2002. The public is too subjective: Public involvement at different levels of health-care decision-making. *Soc Sci Med* 54:1825–37.

Luskin, R. C., J. S. Fishkin, and R. Jowell. 2002. Considered opinions: Deliberative polling in Britain. *Br J Polit Sci* 32:455–87.

Maxwell, J., S. Rosell, and P.-G. Forest. 2003. Giving citizens a voice in healthcare policy in Canada. *Br Med J* 326:1031–3.

McGann, A. 2006. *The Logic of Democracy: Reconciling Equality, Deliberation, and Minority Protection.* Ann Arbor: The University of Michigan Press.

Mouffe, C. 1999. Deliberative democracy or agonistic pluralism. *Soc Res* 66:745–58.

Nease Jr., R. F., and W. B. Brooks. 1995. Patient desire for information and decision-making in health care decisions: The Autonomy Preference Index and the Health Opinion Survey. *J Gen Int Med* 10:593–600.

Nelson, W. L., P. K. J. Han, A. Fagerlin, M. Stefanek, and P. A. Ubel. 2007. Rethinking the objectives of decision aids: A call for conceptual clarity. *Med Decis Mak* 27:609–18.

O'Connor, A. M. 1995. Validation of a decisional conflict scale. *Med Decis Mak* 15(1):25–30.

O'Connor, A. M., C. Bennett, D. Stacey, M. J. Barry, N. F. Col, K. B. Eden et al. 2007. Do patient decision aids meet effectiveness criteria of the international patient decision aid standards collaboration? A systematic review and meta-analysis. *Med Decis Mak* 27:554–74.

O'Connor, A. M., C. L. Bennett, D. Stacey, M. Barry, N. F. Col, K. Eden et al. 2009. Decision aids for people facing health treatment or screening decisions. *Cochrane Database Syst Rev* 3:CD001431.

O'Connor, A. M., H. Llewellyn-Thomas, and A. B. Flood. 2004. Modifying unwarranted variations in health care: Shared decision-making using patient decision aids. *Health Aff* VAR:63–72.

O'Connor, A. M., D. Stacey, V. Entwistle, H. Llewellyn-Thomas, D. Rovner, M. Holmes-Rovner et al. 2003. Decision aids for people facing health treatment or screening decisions. *Cochrane Database Syst Rev* 2:CD001431.

Ong, L. M. L., J. C. J. M. De Haes, A. M. Hoos, and F. B. Lammes. 1995. Doctor-patient communication: A review of the literature. *Soc Sci Med* 40:903–18.

Perloff, R. M., B. Bonder, G. B. Ray, E. Berlin Ray, and L. A. Siminoff. 2006. Doctor-patient communication, cultural competence, and minority health: Theoretical and empirical perspectives. *Am Behav Sci* 49:836–52.

Polletta, F., and J. Lee. 2006. Is telling stories good for democracy? Rhetoric in public deliberation after 9/11. *Am Soc Rev* 71(5):699–723.

President's Commission for the Study of Ethical Problems in Medicine and Biomedical and Behavioral Research. 1982. *Making Health Care Decisions*. Washington DC: U.S. Government Printing Office.

Przeworski, A. 1991. *Democracy and the Market: Political and Economic Reforms in Eastern Europe and Latin America*. Cambridge, UK: Cambridge University Press.

Rawls, J. 1971. *A Theory of Justice*. Cambridge, MA: Harvard University Press.

Richter, D. L., M. L. Greaney, and R. P. Saunders. 2002. Physician-patient interaction and hysterectomy decision making: The ENDOW study. *Am J Health Behav* 26:431–41.

Robinson, A., and R. Thomson. 2001. Variability in patient preferences for participating in medical decision-making: Implication for the use of decision support tools. *Qual Health Care* 10(1):i34–8.

Roter, D. L., S. Larson, C. S. Fischer, R. M. Arnold, and J. A. Tulsky. 2000. Experts practice what they preach: A descriptive study of best and normative practices in end of life discussions. *Arch Int Med* 160:3477–85.

Roter, D. L., J. Rosenbaum, B. deNegri, D. Renaud, L. DiPrete-Brown, and O. Hernandez. 1998. The effectiveness of a continuing medical education program in interpersonal communication skills on physician practice and patient satisfaction in Trinidad and Tobago. *Med Educ* 32:181–9.

Roter, D. L., M. Stewart, S. Putnam, M. Lipkin, W. Stiles, and T. Inui. 1997. Communication patterns of primary care physicians. *J Am Med Assoc* 270:350–5.

Rubinelli, S., and P. J. Schulz. 2006. Let me tell you why! When argumentation in doctor-patient interaction makes a difference. *Argumentation* 20:353–75.

Ryan, M. 1992. Gender and public access: Women's politics in nineteenth-century America. In *Habermas and the Public Sphere*, ed. C. Calhoun, 259–88. Cambridge, MA: MIT Press.

Salgo v. Leland Stanford Junior University Board of Trustees, 154 Cal. App. 2d 560 317 P. 312d 170 (Court of Appeals of California, First District, Division One 1957).

Sanders, L. 1997. Against deliberation. *Polit Theory* 25(June):247–376.

Schneiderhan, E., and S. Khan. 2008. Reasons and inclusion: The foundation of deliberation. *Soc Theory* 26:1–24.

Schudson, M. 1992. Was there ever a public sphere? If so when? Reflections on an American case. In *Habermas and the Public Sphere*, ed. C. Calhoun, 143–63. Cambridge, MA: MIT Press.

Schudson, M. 1997. Why conversation is not the soul of democracy. *Crit Stud Mass Commun* 14:297–309.

Shapiro, I. 1999. Enough deliberation: Politics is about interests and power. In *Deliberative Politics*, ed. S. Macedo, 28–38. Oxford, UK: Oxford University Press.

Steenbergen, M., A. Bachtiger, M. Sporndli, and J. Steiner. 2003. Measuring political deliberation: A discourse quality index. *Comp Eur Polit* 1(1):21–48.

Stewart, M. 1995. Effective physician-patient communication and health outcomes: A review. *Can Med Assoc J* 152:1423–33.

Sudore, R. L., C. S. Landefeld, E. J. Perez-Stable, K. Bibbins-Domingo, B. A. Williams, and D. Schillinger. 2009. Unraveling the relationship between literacy, language proficiency, and patient-physician communication. *Patient Educ Couns* 75:398–402.

Teutsch, S. M., and M. L. Berger. 2005. Evidence synthesis and evidence-based decision making: Related but distinct processes. *Med Decis Mak* 25:487–9.

Thompson, S. C., J. S. Pitts, and L. Schwankovsky. 1993. Preferences for involvement in medical decision-making: Situational and demographic influences. *Patient Educ Couns* 22:133–40.

Walsh, K. C. 2007. *Talking About Race: Community Dialogues and the Politics of Difference*. Chicago: University of Chicago Press.

Weingarten, M. A., N. Guttman, H. Abramovitch, R. Margalit, D. Roter, A. Ziv, J. Yaphe, and J. M. Borkman. 2010. An anatomy of conflicts in primary care encounters: A multimethod study. *Family Pract* 27:93–100.

Wennberg, J. E., A. M. O'Connor, E. D. Collins, and J. N. Weinstein. 2007. Extending the P4P agenda, part 1: How medicare can improve patient decision-making and reduce unnecessary care. *Health Aff (Millwood)* 26:1564–74.

Weston, W. W. 2001. Informed and shared decision-making: The crux of patient-centered care. *Can Med Assoc J* 165(4):438–9.

Williams, W. 2000. The uneasy alliance of group representations and deliberative democracy. In *Citizenship in Diverse Socieites*, ed. W. Kymlicka and W. Norman, 124–54. Oxford, UK: Oxford University Press.

Young, I. M. 1999. Justice, inclusion and deliberative democracy. In *Deliberative Politics*, ed. S. Macedo, 151–8. Oxford, UK: Oxford University Press.

Zarin, D. A., and S. G. Pauker. 1984. Decision analysis as a basis for medical decision-making: The tree of Hippocrates. *J Med Philos* 9:181–213.

11 Design of Performance Evaluation and Management Systems for Territorial Healthcare Networks of Services

Agostino Villa and Dario Bellomo

CONTENTS

INTRODUCTION TO EUROPEAN HEALTHCARE SYSTEMS

In the 16 countries in Europe where the euro is used as currency at present, country borders can be crossed without passport, and political, economic, and social strategies are defined in common by governments. But several obstacles for a full integration of peoples still survive. How is it possible to receive health care in the European Union (EU)? What are the main differences in the healthcare services provided by the EU countries? What are the similarities? To such questions, a recent analysis (see http://www.ilbisturi.it, in Italian) has presented the most significant characteristics of the European healthcare systems, according to information collected from the European Parliament and the organization for economic cooperation and development (OECD).

Some results of this analysis can be of real interest. Considering expenditures for health care, the country with the highest rate for public investments is Germany, with about 8% of the gross national product (GNP), whereas Greece and Sweden reach the highest private investment rate, over 40%. Italy is the country with the largest number of physicians, one for every 200 persons. In turn, the UK has the largest number of beds, one for every 400 persons. The Austrian and Danish healthcare organizations are characterized by a strong decentralization of responsibility at the regional and local level. In Finland, until 1994 each person could have free access to healthcare services, but since then municipalities can decide criteria for each person to contribute to the service costs. In France the healthcare system is based on two principles: the freedom of physicians for prescriptions, and the obligation for each citizen to have an assurance whose cost is proportional to the individual's revenue. In all European countries the healthcare service is managed by a public institution, with some organizational differences: in Greece, the state attributes to each citizen a sanitary personal book to be exhibited to doctors and hospitals; in Ireland healthcare services are managed by eight regional bodies, and in Luxemburg by nine regional bodies. With regards to the coverage of costs by either public or private assurances, the Netherlands laws ask persons with high incomes to have private contracts, as do Spain, Germany, Belgium, Austria, Denmark, Italy, Portugal, and Sweden.

Describing the Italian organization in more detail, the basic unit is the Local Healthcare Agency (LHA; or Azienda Sanitaria Locale [ASL]), referred to as an area whose population is usually corresponding to a province. All LHAs located in a same region depend on the regional government, which coordinates the healthcare service in terms of political plans and attribution of the annual budget. The charitable system is prevalently public; it covers the entire population with contributions of workers and employers (40%), from general taxation (35%), and with contributions of private citizen and assurances (25%). Funding installments are attributed to the different Italian regions, depending on their resident populations. In turn, each regional administration subdivides funds among the LHAs accordingly.

Each person, either Italian citizen or person coming from outside and provided with regular immigration permission, is registered at the National Sanitary Service (SSN) and in the register of the LHA of the municipality where he lives. Each registered person can receive several types of fundamental healthcare services free of any costs, while some other services require partial contributions, as well as the purchase of drugs.

Now, we are in the presence of some critical phenomena: the growing aging of population in almost all countries of "Old Europe" is making the number of "end-users" of healthcare services larger and larger; the explosion of the offer of diagnostic and therapeutic technologies, more and more sophisticated, enlarges costs; and the increasing demand for health (well-being) will transform environmental and relational discomforts into needs for additional healthcare services. All these forces—which in principle are positive for humans—risk undermining the ethical and civil pillars of equity and the universality of healthcare service organizations. This implies that careful interventions must be designed so as to manage these deep changes by supporting some actions of good management, including education and training of good professional operators, development of high qualifications, and rewarding of merits.

To this aim, an effective innovation or "re-engineering" of the national healthcare service organization must satisfy two opposite needs: on one side, to keep the healthcare service costs as low as possible, but on the other, to assure to every person a satisfactory level of good health. One of the most important actions to be taken, and surely the first one, is to design and apply a monitoring system that could allow identification of critical points and potential problems or drawbacks of the system, which could be a real support for managers, both regional and local. This chapter, therefore, aims to propose a design method and utilization suggestions for a good performance evaluation structure.

ADDRESSING THE CONCEPTUAL MODEL OF A REGIONAL HEALTHCARE SYSTEM

As outlined in the introduction, there are substantial differences between the healthcare system organizations in the different European countries. Looking at the national laws and rules concerning health care, the extent of the national healthcare provisions may have an influence on how individuals in these countries define their health and may also influence their attitudes towards care seeking. One would expect high aspirations regarding health (world health organization (WHO) definition) and high expectations regarding care in countries with systems that have abundant supply and generous social security: this phenomenon is usually called *medicalization*. Alternatively, health concepts that are fundamental (e.g., good health means being able to function or work) and reliance on self-care are to be expected in countries where only a basic level of care is provided. It should be noted, however, that these opinions are by no means determined by features of the healthcare system alone but may stem from deeply rooted and culturally determined views (see the survey modules at http://www.europeansocialsurvey.org).

Regarding the Italian situation, healthcare service is centered on the public organization and the diffused culture of the welfare state, originally based on an idea of wide social security coverage. This explains why the management of healthcare services must be done at the level of each LHA by the managerial chief (usually nominated by the regional government), through interactions with the network of healthcare service centers located in the LHA territory. Such a network could include several different types of services, operated in a number of service centers, such as

- "Hospital service," which is here considered (for the sake of simplicity) to include all diagnostic, chirurgical, and rehabilitation activities to be performed in a complex well-equipped center for health
- "Local consulting," "health status testing" centers, and so on, which are intermediate services to which patients are addressed for better diagnosis, but from which they could be addressed to hospitals
- "Family doctors" and "specialists," which are input services to which patients refer for diagnosis and prescription of therapy

These types of services can be viewed as "nodes" in a region, as shown in Figure 11.1, with reference to the area of the LHA of the province of Asti. The managerial chief of an LHA is charged with managing all personnel and resources operating in the agency, by planning the service capacity of the healthcare service network through

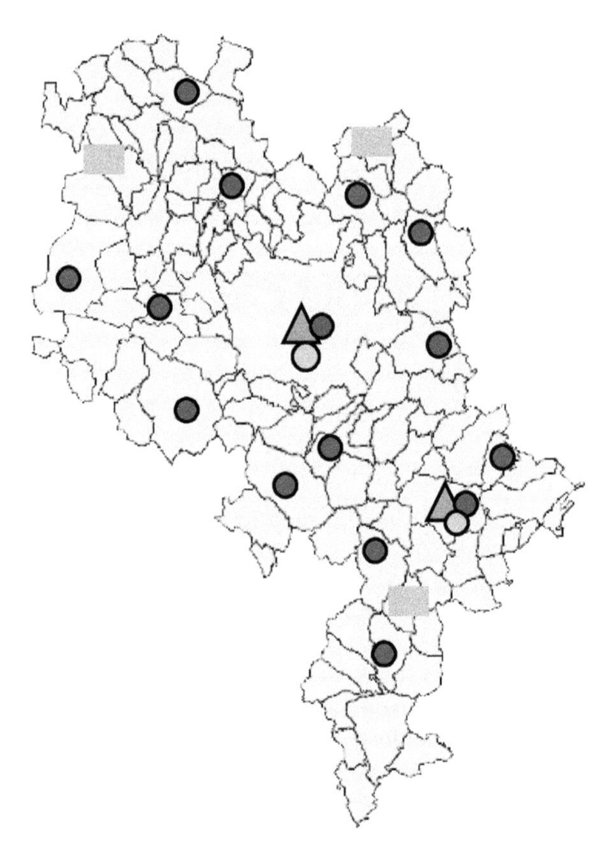

FIGURE 11.1 Illustration of the ASL-AT territory showing all local municipalities and some types of healthcare service centers; the central municipality is the town of Asti; △ denotes a hospital; ● denotes a health status testing lab; ○ denotes a center with healthcare service resources.

attribution of annual funding installments to each service center. Each center, in turn, is managed by a local center director, who has the commitment to provide the services to patients through a good utilization of the centers' operators and resources, according to the budget obtained by the LHA.

Funds are annually programmed at the regional level, thus deciding the funding installments for the about 20 LHAs in the Piemonte region. In an LHA, the managerial chief attributes funding rates to each center in front of requests for new positions, resources, machines, tools, and drugs, planned by the center director. The needs for new resources at a service center are estimated from the past resources utilization, not from the expected service demands (which cannot be easily estimated with sufficient accuracy). This makes each LHA a self-referential management system, with the effect of an excessive cost of the whole healthcare system in terms of GNP percentage, owing both to the "open-loop" planning of funds, and to the wrong idea of a service not "person-centered" but "to-person-supplied" (patients are often considered "passive receivers" who could only give very rough and instinctive evaluations of the received services). This brief outline of the LHA structure confirms that, in the Italian healthcare culture, this service is a provision that depends to a large extent on government policy. Individual citizens have little influence on the quality and accessibility of the health care in their country. In turn, the demand and utilization of care depend on the population's habits and social and cultural factors. In particular, utilization of medicines is beyond government control, thus showing a great increase of cost.

The approach adopted in this chapter comes from a revision of the basic concept of LHA scope, which is here stated as "the maintenance of the health status of the LHA area population" (D. Bellomo 2010, p. 1). Let us better clarify this concept through some reasoning steps:

1. The health maintenance actions occur through the *cooperative interaction* of the two *agents* of the healthcare system: the *persons* living in the area and the *healthcare service* (HS) *providers*. As shown in the schema of the person-to-HS provider interaction (Figure 11.2), a person living in an LHA territory asks for the consultation of an HS provider or medical doctor usually when a vague, not well understood symptom of an illness is felt. The medical encounter is always one of a social and cultural distance between doctor and person asking for care. The perceived distance may have an effect on how satisfied people are. This distance has to be surmounted through a cooperative analysis of the symptom, which allows the doctor to identify a diagnosis. At that time, the person becomes a "patient" of the healthcare system, to whom the HS provider prescribes a therapy. Depending on the therapy effects, two complementary evaluations of the

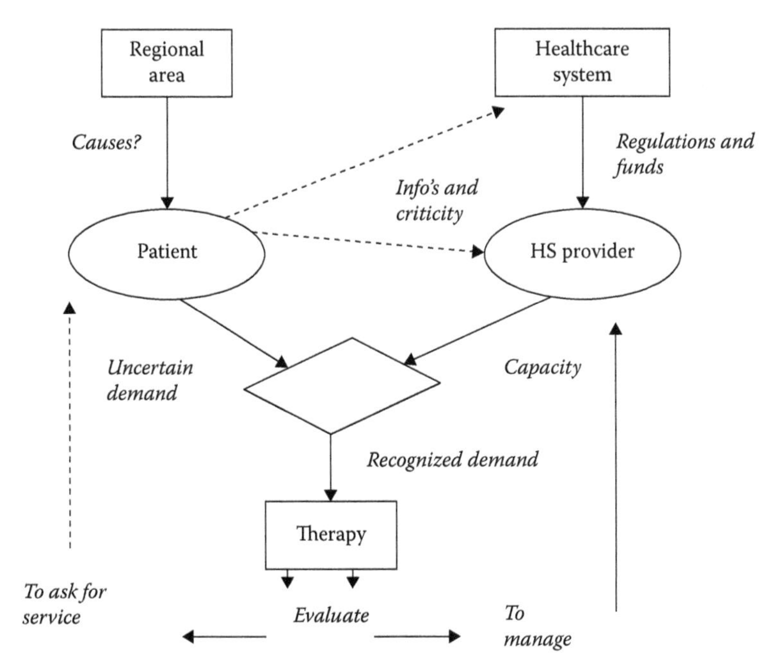

FIGURE 11.2 Schema of the person-to-HS provider interaction.

healthcare system performance can be drawn: on the side of the patient, the evaluation either of the HS provider ability or of some critical issues of the healthcare system; on the side of the HS provider, the evaluation of the service effectiveness and efficiency.

2. To assure good health maintenance action, the monitoring of patient-to-doctor interactions must be sufficiently accurate to discriminate (a) if new investments are necessary (to improve the service capacity); and (b) if the utilization of existing resources must be increased (to reduce/avoid system critical issues). To discriminate between the above two situations, answers to the following questions have to be found: Is the healthcare system able to supply
   i. An *effective* service, that is, is it able to give correct answers to the person's demands?
   ii. An *efficient* service, that is, with a high utilization of resources and a quick answer to a person's questions?
   iii. A *convenient* service, that is, convenient for the population, and without waste of money and time?
3. To answer these questions (which clearly identify the scope of any LHA managerial chief), it is mandatory to have the following information:
   i. A correct knowledge of the *demand* of healthcare services from persons, as it is distributed over the territory and as it is generated from the health status of the population
   ii. A clear knowledge of the *offer* of healthcare services by HS providers and service centers of the LHA, in terms of capacity and formality

   The availability of these data is crucial condition to apply statistical analysis of healthcare utilization, thus paying attention to social and cultural factors and allowing cross-country comparisons from a political perspective.

4. Assuming the ability to collect said information, the problem is how to *compare demand and offer* of healthcare services in terms of a *measurable difference* between these two elements, because
   i. The *demand* is a distributed function with two variables (space and time) and many-valued (typology = pathology; entity = seriousness; person attributes such as gender, age, etc.). In addition, the demand is identified only from the interaction of the two agents: one agent is asking for service without a complete knowledge of what to ask, and the other agent is trying to supply the best service, with the need to acquire a sufficient knowledge on the person's demand, in the same time.
   ii. The offer is a "concentrated function" (at each service center), piecewise constant (typology and capacity can only have sudden variations, owing to investments).

   A direct comparison of the two functions *demand* and *offer* is not only impossible but also lacking in any real meaning; both of them are the result of the interaction of the two agents, the *demand* being identified by the first phase of the interaction, that is, the diagnosis phase, while the *offer* is identified by both the diagnosis and the therapy phases.

5. As a consequence, *demand* and *offer* of healthcare services must be measured through *performance indicators*, whose estimation and use is proper of each one of the two agents. More specifically
   i. The estimation of the *demand*, in terms of (a) how it is generated, and (b) how it is distributed over the territory, is a proper action of the HS providers, through a statistical analysis of the population attributes
   ii. The estimation of the *offer* can be done according two different viewpoints:
      - Estimate how it is generated by the HS providers; this estimation has to be done by the HS provider itself, by evaluating its *offer* capacity and formality.
      - Estimate how it is received by the patient, in terms of *perceived quality* of the service.
6. Based on these concepts, a scheme of the feedback of information, coming from both HS providers and patients and to be used for improving the service *offer*, as outlined in Figure 11.2, can now be stated such as to show, on one side, the different service control and planning actions that must be done by the HS providers, and on the other side, the service quality evaluation done by patients. Figure 11.3 describes in detail the two-layer hierarchy of the LHA management organization.

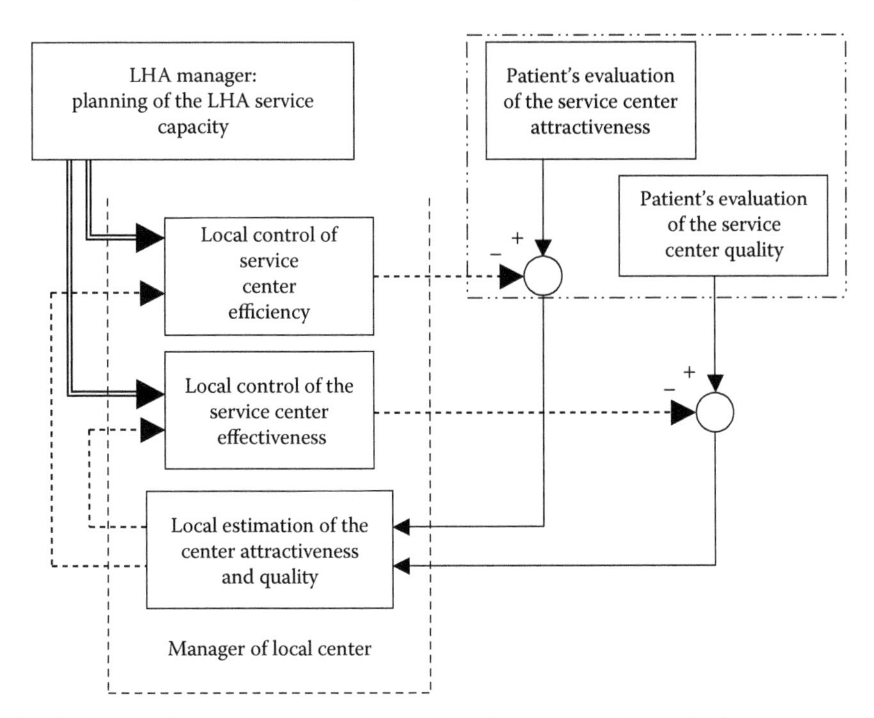

FIGURE 11.3 The two-layer hierarchy of the LHA management organization.

7. In practice, the two *agents* acting in any healthcare system can be detailed by considering as HS providers the following two professionals:
   i. The LHA *manager*, with the scope of planning the service capacity of the healthcare service centers in the LHA area
   ii. The set of healthcare service *operators*, each one being committed to provide a given healthcare service within a LHA center, by controlling the service quality level

   while the third actor of the healthcare system is the set of *end-users*—the *person/patients*—each one asking for a given healthcare service, and able to give a personal evaluation of the quality level of the received care. While the LHA chief has the objective of planning the service capacity of all LHA centers, the HS operator's objective is to assure the best possible efficiency and effectiveness of the functionality of an HS center. Regarding the latter, his or her decision must be based on an estimation of the HS center attractiveness and quality, as it results from both its own evaluation of the service offer and the patient's evaluation of the received service.
8. The real design and development of a performance evaluation and management systems for territorial healthcare networks of services, according to the scheme represented in Figure 11.3, requires finding preliminarily a clear answer to the following questions (usually approached in the industrial frame, but frequently unknown in the healthcare environment):
   i. What could be the meaningful numerical data to be used?
   ii. What should be the most significant performance indicators to be estimated?
   iii. What should be the results to be obtained?
   iv. What should be the models and methods to be used for both estimation and management purposes?

   To answer these questions is the scope of the next sections.

LHA PERFORMANCE EVALUATION BY USE OF INDUSTRIAL MANAGEMENT METHODS

The four questions stated previously summarize the steps that must be followed in designing a performance evaluation procedure for a territorial healthcare service. There can be found intuitive justifications for such a set of questions; the following items refer to the first three questions:

1. First, healthcare service is perhaps the system where to collect data of care services, either requested by patients or provided by operators, is more difficult than in any other service, because such a service is needed by persons who have no clear knowledge of their own needs nor of the necessary service to be received; this has been mentioned before as the social and cultural distance between doctor and patient.

2. Second, the same concept of "healthcare service performance" and of how to measure if an offered service action has been really performable or not is not evident; then, to select meaningful performance indicators is critical, thus reflecting in a difficult evaluation of delivered services by the LHA managers.
3. Third, an analyst of a healthcare system should be extremely careful in stating which type of analysis results he is searching for: even if it is evident that the final result should be an improvement of the healthcare service quality in front of a reasonable compression of costs required, no clear evidence exists about the effects of service modifications which could satisfy both goals.

Meaningful Numerical Data to be Used for LHA Performance Evaluation

An LHA manager as well as an HS operator should have at her or his disposal a model describing the patient's choices, in order to be able to modify the service to be supplied accordingly ("control" action). The necessary information to approach such a local management task could also be collected through questionnaires, but it is well known that data collected through questionnaires suffer several sources of uncertainty (see Walonick 1993).

A different approach, based on industrial experience, consists of estimating the patient flows over the network of service centers by formally modeling the preferences of patients and their dynamics over the service network; this is the line adopted in the research project on which the present report is based (see the acknowledgments at the end of this chapter).

Since it has been experienced that the LHA performance depends on the service programming actuated by the LHA managers and also that it is locally evaluated by patients (who derive their respective choice accordingly), the block diagram in Figure 11.3 suggests three parallel lines of analysis:

1. Analysis of the main characteristics of the LHA territory, so as to know
   a. The main characteristics of the "sources" from which the demands for health are generated, namely the distribution of the population and its main attributes, such as age, employment, study level, and territory attributes, as well as main industrial and servicing sites, main transportation directions, and so on
   b. An estimation of the LHA capacity to supply services through its centers, depending on the amount of resources there allocated
   c. An estimation of the capability of LHA centers to attract patients in terms of a gravity model of the patient flows towards such LHA centers, driven by the service capacity planned at the centers themselves

Note that this first analysis line is a crucial monitoring task for the LHA central manager, because by this analysis the LHA manager could get diverse information on the type of potential patients to whom the managed LHA should assure the required care.

2. Analysis of the performance of each LHA center according to the point of view of the local *service providers* (i.e., the local center managers and operators), by which they obtain estimations of
   a. The level of utilization of the center resources, thus having a measure of the planned efficiency
   b. The level of service "quality offered" by a center, thus having a measure of the planned effectiveness

This second analysis line is the basic monitoring task for a local center manager. This one has to assure the day-by-day care to persons in front of their demands, as soon as possible and as effectively as possible, having cleared patient-to-doctor communication problems.

3. Analysis of the patient flows by using the patients' database of the LHA in order to evaluate how patients select different services and different centers, thus giving indirect (but clear) information on their own preferences

These patient flow data will allow the analyst to recognize, on one hand, how patients evaluate the service centers, their personnel, and the supplied service activity, and on the other hand, to estimate how patients usually follow suggestions of LHA personnel, as doctors, specialists, and lab professionals. These estimations will be evident "measures" of the capability of the LHA system to inform, address, and follow the patients' utilization of the healthcare service supplied.

Depending on the approach here outlined, based on the utilization of "objective data" (not on information obtained by questionnaires) the principal database is the very large table of data of elementary services received by all patients from any operator acting in an LHA center. "Elementary service" means, for example, an individual diagnosis given to a patient by a family doctor, as well as a specific blood test at the consulting room, or an elemental intervention at a hospital department or at a first aid station.

Referring to the Asti LHA, the database for the year 2008 (actually at disposal for the research) includes more than 3 million records, one for each elementary service supplied to patients. From the principal database, for the purpose of simplifying the analysis, a set of working tables has been created:

1. A table of attributes of the elementary services, including
   a. The elementary service code and its date
   b. The service end-user, that is, the patient
   c. The service supplier, that is, the LHA center, and the type of supplied service
   d. The operator who prescribed this elementary service, and the prescription date

   This first table has a number of records as the principal database, but each one with only four data.

2. A table of attributes of the LHA centers, which specifies
   a. The center's place (where it is located)
   b. The number of different types of services which can be supplied there
   c. The resources at disposal

   This second table summarizes all information on the service centers of the LHA, thus giving a description of the LHA structure.

3. A table of attributes of each elementary service to be supplied, in terms of
   a. Service characteristics (type and operations required)
   b. Cost for the LHA (industrial cost)
   c. A ticket to be paid by the patient

   This table is the description of how the LHA can offer answers to the people's needs for health (thus summarizing the capability of the LHA sanitary organization).

4. Table of attributes of patients, which means data describing each patient, provided that he or she is coded to avoid recognition (age, gender, place of residence, exceptions from service payment, etc.)

   This table includes a record for each person who has been registered at the LHA.

5. Table of attributes of LHA operators, to characterize each doctor and center where patients can ask for prescriptions and diagnostic services, in terms of
   a. Place/ location
   b. Types of elementary services which can be offered

   This table has as many entries as the number of doctors and diagnostic centers in the LHA.

Most Significant Performance Indicators to be Estimated

To make clear which types of performance indicators should be useful for healthcare system management, one must remember that the main tasks of the healthcare service can be synthesized by two objectives, which are the two aspects usually perceived by patients:

a. *Quality*, that is, the effectiveness of the offered services
b. *Efficiency*, for instance, the reduction of the costs of drugs and staff, as well as the reduction of patient waiting times

In such systems the main problems to be approached may be classified as follows (Xiong, Zhou, and Manikopoulos 1994):

1. Dimensioning the system, that is, determining the type and number of resources to provide (staff, rooms, beds, etc.)
2. Understanding the workflow and detecting anomalies such as bottlenecks, waiting times, and so on

3. Improving efficiency, that is, using resources in a better way, by decreasing patients' length of stay, reacting to problems such as staff absence, and so on
4. Studying the system's reactivity with respect to an increased workload

In the literature, different approaches are investigated to improve healthcare services. For instance, Qi et al. (2006), studied hospital management by applying industrial engineering strategies. Moreover, Kumar and Shim (2007) modeled business processes in an emergency department in order to minimize patient waiting times. Furthermore, Choi et al. (2005) and Loh and Lee (2005) proposed and investigated medical information systems and Internet-related technologies. Usually, simulation is employed as a tool for verification and validation of the presented solutions for the improvement of quality and efficiency. Typically, simulation is either carried out by way of a dedicated software (Kumar and Shim 2007) or by Petri net (PN) models, as in Xiong, Zhou, and Manikopoulos (1994) and Choi et al. (2005). Another important aspect that has to be considered and that is specific to healthcare services is clinical risk. Therefore, not only efficiency but also clinical risk, the safety of staff and patients, and the reduction of errors are performance indicators (Taxis, Dean, and Barber 1999).

However, these contributions must be located in the framework of hospital management and organization, and scarce attention in literature is given to the performance of the whole healthcare network. Furthermore, dealing with transferability of industrial performance evaluation concepts, it must be noted that direct transfers are not always successful: systems, concepts, and techniques often need specific interpretation, adaptation, and adjustment for the different application domain (see Taxis, Dean, and Barber 1999). Taxis, Dean, and Barber (1999) briefly highlight some of the contextual issues in adapting established industrial practices for use in healthcare domains. There is generally a good knowledge on the "technical" content and basic building blocks of the approaches. However, much less is known about the essential ingredients for success, what works where, why it works, and how a successful transfer and deployment is accomplished. As a consequence, the following expected targets of a new organization of a public healthcare chain can be considered (Figure 11.4):

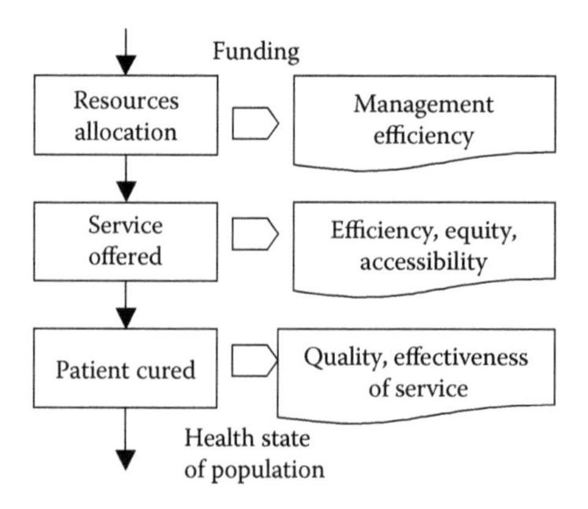

FIGURE 11.4 Scheme of LHA performance evaluation.

- Avoid congestion, minimize queue time, minimize service costs
- Adhere to principles of equity and accessibility of service providers
- Assure a minimum admissible level of service to any person

In practice, referring to the set of database tables reported in the previous section, the performance of the LHA service organization should be evaluated by estimating a number of key performance indicators (KPI). To this end, a specific "dashboard" for LHA managers has been conceived and is now under full development, with the following set of inquiries (here only the most significant are reported, owing to space constraints):

First level inquiring set

A. KPIs referred to the demand of healthcare services from potential patients.

B. KPIs referred to the present offer (supply) of healthcare services by the LHA centers/operators.

C. KPIs referred to how healthcare services are perceived by patients.

Second level A

A-1. How population is distributed over the LHA territory

A-2. How demand for services is generated over the LHA territory

A-3. Comparison of the LHA demand for services with national data

Third level A-1

A-1-1. Population density (i.e., amount of demand)

A-1-2. Average population age (i.e., types of services required)

A-1-3. Birth-rate and old-age rate indices (i.e., types of contacts with the LHA operators)

A-1-4. Distribution of average income per persons, and so on

Third level A-2

A-2-1. Demand density over the territory

A-2-2. Main types of services required

A-2-3. Most frequented centers, for each type of service

Second level B

B-1. How LHA service centers are distributed over the LHA territory

B-2. How the offer (supply) of services is characterized in front of patients' demands

B-3. Comparison of the LHA offer (supply) of services with national data

Third level B-1

B-1-1. Allocation of LHA centers (i.e., local capacity of service)

B-1-2. Area without any service center (i.e., poor offer)

B-1-3. Average distances between towns/villages and service centers (i.e., ease of accessibility)

Third level B-2

B-2-1. Average utilization rate of a center

B-2-2. Average area of use of a service center

B-2-3. Average waiting time at a service center

Second level C

C-1. How patients estimate efficiency and effectiveness of services

C-2. How the innovation/improvement of services could be recognized by patients' satisfaction

Third level C-1

C-1-1. Average accessibility of a center

C-1-2. How attractive a service center is

Preliminary Application of the Performance Evaluation Approach

As mentioned in the section, Meaningful Numerical Data to be Used for LHA Performance Evaluation, the basic idea is to use objective data, which can be collected both from territorial databases, and from data stores at disposal of any LHA. With these assumptions, the performance evaluation procedure has been organized in the following steps:

Step 1. Analysis of the present state of the HS providers in the LHA.

This first step aims to estimate

1. A "map of the LHA population," as it is distributed in the territory under the LHA responsibility, by using local municipalities as demand centers; such a map should give an estimate of the demand for health care directed both inside and outside the LHA centers.
2. A "map of the service quantity," indicating which healthcare services, and the number of resources there employed, are supplied by each LHA provider.
3. A "map of the service-offered quality," in terms of types of healthcare services supplied by each provider. (In this preliminary analysis, it is assumed that the offered quality of a center depends on the variety of services made at disposal of patients.)
4. A "map of providers' accessibility" in terms of distances of the center from each municipality (considered as an "origin" of potential patients).
5. A "map of present HS providers' efficiency," in terms of average service times, types of service procedures, and utilization of resources and personnel.

These five "maps" define the database on which the performance evaluation is based. Typically, what will be derived is an estimation of how the HS providers of the LHA are operating by taking into account the employed resources. Then, the derived estimations will account for the point of view of the LHA managers: no specific use of patients' opinions or data is involved. But it must be noted that this second set of data concerns another parallel estimation, the *patient-perceived quality* of services. In the approach here considered, this second line of performance evaluation, based on qualitative and subjective data, could only be used—if necessary—for recognizing critical points of the LHA service.

Step 2. Evaluation of the attractiveness of each HS provider.

The estimation of the HS provider attractiveness comes from the use of the first map in the list by computing the number of patients who "arrived" to a given provider within a given time period. On the other hand, map 4 allows estimation of how many patients would prefer a given provider, depending on provider accessibility. By comparing the above two estimates, a first evaluation of the provider attractiveness can be derived, by also detecting the greatest flows of patients going to centers outside the LHA area.

Step 3. Comparison between two centers in terms of service quantity and quality. Maps 2, 3, and 5 allow the definition of a comparison procedure concerning the services offered by two (or more) providers in the LHA, thus describing a "service profile," one for each provider. Each service profile could be obtained by reporting the provider characteristics contained in the three mentioned "maps." From a number of service profiles, an "optimal" one can be identified, thus having a benchmark with which all providers could be compared.

Step 4. Evaluation of the employed resources utilization. The estimation of the employed resources' utilization comes from the use of maps 1 and 5. Such an estimation can be derived from the following considerations:

- Definition of a relation linking the service rate both to the number of patients and to the employed resources, by taking into account the standard service times
- Definition of a procedure to compute the necessary service rate such as to assure that the patients' queue be shorter than a given target
- Comparison of the two results

Step 5. Analysis of the criteria to address patients from provider to provider. The addressing of patients from HS provider to another has to be estimated by means of a model able to identify the critical parameters of a provider, in terms of offered services, and the links of the provider with upstream ones. A supply chain model can be used for recognizing such addressing criteria.

In each step of the healthcare performance evaluation, the selection of theoretical models derived from analogy with industrial management/evaluation methods is under application. So, referring to step 2, *gravity and network decomposition models* (as in Iwainsky et al. 1986) are under application, while step 3, as well as step 4, are approached by using *quality control* methods (Franceschini 1998), and in step 5, *production flow analysis* models (Burbidge 1989) are accounted for.

PRELIMINARY RESULTS

The present state of development of the Italian Research Project of National Interest (PRIN) project has reached the definition of the performance evaluation approach, the main analysis tools to be applied, and some preliminary applications of the following procedures: tools to analyze the territory and to obtain statistical distributions of territorial data; gravity models to estimate how patients should choose service centers depending on the center's data; and models to estimate the patient flows on the basis of the LHA patient database and to estimate how they follow some diagnostic and therapeutic "route" among LHA services. Preliminary results are going to be available based on data of the LHA of the Asti healthcare agency. Referring to the list of questions of the dashboard mentioned in the previous section, some useful considerations, even if preliminary, can be drawn.

The main attributes of the Asti area population allow a characterization of the potential demand for healthcare: the average age of the resident population is close to 48 years (see Figure 11.5), with a significant average percentage of persons to whom exemption from paying care costs is assigned, either due to old age or due to chronic diseases (see Figure 11.6). These two data together with the distribution of the population density over the 108 municipalities included in the Asti area (see Figure 11.7), all very small towns except the central town of Asti, all located in a hilly province with a few principal directions for communication (see Figure 11.8), give some preliminary but already clear suggestions to the LHA manager.

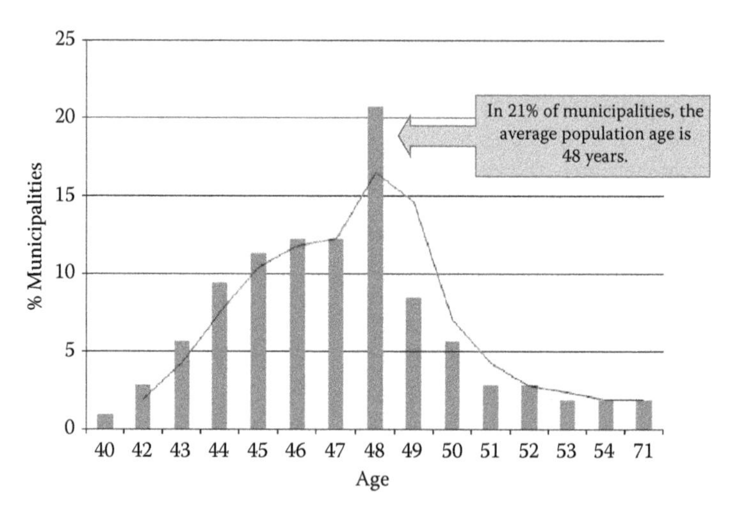

FIGURE 11.5 The age distribution of the resident population.

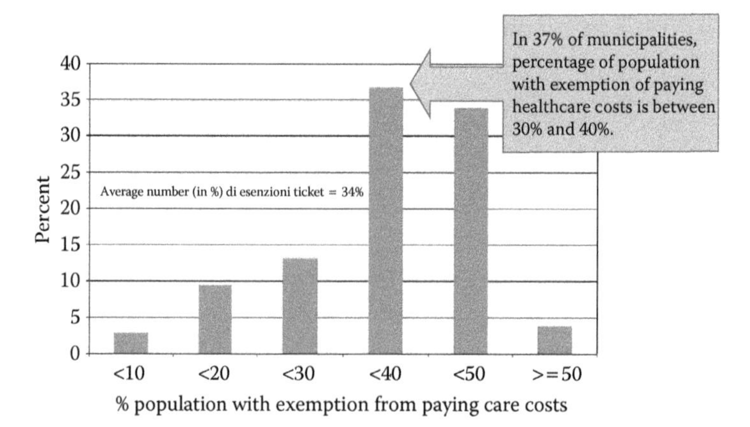

FIGURE 11.6 Percentage of persons to whom exemption from paying care costs is assigned, either due to old age or due to chronic diseases.

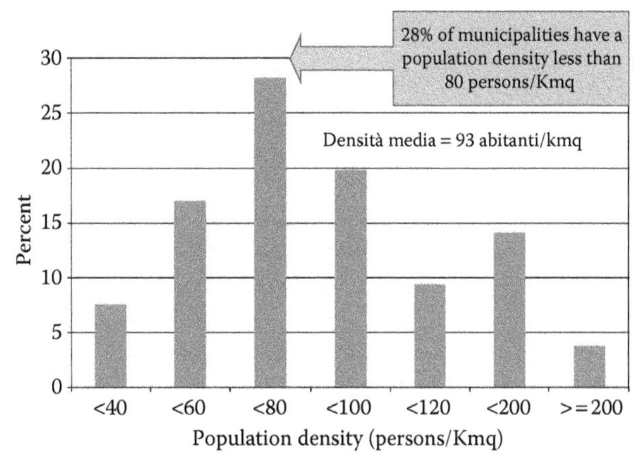

FIGURE 11.7 The population density over the 108 municipalities included in the Asti area.

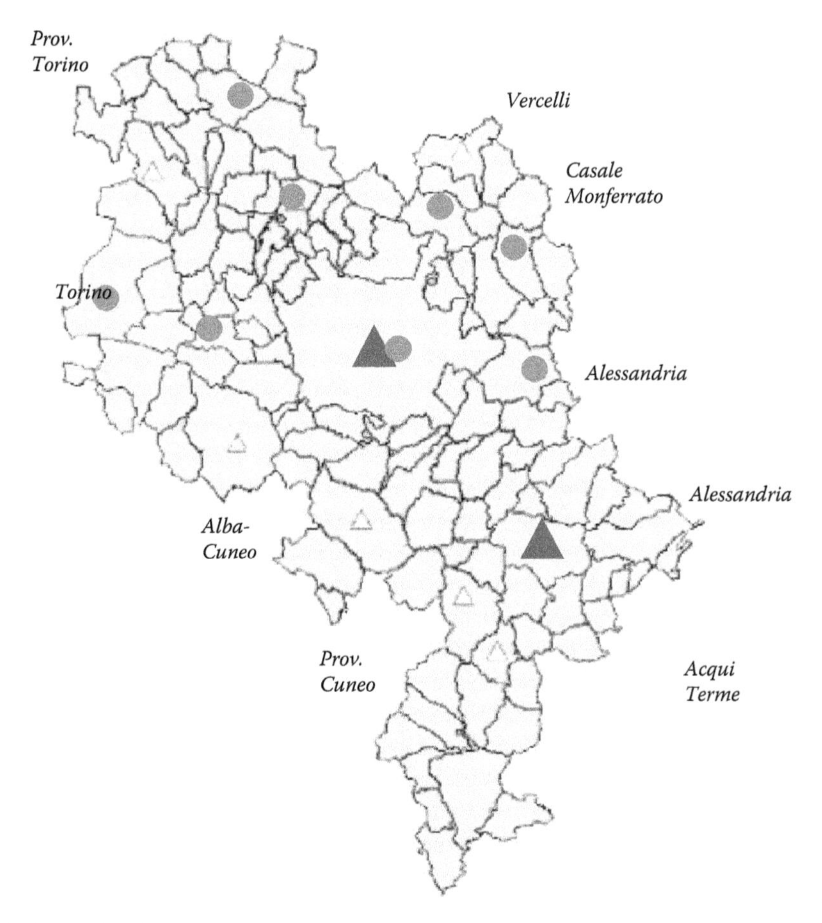

FIGURE 11.8 The principal directions for communication in the Asti area.

FIGURE 11.9 The Asti LHA municipalities with either an absence or only part-time presence of medical doctors (marked by a solid point).

The first hint is to analyze the distribution of the HS centers in the municipalities, in order to recognize zones where a few services are available within a short distance. Figure 11.9 illustrates the municipalities with either an absence or only part-time presence of medical doctors. This figure gives an indirect but evident measure of the HS center's accessibility.

Another hint results from the analysis of flows of patients demanding for healthcare from their municipalities toward the HS centers (here not represented owing to space limits; a sequence of figures, one for each main HS center illustrated in Figure 11.1, would be necessary). This representation offers a clear view of the attractiveness of each HS center, thus giving suggestions about future funding installations.

The current development of the PRIN project also evaluates the utilization of HS centers, depending on the resources assigned there; the goal is to estimate costs and efficiency, besides the effectiveness indices previously mentioned. Results concerning this stage are expected soon, and the recent advancements of the research as well as the more recent results can be seen on the project website (http://www.lep.polito.it/prinsalute).

ACKNOWLEDGMENTS

This work was developed under the Italian Research Project of National Interest "Innovation of Healthcare Service Networks by Using Supply Chain Management Techniques," Ministry of University and Research, chaired by Prof. A. Villa (http://www.lep.polito.it/prinsalute). The author wishes to thank the Azienda Sanitaria Locale-Asti (ASL-AT) general director for having supported this research and for the permission of using the complete database of the agency.

REFERENCES

Bellomo, D. 2010. Internal report, ASL-AT (in Italian).

Burbidge, J. 1989. *Production Flow Analysis for Planning Group Technology*. Oxford, UK: Clarendon Press.

Choi, S. S., M. K. Choi, W. J. Song, and S. H. Son. 2005. Ubiquitous RFID health care systems analysis on PhysioNet grid portal services using Petri nets. In *Proceedings of the 5th IEEE Intern. Conference on Information, Communication and Signal Processing (ICICS)*, Dec. 2005, Bangkok, 1254–8.

Franceschini, F. 1998. Quality Function Deployment—A Design Tool to Link Quality and Innovation, (in Italian), Milano: Ed. Il Sole 24 Ore.

Iwainsky, A., E. Canuto, O. Taraszow, and A. Villa. 1986. Network decomposition for the optimization of connecting structures. *Networks* 16:205–35.

Kumar, A., and S. J. Shim. 2007. Eliminating emergency department wait by BPR implementation. In *Proceedings of the 2007 IEEE Intern. Conference on Industrial Engineering and Engineering Management (IEEM)*, Dec. 2007, Nanyang, 1679–83.

Loh, P. K. K., and A. Lee. 2005. Medical informatics system with wireless sensor network-enabled for hospitals. In *Proceedings of the 2005 International Conference on Intelligent Sensors, Sensor Networks and Information Processing*, Melbourne, 265–70.

Qi, E., G. Xu, Y. Huo, and X. Xu. 2006. Study of hospital management based on hospitalization process improvement. In *Proceedings of the 2007 IEEE Intern. Conference on Industrial Engineering and Engineering Management (IEEM)*, Dec. 2007, Nanyang, 1679–83.

Taxis, K., B. Dean, and N. Barber. 1999. Hospital drug distribution systems in the UK and Germany—a study of medication errors. *Pharm World Sci* 21:25–30.

Walonick, D. S. 1993. Everything you wanted to know about questionnaires but were afraid to ask. http://www.statpac.com/research-papers/questionnaires.htm.

Xiong, H. H., M. C. Zhou, and C. N. Manikopoulos. 1994. Modeling and performance analysis of medical services systems using Petri nets. In *Proceedings of the IEEE International Conference on Systems, Man and Cybernetics*, San Antonio, TX, 2339–42.

12 Decision Making in Healthcare System Design

When Human Factors Engineering Meets Health Care

Pascale Carayon and Anping Xie

CONTENTS

INTRODUCTION

Health care is in great need of redesign (Reid et al. 2005). According to the Institute of Medicine, between 44,000 and 98,000 people die of preventable medical errors every year in the United States (Kohn, Corrigan, and Donaldson 1999). A 2003 study by RAND shows that only 55% of people receive recommended care for either acute, preventive or chronic care (McGlynn et al. 2003). These performance gaps and inefficiencies in health care have been related to the way healthcare systems and processes are designed (Donabedian 1978; Shortell et al. 1994). In order to redesign healthcare systems and processes and, therefore, improve quality of care and patient safety, a human factors systems approach is necessary; this systems approach should address the complexity of healthcare work systems and processes, and give attention to the people dimension (e.g., the needs and characteristics of patients and healthcare professionals and workers) (Carayon et al. 2006).

Applying a human factors systems approach to healthcare system design is neither straightforward nor spontaneous. It is critical to understand how this approach fits or conflicts with the current healthcare culture. The healthcare culture influences how healthcare professionals apply their expertise to conceptualize a problem, to evaluate a solution, and to support their ideas; however, it may also lead to intellectual centrism, which limits interdisciplinary collaboration in providing and improving care (Banerjee and Chiu 2008). Therefore, understanding the similarities and differences between the healthcare culture and the human factors engineering culture can help in integrating the human factors engineering ideas, concepts, and methods into health care.

In this chapter, we present our human factors systems approach to healthcare system design. We then describe the main characteristics of the current healthcare culture. We use the example of checklists to illustrate the fit or conflict between the human factors systems approach and the current healthcare culture. We conclude with recommendations for facilitating and supporting the implementation of a human factors systems approach in health care that will help on the road to healthcare redesign and transformation.

SYSTEMS APPROACH TO HEALTHCARE DESIGN

The 2001 report by the Institute of Medicine (IOM) on "Crossing the Quality Chasm: A New Health System for the 21st Century" stressed the importance of system design in creating safe high-quality healthcare environments (Institute of Medicine Committee on Quality of Health Care in America 2001). This report identifies the poor organization and design of the healthcare delivery system as a major problem: "The prevailing model of healthcare delivery is complicated, comprising layers of processes and handoffs that patients and families find bewildering and clinicians view as wasteful" (p. 28). The IOM committee recommends the application of engineering concepts and methods to improve the design of care processes and therefore improve patient safety and other quality of care outcomes. Others have also recommended a shift in the approach to healthcare system design. According to Shortell and Singer (2008, p. 445), "although creating a culture of safety is important, creating a culture of systems is a more fundamental challenge." A substantially broader vision with a real adoption of a systems approach needs to be applied to the design of healthcare systems and processes (Shortell and Singer 2008). In this section, we present our systems approach to healthcare system design and describe the underlying cultural values to this approach.

SEIPS Model of Work System and Patient Safety

Given the complexity of the healthcare system, it is important to adopt a systems approach to the analysis and redesign of healthcare systems (Vincent 2004). Our systems approach, that is, the SEIPS (Systems Engineering Initiative for Patient Safety) model of work system and patient safety, is based on the work system model developed by Smith and Carayon-Sainfort (1989) and the structure-process-outcome (SPO) model of Donabedian (1978). According to the work system model developed by Carayon and Smith (Carayon and Smith 2000; Smith and Carayon 2000; Smith and Carayon-Sainfort 1989), various people (e.g., physician, nurse, patient) perform or are involved in a range of care tasks (e.g., nurse administering medication, patient reviewing medication instructions) using various tools and technologies (e.g., physician using a stethoscope, nurse documenting a medication administration in the electronic medication administration record) in a physical environment (e.g., patient exam room, operating room) under certain organizational conditions (e.g., coordination between primary care physician and specialist, work schedules). Donabedian's SPO model links the structure and processes of care to subsequent patient outcomes (Donabedian 1988). By integrating these two models, the SEIPS model highlights how work system design (structure) is linked to patient safety (outcome) through care processes. In addition, because the SEIPS model is anchored in the human factors and ergonomics discipline, the outcomes of the model include not only patient outcomes (e.g., patient safety) but also employee and organizational outcomes (e.g., worker safety and well-being). See Figure 12.1 for a graphical representation of the SEIPS model.

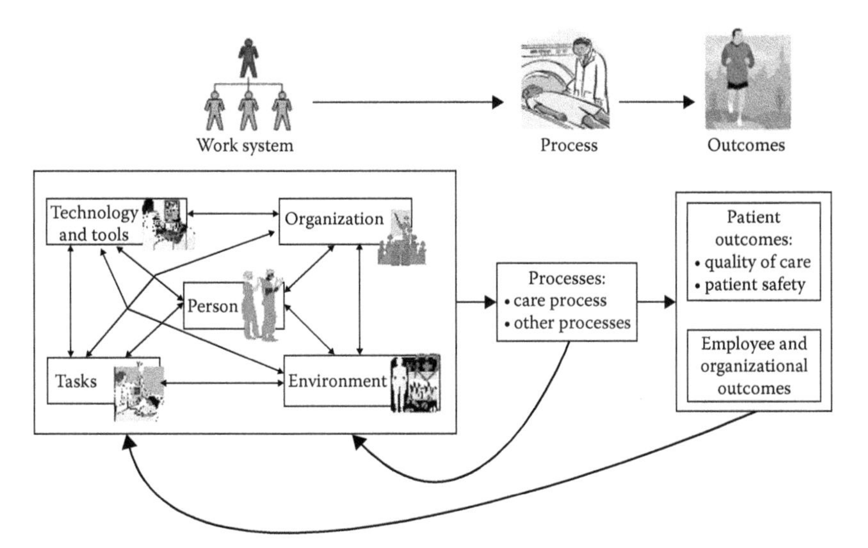

FIGURE 12.1 SEIPS model of the work system and patient safety. (From Carayon, P. 2006. *Appl Ergon* 37:525–35. With permission.)

Core Values of Human Factors Systems Approach

The human factors systems approach to healthcare system design is based on underlying assumptions and values. In this section, we discuss four core values of the human factors systems approach: (1) people at the center of the system, (2) systems thinking, (3) continuous improvement, and (4) balancing multiple objectives.

People at the Center of the System

According to the discipline of human factors and ergonomics, the design of systems has to be compatible with the needs, abilities, and limitations of people. In the SEIPS model, the individual is at the center of the work system. The work system should be designed to enhance and facilitate performance by the individual and to reduce and minimize the negative consequences on the individual (such as stress, fatigue, and burnout) and the organization (for example, organizational performance and turnover).

Three categories of human characteristics need to be considered: (a) physical characteristics, (b) cognitive characteristics, and (c) psychosocial characteristics. The dimensions and layout of a hospital patient room may limit physical access to the patient, and therefore negatively affect the ability of the healthcare professional to provide care; this is an example of the role of physical characteristics in patient safety and healthcare quality. When the cognitive abilities of a patient who is waiting for his surgery are affected, the patient may not be able to fully understand the instructions or questions of the healthcare professional; this may be critical if the question to the patient is about checking the side of surgery. Therefore, cognitive characteristics of various people (e.g., patients) in the system can play a role in patient safety. One example of psychosocial characteristics of importance to healthcare quality and safety is the ability of healthcare professionals to communicate and work in teams. These examples of physical, cognitive, and psychosocial characteristics underline the importance of considering the multiple needs, abilities, and limitations of people involved in a healthcare system. The physical, cognitive, and psychosocial characteristics of the people at the center of the system are varied and influence each other.

It is also important to recognize that a healthcare work system includes a variety of people: healthcare professionals (e.g., physicians, nurses, pharmacists, technicians, and maintenance personnel), patients, and their families. Each individual has physical, cognitive, and psychosocial needs, abilities, and limitations that should be considered when designing or redesigning a healthcare system or process. Designing a healthcare system or process for multiple stakeholders can, however, be challenging; the needs of one group may conflict or may not fit with the needs of another group. For instance, the physical layout of an intensive care unit can affect the ability of nurses to monitor patients: if the ICU has an open layout, it is easier for nurses to monitor patients and quickly intervene if necessary. On the other hand, an open layout may limit patient and family privacy. In this example, the design of the ICU physical layout requires a balance between the needs of nurses and those of patients and their families.

Systems Thinking

The balance theory and the associated work system model emphasize that a workplace, process, or job is more than the sum of the individual components of the system (Carayon 2009; Smith and Carayon 2000; Smith and Carayon-Sainfort 1989). The interactions among the various elements of the system produce results that are greater (or lesser) than the additive aspects of the individual parts of the system. It is the way in which the system elements relate to each other that determines the potential for the system to produce positive results. Thus, work system improvements must consider and accommodate all elements of the work system and their interactions. A major advantage of the balance theory is that it does not highlight any single factor or a small set of factors. Rather, it examines the design of work systems from a holistic perspective to emphasize the potential positive elements in a work system that can be used to overcome the adverse aspects.

In health care, "quality problems occur typically not because of a failure of goodwill, knowledge, effort, or resources devoted to healthcare, but because of fundamental shortcomings in the ways care is organized" (Institute of Medicine Committee on Quality of Health Care in America 2001, p. 25). To improve the quality and safety of care, a systems approach, which focuses on the structure of health care, healthcare processes, and the interactions among work system elements, should be applied for healthcare system design. Those improvement activities need to consider the work systems and the interactions between work system elements: a change in one element of the work system can affect other elements of the work system. Therefore, any project aimed at improving healthcare systems and processes needs to consider the system interactions by anticipating how the change may affect other system elements. In the case of CPOE (computerized provider order entry) implementation in hospitals, a number of so-called "unintended consequences" have been identified (Ash et al. 2007, 2009). One unintended (unanticipated and undesirable) consequence of CPOE identified by Ash and colleagues (2007, 2009) was the increase in physical ergonomic problems and subsequent musculoskeletal discomfort and disorders. Many of these unintended consequences may have been anticipated if the CPOE planning and implementation phases had been attentive to the system influences of the CPOE change. The leaders of system improvement activities need to be aware of system influences of the changes being proposed and implemented.

Continuous Improvement

System design is an activity that extends over time, and continues beyond system implementation and throughout use (Carayon 2006; Clegg 2000). Three phases of system development can be distinguished: system design, implementation of change, and continuous adaptation and improvement (Carayon 2006; Meister and Enderwick 2001). The various phases of system design and changes influence the design of individual system elements and produce changes in other connected elements, thereby creating a continuous movement towards adaptation and improvement.

Feedback has been emphasized as an important organizational design element in the healthcare literature (Evans et al. 1998; McDonald et al. 1996). The importance of feedback in managing the change process is echoed in the literature on quality management (see, for example, the plan-do-check-act cycle proposed by Deming [1986]). Improving quality of care and patient safety should be considered as a continuous process, not as a single project; this concept of continuous change as opposed to discrete change has been discussed by Weick and Quinn (1999) and examined in the context of smart IV pump implementation by Carayon and colleagues (2008).

The SEIPS model specifies feedback loops from process to work system and from outcomes to work system (see Figure 12.1). The feedback loop that goes from the outcomes to the work system represents a pathway to redesigning the work system. Information is collected on outcomes and can be used to identify problems with the way the work system is designed. This information can then be used to redesign the work system. The second feedback loop goes from the process to the work system and indicates a two-way interaction between the work system and the process: the work system influences how the care process is carried out and the actual care process may lead to changes or adaptation in the work system. An example of this adaptive mechanism is work-around: the work system puts constraints on how the care process can be efficiently and effectively carried out; in order to achieve the care objectives, the healthcare professional has to work around some of the work system obstacles. This adaptive mechanism shows how the care process may lead to changes in the work system.

Balancing Multiple Objectives

According to the International Ergonomics Association (2000), human factors systems design is concerned with the optimization of both human well-being and overall system performance. The SEIPS model emphasizes the linkages between patient outcomes and employee/organizational outcomes (see Figure 12.1). A healthy work organization must have good organizational outcomes and a healthy and safe workforce, and provide high quality safe patient care (Sainfort et al. 2001).

Woods (2006) contrasts two groups of organizational objectives: (1) acute goals (e.g., production goals), and (2) chronic goals (e.g., safety). The acute goals happen in the short term, whereas the chronic goals require long-term investment and attention. There can be a conflict between the acute and chronic goals, which often leads to attention to short-term issues at the expense of long-term issues; examples of this process include the Columbia accident, which was the result of "production pressures" eroding the organizational identification and control of safety risks (Woods 2006). In health care, the acute goals include the delivery of efficient, effective, timely care; the chronic goals include patient safety, patient-centered care and individual outcomes such as worker safety, well-being, and satisfaction. These system objectives need to be balanced out: too much attention to one objective may lead to negative impact on another objective, especially if the former objective is an acute goal and the latter objective is a chronic goal.

HEALTHCARE CULTURE

According to Schein (1992), culture can be discussed at three distinct levels: artifacts and behaviors, espoused values, and assumptions. The first level includes any tangible or verbally identifiable elements, which are easy to observe but difficult to interpret. For instance, patterns of behavior or norms can be easily extracted, while it is difficult to understand the meaning of these activities. At the second level are conscious values and beliefs that members of a cultural group espouse and profess. These values represent an ideal aspiration and can be articulated. The third level consists of driving elements of a culture. These underlying assumptions are usually developed over a long period of time, and can hardly be seen and identified. Schein's model has been proposed as a framework for understanding organizational culture (Schein 1992, 1999). In this section, we use Schein's model to describe the cultural values of health care.

The healthcare culture extends beyond regional and organizational differences; it can be considered as a professional culture, that is, a network of distributed knowledge (ideas, beliefs, practices, and values) that is produced, shared, and reproduced among a collection of interconnected individuals (Banerjee and Chiu 2008). Healthcare culture can be further divided into subcultures defined by different professions such as medicine, nursing, and administration (Carroll and Quijada 2007). Our discussion of healthcare culture is focused on the level of espoused values, and four core values of health care are described: (1) scientific inquiry, (2) individual responsibility for patient care, (3) autonomy, and (4) excellence (Carroll and Quijada 2007; Smith and Bartell 2007). These values, which were originally considered to be peculiar to the medical culture, are now shared by most healthcare professional subcultures.

Scientific Inquiry

Scientific inquiry or evidence-based practice in health care emphasizes the integration of scientific evidence with clinical expertise and patient values in the decision making of individual patient care, and calls for standardization of best practices. As the 2001 IOM report asserted, "the best care results from the conscientious, explicit, and judicious use of current best evidence and knowledge of patient values by well-trained, experienced clinicians" (p. 76). The use of clinical practice guidelines is an example of implementing scientific inquiry. By synthesizing the best evidence, clinical practice guidelines provide professionals with formal recommendations about appropriate and necessary care for specific clinic circumstances (Lohr, Eleazer, and Mauskopf 1998). They are expected to reduce undesirable variations in clinical practices and improve healthcare outcomes.

To incorporate scientific inquiry into practice, both individual professionals and the organization have to be well prepared. At the individual level, professionals not only need to know how to diagnose and treat patients but also how to acquire research skills that are not part of traditional medical or other healthcare training (Evidence-Based Medicine Working Group 1992). Specifically, professionals should be able to formulate a clear clinical question from that problem, to search for the relevant information from the best possible published or unpublished sources, to evaluate that evidence for its validity and usefulness, and to implement the appropriate findings (Davidoff 1999). At the organizational level, evidence-based policies, the development of evidence-based practice, and the promotion of a culture of evidence need to be implemented and promoted.

Individual Responsibility

As Park et al. (2007) indicated, healthcare professional responsibility consists of four major factors: patient care, education, self, and collegial relationships. We focus our discussion on the responsibility to patient care, which is at the center of professional responsibility. Traditional medical ethics emphasized the relationship between individual physicians and individual patients (Pellegrino 1982). Physicians worked as individual practitioners in an isolated "cottage industry" with less sophisticated interventions. Good medicine depended on physician training and character. Patients trusted physicians to apply their knowledge and skills for the patients' best interest, while physicians acted as the sole arbitrator in the decision making of all matters relevant to health and well-being (Kenny, Mann, and MacLeod 2003; Mathews and Pronovost 2008; Shortell et al. 1998). In this case, professional responsibility was viewed as an individual matter, stressing individual credit for success and individual blame for failure (Newton 1982).

Individual responsibility is expected to be untenable in today's health care, which is characterized by increased complexity (Pellegrino 1982). On the patient side, the complexity of health care increases when patients face multiple interrelated problems that are ambiguous (Molleman et al. 2008). On the provider side, scientific and technological growth leads to a considerable increase in advanced and comprehensive diagnoses and treatments (Wilson, Holt, and Greenhalgh 2001). The complexity of health care results in increased differentiation and specialization, and requires professionals from multiple disciplines to cooperate. However, in many cases, healthcare complexity leads to consultation rather than true multidisciplinary teamwork, and professionals still hold strong beliefs in individual responsibility. For instance, Molleman et al. (2008) interviewed members of two clinical teams about the relationship between healthcare complexity and cooperation. Most of the interviewees believed that they were able to deal with situations involving multiple problems themselves, and considered this to be a basic aspect of their job. The ethics of individual responsibility are deeply rooted in the traditional clinical training and closely related to professional autonomy, which can limit the shift of professional responsibility from individual to collective (Molleman et al. 2008).

Autonomy

Autonomy has been cited by Mathews and Pronovost (2008) as the ability to control and self-monitor one's technical work. In health care, it means control over (1) the nature and volume of tasks, (2) the acceptance of patients, (3) diagnosis and treatment, (4) the evaluation of care, and (5) other professionals (Schulz and Harrison 1986). Autonomy corresponds to the development of professional responsibility. Traditionally, because of the authoritative relationship between individual physicians and individual patients, autonomy was regarded as the central aspect of physician professionalism (Carroll and Quijada 2007; Freidson 1970). Along with increased healthcare complexity, specialization and cooperation are believed to accompany weakened autonomy. However, due to the lack of true multidisciplinary teamwork, professionals tend to immerse themselves in the knowledge and culture of their own group and have few opportunities to interact with other disciplines (Hall 2005; Hall and Weaver 2001). Therefore, physician autonomy is even strengthened. Meanwhile, autonomy also becomes part of other healthcare professional subcultures (Carroll and Quijada 2007). For instance, nurses may be considered incompetent if they cannot solve problems on their own and need to ask their supervisors for help (Tucker and Edmondson 2003).

The development of professional autonomy is demonstrated by the study by Leipzig et al. (2002). They investigated the role of physician in multidisciplinary teams by surveying students in medicine, nursing, and social work. Seventy-three percent of medical students, in contrast to 44% of nursing students and 47% of social work students, believed that the primary purpose of a team was to assist physicians in achieving treatment goals for patients. Besides, 80% of medical students, in contrast to 40% of nursing students and 38% of social work students, agreed that the physician had the right to change the team's patient care plans without the consent of the team and had the final word on team decisions.

Excellence

Excellence means making conscientious effort to be the best at what one does and being recognized for it (Carroll and Quijada 2007). Christmas et al. (2008) described clinical excellence in several distinctive dimensions, including communication and interpersonal skills, professionalism and humanism, diagnostic acumen, skillful negotiation of the healthcare system, knowledge, taking a scholarly approach to clinical practice, and having passion for clinical medicine. Holding the responsibility to reduce medical error, increase patient safety, minimize overuse of healthcare resources, and optimize the outcomes of care, healthcare professionals essentially seek to be excellent. Due to the long-standing physician-patient relationship, professionals are also anticipated to be excellent by their patients. Moreover, the value of excellence echoes the emphasis on role models in the training of healthcare professionals.

A difficult question related to the value of excellence is the evaluation of clinical performance. It is important to decide what the proper measures of clinical performance are (e.g., patient outcomes, care processes, or patient satisfaction) and how to measure them accurately. Another difficult question is the relationship between excellence and standardization. Does excellence contradict standardization? Or does being excellent require professionals to follow standards?

WHEN HEALTHCARE SYSTEM DESIGN MEETS THE HEALTHCARE CULTURE

Table 12.1 summarizes the core values of the healthcare culture and the human factors systems approach to healthcare system design. To apply the human factors systems approach to healthcare system design, it is critical to understand how this approach fit or conflict with the current healthcare culture. In this section, we discuss the values (1) that fit between the two cultures, (2) that conflict with each other, and

TABLE 12.1
Cultural Values in Health Care and Human Factors Engineering

Cultural Values in Health Care
1. Scientific inquiry
2. Individual responsibility
3. Autonomy
4. Excellence

Cultural Values in Human Factors Engineering
1. People at the center
2. System thinking
3. Continuous improvement
4. Balancing multiple objectives

(3) that need to be adapted. Finally, we use the example of checklists to illustrate the fit or conflict between the human factors systems approach and the current healthcare culture.

Values That Fit between the Two Cultures

In health care, pursuing excellence can be considered a continuous improvement process. First, clinical education takes time; it is necessary for professionals to spend a long time to accumulate the knowledge, skills, and experience that are prerequisites for being excellent. Second, excellence is a continuous-change objective rather than a static state. To keep themselves excellent, professionals have to be committed to lifelong learning and maintain their medical knowledge and clinical skills (ABIM Foundation, ACP-ASIM Foundation, and European Federation of Internal Medicine 2002). Third, feedback is important for professionals to learn how far they are from being excellent. Professionals adapt themselves according to the feedback. Finally, when pursuing excellence becomes a shared value among professionals, it may create an atmosphere of competition, which may promote continuous improvement.

Values That Conflict with Each Other

There are many ways that healthcare culture can conflict with the cultural values of a human factors systems approach. Healthcare culture focuses on the performance of practitioners, while the human factors system approach focuses on the conditions under which individuals work. Healthcare culture emphasizes individual knowledge and skills, while the human factors system approach emphasizes team collaboration and communication. Healthcare culture tends to blame individual clinicians for forgetfulness, inattention, carelessness, and poor productivity, while the human factors system approach attempts to build defenses to prevent errors or mitigate their effects proactively. If errors occur, the predominant healthcare culture assumes that the problem must be due to a lack of competence or carelessness of the healthcare professional. Therefore, the response to errors is that individuals need to be trained more and better, be alerted to the need to attend to safety and follow rules, and be motivated to be careful, and individuals are punished if they err.

Finally, the healthcare culture tends to focus on the relationships between process and health-related outcomes, while the human factors system approach focuses on the interaction of the individual and work surroundings including structure, process, and individual and system outcomes (see Figure 12.1).

Values That Need to Be Adapted

Scientific inquiry is strongly recommended to support clinical decision making (Institute of Medicine Committee on Quality of Health Care in America 2001). It is expected to result in better and more efficient care, improved healthcare outcomes, better-educated patients and clinicians, a scientific base for public policy, a higher quality of clinical decisions, and better-coordinated research activities (Timmermans and Berg 2003). However, implementing scientific inquiry in practice is highly controversial. While proponents believe that scientific inquiry will reduce variation in clinical practices and increase credibility of medicine, opponents consider scientific inquiry as a threat to professional autonomy (see section on Autonomy) and the value of excellence (see section on Excellence) because it discourages individual innovation in patient care and healthy competition among practitioners (Timmermans and Mauck 2005). The key point to bridge the gap between these opposing viewpoints is not arguing whether scientific inquiry is beneficial to clinical decision making. Instead, it is important to figure out what evidence is appropriate and how the evidence can be used to guide clinical practice. Barriers to the use of evidence include factors related to practitioners (e.g., unawareness, disagreement), tasks (e.g., task ambiguity, high workload), organization (e.g., responsibility ambiguity, ineffective communication), and technologies (e.g., lack of necessary supplies) (Cabana et al. 1999; Gurses et al. 2008). From the perspective of systems thinking, a multifaceted approach that reaches out to multiple stakeholders and addresses different components of the work system is recommended to support the implementation of scientific inquiry (Timmermans and Mauck 2005).

Individual responsibility, which stresses the importance of training, is necessary for professionals to improve their performance, but not sufficient to lead to a safe environment for patients. The cultural value of individual responsibility holds the assumption that errors are due to lack of competence or carelessness of professionals. In order to achieve improvements in patient safety, we need to move from a culture of blaming individuals for errors to a culture in which errors are treated as signs and opportunities to improve the system (Institute of Medicine Committee on Quality of Health Care in America 2001). Besides, health care is a complex work system, in which errors result from complex causes rather than from only the failure of individual professionals. Therefore, it is important to evaluate the many contributors to errors and to use this knowledge to design healthcare systems. In addition, due to the complexity of health care, individual professionals are no longer able of delivering complex diagnostic and therapeutic interventions on their own (Mathews and Pronovost 2008); cooperation among professionals from different disciplines is required. However, true multidisciplinary teamwork cannot be achieved until responsibilities and accountabilities are shared among team members. Important questions about collective responsibilities include (1) how to allocate responsibility in collective decisions, (2) how to make collective decisions, (3) where one's moral responsibility lies in the face of conflict, and (4) how to resolve conflict within the collectivity or team (Park et al. 2007; Pellegrino 1982).

Autonomy is a cultural strength in health care when it emphasizes professional competence. All professionals should commit themselves to lifelong learning and to maintaining knowledge and skills that are necessary for the provision of quality care (ABIM Foundation, ACP-ASIM Foundation, and European Federation of Internal Medicine 2002). However, when professionals believe that they can do everything themselves and refuse advice and help from others, autonomy becomes a weakness. As discussed previously, autonomy may conflict with the idea of scientific inquiry and standardization, which is considered an effective way to improve quality and reduce costs of care (Mathews and Pronovost 2008). Autonomy can also be a barrier to teamwork since it can foster a hierarchical power structure (e.g., physician autonomy) and lead to interprofessional conflict (e.g., autonomy emphasized by different professions) (Brown et al. 2011; Hall 2005). To promote the strengths of autonomy and avoid its weaknesses, Carroll and Quijada (2007) suggest a shift of the assumptions underlying autonomy. For instance, the assumption that asking for help is a fault has to be changed into that asking for help is what a responsible and caring professional does. Furthermore, a balance between professional autonomy and patient autonomy is recommended to encourage information exchange and negotiation between patients and healthcare professionals (Quill and Brody 1996).

Checklists as an Example

In this section, the design, implementation, and operational use of checklists provides an example to illustrate the differences between the healthcare culture and the human factors systems approach to healthcare system design; we then discuss how the healthcare culture should be adapted.

Although the idea of using checklists to reduce errors and failures and ensure consistency and completeness in carrying out a task is not new, it was not until recently that evidence showed the role of checklist in improving quality of care and patient safety. In 2001, Pronovost and his colleagues began to implement a simple checklist in ICUs to reduce the incidence of catheter-related bloodstream infections. In 2006, they published their results of a study conducted in 108 ICUs throughout the entire state of Michigan; the study was published in the *New England Journal of Medicine* and showed a reduction of catheter-related bloodstream infection rate from 2.7 per 1000 catheter-days at baseline to 0 at 3 months after the implementation of the checklist (Pronovost et al. 2006).

The use of checklists in health care seems to be a paradox. While checklists can bridge the gap between best evidence and best practice, practitioners in many healthcare settings fail to use them. Some practitioners even ignore or berate others who remind them to use checklists. One of the main reasons why practitioners refuse to use checklists is that the use of checklists conflicts with practitioner autonomy. According to Mathews and Pronovost (2008), two types of conflict can be distinguished: (1) practitioners are reluctant to yield decision-making power and to relinquish their autonomy and (2) practitioners believe that autonomy can buffer patients from ill-advised and poorly researched standards. The human factors systems approach to healthcare system design can help to resolve both types of conflict and to foster the use of checklists.

When practitioners consider checklists to be a threat to their autonomy, strategies that facilitate teamwork and improve communication are required to improve practitioners' adherence to the use of checklists and evidence-based guidelines. For instance, Berenholtz et al. (2004) found that without nursing intervention, physicians skipped at least one step in the checklist in more than a third of the procedures. However, in many cases, physicians perceive nursing intervention or teamwork as a threat. Neily et al. (2010) implemented a multifaceted intervention to promote teamwork and reduce surgical mortality. Besides the checklist, their intervention included teamwork training, ongoing coaching, and tools that supported teamwork. The checklist acted as a tool to trigger operating room communication rather than as a simple memory aid; in light of the SEIPS model (see Figure 12.1), the checklist is a tool that requires changes in other elements of the work system.

When practitioners do not trust checklists, strategies should focus on the design, implementation, and continuous adaptation of checklists. A well-established evidence base is critical for the design and use of checklists. What is equally important is evidence that shows the effectiveness of checklists. One challenge of designing and using checklists is the complexity of health care. Considering the diversity of patients, practitioners must be prepared for unpredictable events that checklists are unsuited to address (Gawande 2010). In addition, practitioners as the users of checklists should actively participate in the development, implementation, and adaptation of checklists (Carayon 2006; Smith and Carayon 1995).

Practitioners may not use checklists for various complex reasons. Therefore, all strategies discussed should be considered when planning the design, implementation, and operational use of checklists in health care. These strategies correspond to the human factors systems approach to healthcare system design (see Figure 12.1).

RECOMMENDATIONS FOR DECISION MAKING IN HEALTHCARE SYSTEM DESIGN

We have described the characteristics of the healthcare culture and the cultural values of the human factors systems approach to healthcare system design, and have highlighted similarities and differences between the two cultures. Given the critical role of the human factors systems approach in improving healthcare delivery, we need to identify ways that the two cultures can work together in creating solutions and implementing system changes for better quality and safety of care. In this section, we propose three recommendations for bridging the gaps between the two cultures: (1) awareness of cultural differences, (2) collaboration between health care and human factors engineering, and (3) biculturalism.

Awareness of Cultural Differences

Any attempt to apply a systems perspective to healthcare delivery must take into account the current healthcare culture. However, as we discussed in the section on Values That Conflict with Each Other, there are potential conflicts between the cultural values of health care and the cultural values of the human factors systems approach. Therefore, it is first necessary for the various parties involved in a system redesign effort or improvement activity to be aware of these cultural differences.

To our knowledge, this chapter is the first attempt at clarifying and comparing the cultural values of the two professional environments of health care and human factors engineering. Further effort to describe the similarities and differences in cultural values between health care and human factors engineering should be encouraged. As indicated in the section on Healthcare Culture, health care has many different professional cultures. For instance, there is extensive research showing cultural differences between the medical culture and the nursing culture; see, for instance, the research comparing perceptions and attitudes of ICU physicians and nurses (Thomas, Sexton, and Helmreich 2003). The human factors engineering discipline has grown significantly in the past 20 years, and subdisciplines have emerged, such as cognitive ergonomics and macroergonomics. In this chapter, we have described cultural values of the healthcare and human factors engineering cultures; further work needs to describe the subcultures within each of these cultures.

Since 2004, the SEIPS group at the University of Wisconsin-Madison has conducted a week-long course on human factors engineering and patient safety (http://cqpi.engr.wisc.edu/shortcourse_home). This course provides basic information about human factors engineering and describes a range of applications of human factors engineering to health care and patient safety, such as proactive risk assessment and design and implementation of health information technology. A key message provided to the short course participants at the beginning of the short course is "By the end of the week, you will look at the world differently." This statement shows that a major objective of the short course is to initiate a process of "acculturation": the short-course participants learn about human factors engineering theories, concepts, and methods and are, therefore, exposed to a different culture. Through this training mechanism, healthcare professionals become aware of the cultural values of human factors engineering. It is also important for human factors professionals and researchers to be aware of these cultural values of health care when engaged in learning and teaching activities.

An increasing number of human factors professionals and researchers are involved in health care and patient safety. In order to be effective and have significant impact on healthcare system design, they need to be aware of their own cultural values and the similarities and differences between their cultural values and those of health care.

Collaboration between Health Care and Human Factors Engineers

Awareness of cultural similarities and differences between health care and human factors engineering is an important step in the adoption of a systems approach to healthcare system design, but it is not sufficient. Integration of human factors engineering in health care will require increased communication and collaboration between the two groups. The design of healthcare systems and processes needs to involve the collaboration of human factors engineers with healthcare professionals (Carayon 2006; Rasmussen 2000). Integration of human factors engineering knowledge with healthcare expertise is essential and can be achieved by the participation of human factors engineers in healthcare improvement teams. The development of multidisciplinary improvement teams within healthcare organizations takes advantage of system-level work redesign to provide effective health care (Smith and Bartell 2007). The involvement of a human factors engineer in a multidisciplinary improvement team will increase interactions and information flow between the team members and enhance collaboration in analyzing and solving complex patient safety and quality of care problems.

The human factors engineer can play various roles in the multidisciplinary improvement team: facilitator, consultant or technical advisor, educator, or leader. The role of the human factors engineer must be flexible depending on the needs of the improvement activities and the teamwork dynamics. In addition, the human factors approach to healthcare system design can successfully be disseminated if human factors engineering can be shared and "given away" (Carayon 2010). This learning and adaptation process develops over time as the role of the human factors engineer changes from one of leading and facilitating the improvement activity to one of consultant and advisor. This process from external regulation to internal regulation by the team members has been described in the context of participatory ergonomics projects (Haims and Carayon 1998), and should be implemented in multidisciplinary healthcare improvement teams.

Henriksen (2007) suggests that human factors researchers and professionals can play a leadership role not only in improvement teams but in broader endeavors aimed at designing high-quality, safe healthcare systems and transforming health care. A few healthcare organizations have hired human factors engineers to lead their patient safety efforts; this may be a sign of the increasing leadership role of human factors engineering in patient safety.

Biculturalism

The involvement of human factors engineers in quality of care improvement activities can help in bridging the cultural divide between the two disciplines because access to human factors engineering knowledge is easier: health care has access to human factors theories, concepts, and methods through the involvement of human factors engineers. But there may still be situations of "intellectual centrism" where intellectual inputs from one discipline are ignored by the other discipline. The third strategy, biculturalism, addresses this deep cultural problem.

Biculturalism is a key to navigating domains where knowledge from seemingly dissimilar disciplines is recruited for solving problems. A *bicultural* is a person who possesses insider knowledge of two cultural knowledge traditions (Benet-Martinez et al. 2002). Biculturalism has been proposed as a method for enhancing creativity and productivity in idea development, problem solving, and implementation of innovation at the boundary between the research and development function and the marketing function (Banerjee and Chiu 2008).

In our context of improving healthcare delivery by using a human factors systems approach, the biculturals are likely to be healthcare professionals (e.g., a physician or a nurse) who are trained in human factors engineering. Biculturalism emphasizes the integration of and balance and switch between different cultures. The healthcare biculturals will be able to engage in "cultural frame-switching" between their healthcare culture and the human factors engineering culture. Research on biculturals in the context of national cultures, in particular with groups of European Americans or Asian Americans, has demonstrated this process of frame-switching (Benet-Martinez et al. 2002). The cultural frame-switching process allows the bicultural to rely on his/her primary culture (i.e., health care) or his or her secondary culture (i.e., human factors engineering) depending on the demands of the situation (Banerjee and Chiu 2008). Healthcare biculturals can help to generate solutions to complex quality of care and patient safety problems because they have access to a broader range of knowledge and can call on these different bodies of knowledge during problem solving.

There are probably very few healthcare biculturals, that is, individuals who have deep knowledge in a healthcare discipline and human factors engineering. There may be situations where the healthcare biculturals are isolated or not supported by their colleagues or their organization. An organizational environment that fosters multidisciplinary exchange and learning is necessary for healthcare biculturals to provide input and suggestions and be effective contributors to healthcare system design. In addition, effective biculturals are more likely to be people with cognitive flexibility (Tadmor and Tetlock 2006): they can easily and quickly switch between the healthcare culture and the human factors engineering culture.

Healthcare professionals who gain deep knowledge of human factors engineering can become healthcare biculturals and are likely to benefit from bicultural effectiveness training (Szapocznik et al. 1986). This training can help the healthcare biculturals in recognizing cultural differences and developing strategies to further integrate human factors engineering in healthcare system design. The training may also help the biculturals in integrating the two cultures, therefore, reducing the likelihood of perceiving the two cultures as contradictory and conflicting (Tadmor and Tetlock 2006).

CONCLUSION

There is increasing recognition among healthcare professionals, leaders, and researchers that the human factors systems approach can play a major role in healthcare transformation. The integration of the human factors systems approach into health care is faced with numerous challenges, including potential cultural conflicts. In this chapter, we have highlighted the sources of these cultural conflicts and have proposed a series of recommendations for moving forward in this integration process.

In order to enhance the integration of human factors engineering in healthcare system design and patient safety, we proposed three recommendations aimed at increasing the awareness of cultural similarities and differences between the healthcare culture and the human factors systems approach, at increasing the involvement of human factors engineers in healthcare improvement activities, and at training healthcare biculturals or healthcare professionals with a deep knowledge in human factors engineering. The implementation of these recommendations may be facilitated as healthcare organizations face continued pressures to improve patient safety and recognize the need for integrating human factors engineering in their patient safety efforts.

REFERENCES

ABIM Foundation, ACP-ASIM Foundation, and European Federation of Internal Medicine. 2002. Medical professionalism in the new millennium: A physician charter. *Ann Intern Med* 136(3):243–6.

Ash, J. S., D. F. Sittig, R. Dykstra, E. Campbell, and K. Guappone. 2009. The unintended consequences of computerized provider order entry: Findings from a mixed methods exploration. *Int J Med Inform* 78(Suppl. 1):S69–76.

Ash, J. S., D. F. Sittig, R. H. Dykstra, K. Guappone, J. D. Carpenter, and V. Seshadri. 2007. Categorizing the unintended sociotechnical consequences of computerized provider order entry. *Int J Med Inform* 76(Suppl. 1):S21–27.

Banerjee, P. M., and C. Y. Chiu. 2008. Professional biculturalism enculturation training: A new perspective on managing R&D and marketing interface. In *Current Topics in Mangement*, ed. M. A. Rahim, 145–59, Vol. 13. New Brunswick, NY: Transaction Publishers.

Benet-Martinez, V., J. Leu, F. Lee, and M. W. Morris. 2002. Negotiating biculturalism—Cultural frame switching in biculturals with oppositional versus compatible cultural identities. *J Cross Cult Psychol* 33(5):492–516.

Berenholtz, S. M., P. J. Pronovost, P. A. Lipsett, D. Hobson, K. Earsing, J. E. Farley et al. 2004. Eliminating catheter-related bloodstream infections in the intensive care unit. *Critical Care Medicine* 32(10):2014–20.

Brown, J., L. Lewis, K. Ellis, M. Stewart, T. R. Freeman, and M. J. Kasperski. 2011. Conflict on interprofessional primary health care teams—can it be resolved? *J Interprof Care* 25(1):4–10.

Cabana, M. D., C. S. Rand, N. R. Powe, A. W. Wu, M. H. Wilson, P. A. Abboud et al. 1999. Why don't physicians follow clinical practice guidelines? A framework for improvement. *J Am Med Assoc* 282(15):1458–65.

Carayon, P. 2006. Human factors of complex sociotechnical systems. *Appl Ergon* 37:525–35.

Carayon, P. 2009. The balance theory and the work system model... Twenty years later. *Int J Hum Comput Interact* 25(5):313–27.

Carayon, P. 2010. Human factors in patient safety as an innovation. *Appl Ergon* 41(5):657–65.

Carayon, P., A. S. Hundt, B.-T. Karsh, A. P. Gurses, C. J. Alvarado, M. Smith et al. 2006. Work system design for patient safety: The SEIPS model. *Qual Saf Health Care* 15(Suppl. I):i50–8.

Carayon, P., and M. J. Smith. 2000. Work organization and ergonomics. *Appl Ergon* 31:649–62.

Carayon, P., T. B. Wetterneck, A. S. Hundt, S. Rough, and M. Schroeder. 2008. Continuous technology implementation in health care: The case of advanced IV infusion pump technology. In *Corporate Sustainability as a Challenge for Comprehensive Management*, ed. K. Zink, 139–51. New York: Springer.

Carroll, J. S., and M. A. Quijada. 2007. Tilting the culture in health care: Using cultural strengths to transform organizations. In *Handbook of Human Factors and Ergonomics in Health Care and Patient Safety*, ed. P. Carayon, 823–32. Mahwah, NJ: Lawrence Erlbaum Associates.

Christmas, C., S. J. Kravet, S. C. Durso, and S. M. Wright. 2008. Clinical excellence in academia: Perspectives from masterful academic clinicians. *Mayo Clin Proc* 83(9):989–94.
Clegg, C. W. 2000. Sociotechnical principles for system design. *Appl Ergon* 31(5):463–77.
Davidoff, F. 1999. In the teeth of the evidence: The curious case of evidence-based medicine. *Mt Sinai J Med* 66(2):75–83.
Deming, W. E. 1986. *Out of the Crisis*. Cambridge, MA: MIT Press.
Donabedian, A. 1978. The quality of medical care. *Science* 200:856–64.
Donabedian, A. 1988. The quality of care: How can it be assessed? *J Am Med Assoc* 260(12):1743–8.
Evans, R. S., S. L. Pestotnik, D. C. Classen, T. P. Clemmer, L. K. Weaver, J. F. J. Orme et al. 1998. A computer-assisted management program for antibiotics and other antiinfective agents. *New Engl J Med* 338(4):232–8.
Evidence-Based Medicine Working Group. 1992. Evidence-based medicine. A new approach to teaching the practice of medicine. *J Am Med Assoc* 268(17):2420–5.
Freidson, E. 1970. *The Profession of Medicine: An Essay in the Sociology of Applied Knowledge*. New York: Dodd Mead.
Gawande, A. 2010. *The Checklist Manifesto: How to Get Things Right*. New York: Henry Holt and Co.
Gurses, A. P., K. L. Seidl, V. Vaidya, G. Bochicchio, A. D. Harris, J. Hebden et al. 2008. Systems ambiguity and guideline compliance: A qualitative study of how intensive care units follow evidence-based guidelines to reduce health care-associated infections. *Qual Saf Health Care* 17(5):351–9.
Haims, M. C., and Carayon, P. 1998. Theory and practice for the implementation of 'in-house', continuous improvement participatory ergonomic programs. *Appl Ergon* 29(6):461–72.
Hall, P. 2005. Interprofessional teamwork: Professional cultures as barriers. *J Interprof Care* 19(Suppl 1):188–96.
Hall, P., and L. Weaver. 2001. Interdisciplinary education and teamwork: A long and winding road. *Med Educ* 35(9):867–75.
Henriksen, K. 2007. Human factors and patient safety: Continuing challenges. In *Handbook of Human Factors and Ergonomics in Health Care and Patient Safety*, ed. P. Carayon, 21–37. Mahwah, NJ: Lawrence Erlbaum Associates.
Institute of Medicine Committee on Quality of Health Care in America. 2001. *Crossing the Quality Chasm: A New Health System for the 21st Century*. Washington, DC: National Academy Press.
International Ergonomics Association (IEA). 2000. *The Discipline of Ergonomics*. http://www.iea.cc (accessed August 22, 2004).
Kenny, N. P., K. V. Mann, and H. MacLeod. 2003. Role modeling in physicians' professional formation: Reconsidering an essential but untapped educational strategy. *Acad Med* 78(12):1203–10.
Kohn, L. T., J. M. Corrigan, and M. S. Donaldson, eds. 1999. *To Err Is Human: Building a Safer Health System*. Washington, DC: National Academy Press.
Leipzig, R. M., K. Hyer, K. Ek, S. Wallenstein, M. L. Vezina, S. Fairchild et al. 2002. Attitudes toward working on interdisciplinary health care teams: A comparison by discipline. *J Am Geriatr Soc* 50(6):1141–8.
Lohr, K. N., K. Eleazer, and J. Mauskopf. 1998. Health policy issues and applications for evidence-based medicine and clinical practice guidelines. *Health Policy* 46(1):1–19.
Mathews, S. C., and P. J. Pronovost. 2008. Physician autonomy and informed decision making: Finding the balance for patient safety and quality. *J Am Med Assoc* 300(24):2913–5.
McDonald, C. J., J. M. Overhage, W. M. Tierney, G. R. Abernathy, and P. R. Dexter. 1996. The promise of computerized feedback systems for diabetes care. *Ann Intern Med* 124(1S-II):170–4.

McGlynn, E. A., S. M. Asch, J. Adams, J. Keesey, J. Hicks, A. DeCristofaro et al. 2003. The quality of health care delivered to adults in the United States. *New Engl J Med* 348(26):2635–45.

Meister, D., and T. P. Enderwick. 2001. *Human Factors in System Design, Development, and Testing*. Mahwah, NJ: Lawrence Erlbaum Associates.

Molleman, E., M. Broekhuis, R. Stoffels, and F. Jaspers. 2008. How health care complexity leads to cooperation and affects the autonomy of health care professionals. *Health Care Anal* 16(4):329–41.

Neily, J., P. D. Mills, Y. Young-Xu, B. T. Carney, P. West, D. H. Berger et al. 2010. Association between implementation of a medical team training program and surgical mortality. *J Am Med Assoc* 304(15):1693–700.

Newton, L. H. 1982. Collective responsibility in health care. *J Med Philos* 7(1):11–21.

Park, J., S. I. Woodrow, R. K. Reznick, J. Beales, and H. M. MacRae. 2007. Patient care is a collective responsibility: Perceptions of professional responsibility in surgery. *Surgery* 142(1):111–8.

Pellegrino, E. D. 1982. The ethics of collective judgments in medicine and health care. *J Med Philos* 7(1):3–10.

Pronovost, P., D. Needham, S. Berenholtz, D. Sinopoli, H. Chu, S. Cosgrove et al. 2006. An intervention to decrease catheter-related bloodstream infections in the ICU. *New Engl J Med* 355(26):2725–32.

Quill, T. E., and H. Brody. 1996. Physician recommendations and patient autonomy: Finding a balance between physician power and patient choice. *Ann Intern Med* 125(9):763–9.

Rasmussen, J. 2000. Human factors in a dynamic information society: Where are we heading? *Ergonomics* 43(7):869–79.

Reid, P. R., W. D. Compton, J. H. Grossman, and G. Fanjiang. 2005. *Building a Better Delivery System: A New Engineering/Health Care Partnership*. Washington, DC: The National Academies Press.

Sainfort, F., B. Karsh, B. C. Booske, and M. J. Smith. 2001. Applying quality improvement principles to achieve healthy work organizations. *J Qual Improv* 27(9):469–83.

Schein, E. H. 1992. *Organizational Culture and Leadership*. 2nd ed. San Francisco, CA: Jossey-Bass.

Schein, E. H. 1999. *The Corporate Culture Survival Guide: Sense and Nonsense about Culture Change*. 1st ed. San Francisco, CA: Jossey-Bass.

Schulz, R., and R. Harrison. 1986. Physician autonomy in the Federal Republic of Germany, Great Britain and the United States. *Int J Health Plann Manage* 1(5):335–55.

Shortell, S. M., and S. J. Singer. 2008. Improving patient safety by taking systems seriously. *J Am Med Assoc* 299(4):445–7.

Shortell, S. M., T. M. Waters, K. W. Clarke, and P. P. Budetti. 1998. Physicians as double agents: Maintaining trust in an era of multiple accountabilities. *J Am Med Assoc* 280(12):1102–8.

Shortell, S. M., J. E. Zimmerman, D. M. Rousseau, R. R. Gillies, D. P. Wagner, E. A. Draper et al. 1994. The performance of intensive care units: Does good management make a difference? *Med Care* 32(5):508–25.

Smith, M. A., and J. M. Bartell. 2007. The relationship between physician professionalism and health care systems change. In *Handbook of Human Factors and Ergonomics in Health Care and Patient Safety*, ed. P. Carayon, 139–46. Mahwah, NJ: Lawrence Erlbaum Associates.

Smith, M. J., and P. Carayon. 1995. New technology, automation, and work organization: Stress problems and improved technology implementation strategies. *Int J Hum Factors Manuf* 5(1):99–116.

Smith, M. J., and P. Carayon. 2000. Balance theory of job design. In *International Encyclopedia of Ergonomics and Human Factors*, ed. W. Karwowski, 1181–4. London: Taylor & Francis.

Smith, M. J., and P. Carayon-Sainfort. 1989. A balance theory of job design for stress reduction. *Int J Ind Ergon* 4(1):67–79.

Szapocznik, J., A. Rio, A. Perez-Vidal, W. Kurtines, O. Hervis, and D. Santisteban. 1986. Bicultural effectiveness training (BET): An experimental test of an intervention modality for families experiencing intergenerational/intercultural conflict. *Hisp J Behav Sci* 8(4):303–30.

Tadmor, C. T., and P. E. Tetlock. 2006. Biculturalism—A model of the effects of second-culture exposure on acculturation and integrative complexity. *J Cross-Cult Psychol* 37(2):173–90.

Thomas, E. J., J. B. Sexton, and R. L. Helmreich. 2003. Discrepant attitudes about teamwork among critical care nurses and physicians. *Crit Care Med* 31(3):956–9.

Timmermans, S., and M. Berg. 2003. *The Gold Standard: The Challenge of Evidence-Based Medicine and Standardization in Health Care*. Philadelphia: Temple University Press.

Timmermans, S., and A. Mauck. 2005. The promises and pitfalls of evidence-based medicine. *Health Affairs (Millwood)* 24(1):18–28.

Tucker, A. L., and A. Edmondson. 2003. Why hospitals don't learn from failures: Organizational and pschological dynamics that inhibit system change. *Calif Manage Rev* 45(2):55–72.

Vincent, C. A. 2004. Analysis of clinical incidents: A window on the system not a search for root causes. *Qual Saf Health Care* 13:242–3.

Weick, K. E., and R. E. Quinn. 1999. Organizational change and development. *Ann Rev Psychol* 50:361–86.

Wilson, T., T. Holt, and T. Greenhalgh. 2001. Complexity science: Complexity and clinical care. *Br Med J* 323(7314):685–8.

Woods, D. D. 2006. Essential characteristics of resilience. In *Resilience Engineering—Concepts and Precepts*, ed. E. Hollnagel, D. D. Woods, and N. Leveson, 21–34. Burlington, VT: Ashgate.

13 Cultural Factors in the Adoption and Implementation of Health Information Technology

Atif Zafar and Mark R. Lehto

CONTENTS

INTRODUCTION

The U.S. healthcare system is immensely complex and fragmented with multiple providers, multiple sources of funding, and multiple locations of care. There are more than 360,000 care delivery sites in the United States, and often there is little or no communication between them. It has been estimated that in as many as 80% of outpatient encounters, the provider lacks potentially important information about the patient that is hidden away in the files of other providers. What *is* communicated is often incomplete, inaccurate (wrong or out of date), or unclear (i.e., illegible, nonsensical notes). Clinical decisions based on invalid or out-of-date information can have disastrous consequences and result in much inefficiency, including duplication of tests, delays in care delivery, and inappropriate or unnecessary care. All of this can ultimately raise the cost of care and lead to patient harm.

The overuse and underuse of healthcare services and interventions is a particularly pressing concern. There are great disparities in the delivery of health care to needed populations, as documented in sources such as the Dartmouth Atlas Project (http://www.dartmouthatlas.org) and the AHRQ National Healthcare Disparities Report (available at http://www.ahrq.gov). Illustrating this trend, a recent review of one local hospital revealed (1) 30% of treated children receive excessive antibiotics for otitis; (2) 20%–50% of surgical procedures are not necessary; (3) 50% of back pain X-rays are not necessary; and (4) 50% of elderly patients do not get a pneumovax.

Adverse drug events (ADEs) are another consequence of a fragmented healthcare system. They have caught the public eye in recent years, and more and more government grants have gone into cataloging, studying, and preventing them. In one meta-analysis, 84% of ADEs were classified as preventable, and the average settlement cost for litigation in these cases was $4.3 million (Kelly 2001).

The need to improve the delivery of health care in the United States was highlighted in the year 2000, in a seminal report published by the Institute of Medicine (IOM) entitled "To Err is Human: Building a Safer Health System" (Kohn, Corrigan, and Donaldson 2000). This report stated that adverse drug events and medical errors may be responsible for up to 98,000 unnecessary deaths per year in the United States. The key conclusion from this study was that there is no single responsible entity but that this is a *systems-wide* problem. Kohn et al. cited as key contributing factors to this problem the lack of public attention to the issue, failure of appropriate licensure and accreditation, the perceived threat of litigation that lets errors go unreported, the decentralized and fragmented nature of the healthcare system, and the lack of incentives for improved safety. They concluded that there is no "magic bullet" that can reduce the harm done but that a "comprehensive approach" is needed.

The rapidly evolving knowledge of clinical best practices also creates a challenge to continually educate healthcare personnel. In a landmark paper published in the *New England Journal of Medicine*, L'Enfant (2003) wrote that it takes an average of 17 years for known medical best practices to actually be applied to patient care. The volume of research data that must be tracked is enormous. An article in the Harvard Business Review stated that physicians must keep track of "almost 10,000 different diseases and syndromes, 3,000 medications, 1,100 laboratory tests, and many of the 400,000 articles added each year to the biomedical literature" (Davenport and Glaser 2002, p. 107). This is physically impossible for any human being and the problem becomes one of information management rather than knowing everything there is to know.

Information management in health care is complicated by the inherent complexity in the way data are generated, stored, communicated, operated on, represented, and understood. All healthcare processes generate data, and the data are stored in many ways and in many locations, filed away on paper, or within "silos" behind the firewalls of institutions, or as tacit knowledge in someone's mind, which often makes the data *inaccessible* to clinicians and healthcare workers at the time it is needed. This inaccessibility adds to the complexity and fragmentation of health care, because healthcare delivery is designed to be a linear arrangement of tasks, and each task requires knowledge from the previous one. If prior knowledge is not available, the task is not efficient and requires a duplication of subtasks from the previous task in order to *regenerate* the required knowledge.

The field of clinical informatics has taken on the challenge of making this information management process happen through the implementation of health information technology (HIT). The mantra has become one of using HIT to "get the right information to the right person in the right modality at the right time and with the right level of urgency" for effective health care to occur.

HIT Solution

HIT is an umbrella term that encapsulates many technology tools, processes and policies present in health care today (Table 13.1). Examples include electronic medical records, clinical decision support systems, health information exchange networks, computerized provider order entry systems, security and privacy policies, and processes for patients' "protected health information" (HIPAA), personal health records, and ePrescribing systems.

One of the key advantages of HIT is that it can provide for better communication and coordination of care in an increasingly specialized and fragmented healthcare delivery environment. The features and functions of HIT systems can lead to a reduction in duplicate testing, better handoffs between care sites and care providers, point-of-care advice about drug-drug, drug-diagnosis, and drug-lab interactions, reminders about clinical practice guidelines using evidence-based medicine, checking for corollary orders using order sets, and guidance on selection of cost-effective medications and management strategies. Note that these solutions need not be electronic. Informatics is about *information delivery*, in whatever medium best suits the context. This includes paper-based solutions.

The potential benefits of implementing HIT have led to a strong push for adoption in the United States by several groups such as the Institute for Medicine, the National Academy of Engineering, the Leapfrog Group, and the U.S. federal government. The economic benefits have been estimated to be in the order of hundreds of billions of

TABLE 13.1
Levels of Health Information Technology Application

Application Level
- The specific applications that allow healthcare personnel to interact with patient data (including clinical data, appointment histories, medication lists, demographics, and financial information)

Communication Level
- How one clinical information system talks to another
  - How a lab data system talks to an electronic medical record
  - How a pharmacy system talks to a billing system

Process Level
- Processes and procedures for ensuring security of data
- Processes and procedures for ensuring completeness and accuracy of the data

Device/Technology Level
- The actual computer hardware on which these applications work

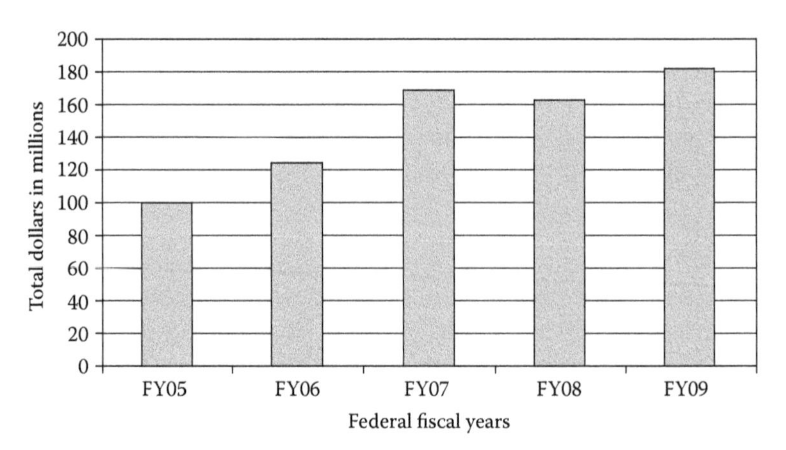

FIGURE 13.1 Annual federal expenditures in millions of dollars for HIT.

dollars, and the U.S. federal government has and continues to spend millions of dollars on the implementation of HIT systems (Figure 13.1). However, despite the large potential advantages of HIT systems, most physicians in general practice still do not use them. For example, one recent study found only 23% of practices in the state of Massachusetts had an electronic health record (EHR) system (Simon et al. 2007).

Barriers to implementation in real-world environments include a variety of technical, cultural, and financial issues. Integration with legacy systems, managing workflow change, and keeping costs affordable continue to be real-world technical and economic issues faced by purchasers of HIT systems. However, the most significant barrier is that successful introduction of HIT into healthcare settings will often require a large-scale effort to bring about cultural change (GAO 2003).

Cultural Barriers and Constraints

Cultural factors such as perceived threats to autonomy; loss of control, autonomy, or status; role and power shifts; privacy concerns; reluctance to learn new skills; perceived depersonalization of patient and provider interactions; or indifference to information technology can influence both the successful use and acceptance of HIT by clinicians and other stakeholders in the healthcare system (Angst and Agrawal 2009; Doebbeling et al. 2006; Holden 2009; Meijden et al. 2003; Moen, Gregory, and Brennan 2007; Shortliffe 2005; Timmons 2001). This issue is quite complex, as it involves multiple stakeholders, competing interests, different professions, and critically important consequences.

The technology acceptance model (TAM) (Davis 1986), which holds that behavioral intent to use a system is determined by *perceived usefulness* and *perceived ease of use*, provides a good starting point for assessing how cultural factors can impact acceptance of HIT. To gain acceptance TAM suggests that it is critical that physicians appreciate the advantages HIT directly offers to them (see Table 13.2), including the types of data collected and made available to practitioners by HIT (see Table 13.3), as well as added capabilities such as those obtained

TABLE 13.2
Key Advantages of Health Information Technology to the Practitioner

IT solutions can provide needed data in the exam room
- Latest lab and test results
- Medication lists
- List of appointments
- Clinic notes and consult recommendations

IT solutions can help with clinical decision support
- Medication conflicts
- Research results and evidence based guidelines
- Clinical knowledge—differential diagnoses and so on

IT solutions can help with prevention and patient education
- Preventive services order sets
- Patient handouts and pamphlets

IT solutions can help with the documentation process
- Macros and templates for rapid documentation activities
- Through advanced data entry methods—speech and handwriting recognition
- Entrance by exception—enter data only if changed
- Automated clinical pathways—decision support
- Trend tracking

IT solutions can help physicians communicate better—with colleagues, specialists, and patients, and coordinate care delivery
- Using telemedicine if in remote/rural sites
- Communicate with home health nurses, nursing homes, and so on
- Using email and other communication channels besides paper
- Help bridge the health disparities gap

IT solutions can help manage busy workflows
- Keep track of patients as they come to a clinic (greaseboard function)
- Help you communicate with the front-office staff more efficiently
  - Order pneumovax/flu shots, tests (EKGs), or medications from the exam room so the nurse is ready to give the shot or do the test when the patient walks out

IT solutions can help improve patient satisfaction
- Improved patient compliance
  - Easy to read and understand written instructions
  - Better medication side effect tracking
- Improved access to and more "personalized" care for the patient and caregivers
- Patient-centered care for high-risk patients, that is, better monitoring

(Continued)

TABLE 13.2 (*Continued*)
Key Advantages of Health Information Technology to the Practitioner

IT solutions can help care for patients at long distance
- Telemedicine tools can help a primary care provider communicate with a specialist long-distance with the patient in the room
- Examples include eleradiology, telecardiology, teledermatology, and so on

IT solutions can help reduce the cost of care
- Help select cost-effective interventions (lab tests, medications, etc.)
- Help bill more effectively and more completely
- Help protect physicians from costly lawsuits by documenting better
- Help provide better time management of healthcare personnel

TABLE 13.3
Types of Clinical Data Made Available to Practitioners by HIT

Type of Clinical Data	Typical Storage Medium
Demographics (name, address, dob)	Registration system (IDX, etc.)
Insurance information	Registration system/billing system
Appointments	Registration system (IDX, etc.)
Lab data	Lab system or paper-based letters or faxes
Radiology images (X-rays, etc.)	Radiology information system (PACs)
Dictated reports (X-rays/path/procedures)	Electronic medical record (EMR), often as paper-based reports in charts
Clinic notes	EMR, often paper-based notes in charts
Consultant notes	Letters and copies of notes in paper chart
EKGs	EMR, but often paper-based
Emergency visits	EMR, but paper ER charts and hospital charts
Prescription histories	EMR, but often in paper-based clinic notes
Allergies	EMR, but often in paper-based charts
Discharge summaries (from hospital)	EMR, but often sent as paper mail or faxes

by having shared data repositories in the form of Health Information Exchange (HIE; see Table 13.4). TAM also suggests that it is important that users of HIT find it easy to use.

A large set of studies have confirmed for the most part the prediction of TAM that acceptance of HIT is strongly related to both perceived usefulness and ease of use (for a comprehensive review, see Holden 2009). Overall, these findings support Shortliffe's observation that (1) many clinicians have a poor appreciation of the value of information technology and the important "role IT could be playing in addressing their fiscal, quality, and organizational challenges," (p. 1226) and (2) "given the many other pressures on today's clinicians and health care workers' relative lack of experience with computing during their training, there can be a reluctance to learn new skills in an area that seems foreign and tangential to medical care" (Shortliffe 2005, p. 1227).

TABLE 13.4
Physician Needs Addressed by Capabilities of Health Information Exchange

Outpatient physicians do not know what happened in the hospital to one of their patients.
- This is addressed by
  - Medication lists
  - Lab and test results
  - Diagnoses and problems
  - Discharge summary

The ER does not know the history of a patient being seen by a primary care provider.
- This is addressed by
  - Clinic notes
  - Medication lists
  - Diagnoses and problems

A specialist does not know what tests were done on a referred patient. This is addressed by
- Referral question, that is, why were they referred?
- Lab and test results
- Radiology and nuclear medicine data
- Medication lists
- Diagnoses

A primary care provider does not know what a specialist did. This is addressed by
- Specialty care clinic notes
- Follow-up recommendations

Other questions regarding usage:
- Was the patient seen in other clinics or in other ERs recently, for what, and what was done?
  - Patients move around a lot
- Which pharmacies are filling the prescriptions?
- What appointments does the patient have that are upcoming or which appointments were missed?

Prevention and surveillance
- Immunization and disease outbreaks

Home health care

Purposive sampling and interviews (e.g., Holden 2009), along with observational studies (e.g., Saleem et al. 2005), have led to the identification of many factors contributing to these cultural barriers, and at the same time also suggest strategies for improving acceptance. In particular, studies have shown that ratings of the perceived usefulness of HIT are strongly related to previous positive experiences with HIT (Wu et al. 2006). Healthcare staff trained in the use of computers also show more favorable attitudes towards the use of computers in the workplace (Wu et al. 2006).

However, attaining acceptance of HIT is a much more complicated issue than simply one of perceived usefulness and ease of use. To gain acceptance from certain groups, a large set of issues that are in some cases entirely unrelated to usefulness and ease of use must be addressed. For example, providers might be concerned that their use or failure to comply with the recommendations of a medical decision support system will increase their risk of liability in legal settings. Medical staff may be similarly suspicious that higher-level administration is more interested in using HIT to track performance and punish nonconformers who fail to meet arbitrary clinical guidelines or standardized practices than in improving the quality and delivery of services. In some cases, tradeoffs must be made to gain acceptance that significantly *interfere* with HIT effectiveness or ease of use. For example, one commonly voiced concern is that patients will feel uncomfortable if the clinician is on the computer entering data while talking to the patient. Because of this cultural issue, some providers avoid using the computer until after the patient encounter is over, which creates numerous problems.

To provide a general picture of how these different cultural factors fit together, we have developed a simple model describing the dynamic, or ever-changing, relationship between acceptance of HIT (AHIT), and three other factors: (1) stakeholder culture, which corresponds to the shared roles, values, norms, perceived threats, and experiences shared by a group of stakeholders; (2) cultural constraints, which corresponds to both HIT design elements to satisfy shareholders and the way shareholders actually use it, and (3) HIT effectiveness and ease of use when these cultural constraints are present.

As shown in Figure 13.2, one part of the AHIT model shows that stakeholder acceptance of HIT is both related to and influences stakeholder culture. That is, acceptance of HIT can lead to either positive or negative changes in stakeholder culture (i.e., changes in perceived roles, needs, values, norms, threats, etc.), and changes in culture can lead to either positive or negative changes in HIT acceptance. The figure also shows that shareholder acceptance can be at different levels of HIT, that is, with the application, process, or device level. In some cases, acceptance is entirely driven by a particular shareholder cultural element, such as provider perception that payers receive most of the benefit. Along these lines, Shortliffe (2005) notes that the latter perception is recognized as a major barrier to HIT implementation, and points

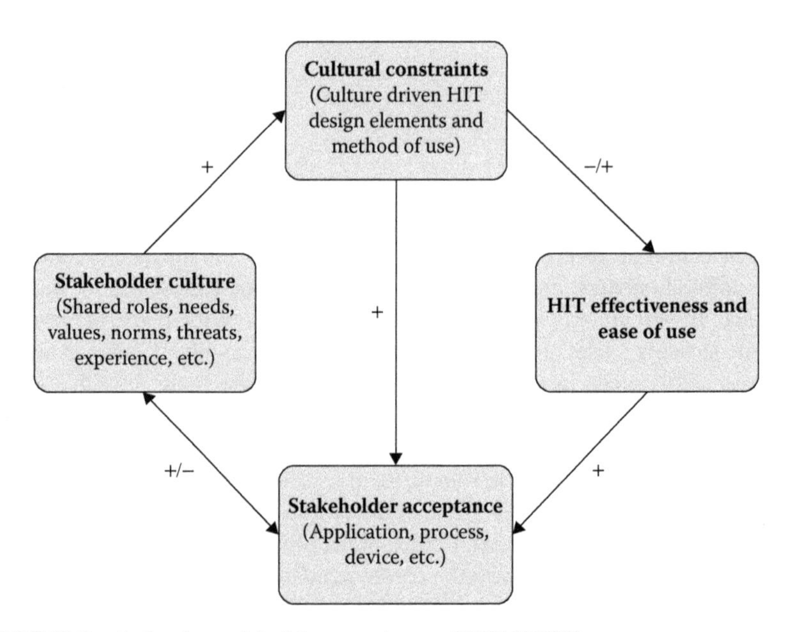

FIGURE 13.2 A simple model of the acceptance of HIT (AHIT).

out one study showing that 89% of the benefits of HIT implementation went to payers and 11% to providers (Johnston et al. 2003).

In other cases, stakeholder acceptance is driven by HIT effectiveness and ease of use. As such, this part of our model is equivalent to the technology acceptance model discussed previously. A third element of our model reflects the fact that stakeholder culture imposes constraints on how the HIT will be designed and used, such as requiring that providers not use the computer while they are interacting with the patient. As shown in Figure 13.2, satisfying these cultural constraints should improve (as indicated with a +) stakeholder acceptance, but might actually reduce HIT effectiveness and ease of use. In some cases, this reduction in effectiveness and ease of use might outweigh the benefit of satisfying the cultural constraint, resulting in a net loss of shareholder acceptance. This in turn, could result in changes in stakeholder culture, leading to pressure to eliminate the cultural constraint, and so on.

Returning to our earlier example of a few sentences ago, involving a clinic where providers were asked to not use the computer while they are interacting with the patient, we actually observed this process unfold over a period of a few years in one of our local clinics. In this case, interns complained intensely because they were required to enter in the data about their visit from memory in a separate computer room often several hours after the patient appointment was finished. The ultimate result was that the interns now have computers in the exam room and can interact with the HIT while interacting with the patient if they choose to do so.

As such, the AHIT model helps illustrate several aspects of what happens when HIT is used in the dynamic, ever-changing culture of health care. Understanding the cultural clashes introduced by the introduction of HIT, and how they unfold over time, requires consideration of the context within which HIT is used, as expanded upon in the next sections.

Complexity Case-Mix Equation and Clinical Workflows

The process of health care is complex and convoluted. Different problems require different specialists and different venues of care. As a consequence, different aspects of the care of a single individual may be spread widely apart in time and space. To illustrate the complexity of healthcare delivery, consider the case scenario for a man 65 years of age illustrated in Table 13.5, for a typical patient seen in primary care settings, with *multiple chronic medical problems*, requiring *day to day management*. As shown in the table, this person is taking multiple medications and will need to make multiple visits to different healthcare facilities to see specialists or for other reasons during the course of a year. This equates to up to 13 ambulatory provider (specialist) visits a year. In addition, the person may have the need for occasional hospitalizations and emergency room visits for exacerbations of chronic diseases. This means that for this single patient, around 19 individual provider visits per year must be coordinated. Each of these provider visits may involve different visit processes, associated with changes in medication (name, dose, format, etc.), tests ordered (labs, X-rays, procedures), and instructions provided to the patient. In addition to all of this, health maintenance screening must be performed in order to prevent chronic disease. This patient would qualify for

TABLE 13.5
Case Scenario for a Typical Patient Seen in Primary-Care Settings

Medical Problems
1. Diabetes
2. Heart disease
3. Hypertension
4. High cholesterol
5. Sleep apnea
6. Obesity
7. Tobacco abuse
8. COPD
9. Chronic low back pain

Medications
1. Glipizide
2. Metformin
3. Atenolol
4. Lisinopril
5. Zocor
6. Albuterol
7. Combivent
8. Ibuprofen

Specialist Visits
1. Primary care provider (PCP; 3–4 times a year)
2. Cardiology (1–2 times a year)
3. Pulmonary (1–2 times a year)
4. Sleep specialist (1 time per year)
5. Nutritionist (1 time per year)
6. Physical therapy (2–3 times a year)

Other Visits
1. Hospitalizations (1–2 times a year)
2. ER and urgent care visits (3–4 times a year)

Types of Acute Care Visits
1. Respiratory illness including bronchitis, sinusitis, pneumonia
2. Chest pain
3. Exacerbation of back pain
4. COPD exacerbation—viral illness or otherwise

Visit Processes
1. Change in medication (name, dose, format, etc.)
2. Tests ordered (labs, X-rays, procedures)
3. Instructions provided to the patient

Screening Tests
1. Colonoscopy (once every 10 years)
2. Prostate screening (once/year)
3. Yearly chronic labs: A1c, HDL, LDL, serum creatinine, urine protein, electrolytes, CBC

the following screening, including a colonoscopy (once every 10 years), prostate screening (once/year), and early chronic labs: A1c, HDL, LDL, serum creatinine, urine protein, electrolytes, CBC.

To adequately deliver these services, the patient and his or her providers will have to keep track of all of this information. The average numbers of lab results may include 30–50 individual lab test results, X-rays and other procedures (e.g., cardiac echo). This information has to be updated for each visit and communicated to the rest of the providers in order to maximize the efficiency of care. If the information is located on paper, as is the case for 95% of outpatient practices in the country, it may be exceedingly difficult to coordinate delivery of this information to the right person at the right time in the right modality. As a consequence of not knowing what has transpired between visits to the same provider, tests are duplicated, medications improperly reconciled, diagnoses missed, and patient harm done.

A second issue is that the workflow in a typical primary care office is immensely complex. Multiple clinical and administrative people have contact with patients during the course of a visit. To get a better understanding of this issue, we have evaluated clinical workflows using time-motion studies and task-analysis (Louthan et al. 2006; Overhage et al. 2001; Zafar and Lehto 2009). In this setting, the typical patient flow was as follows:

check-in clerk → nurse's aid → doctor → nurse → check-out clerk

Some of the tasks performed and their durations during a visit are listed in Table 13.6. One of our observations was that the tasks performed in this overall sequence needed to flow well, or the entire system would stall. To accomplish this

TABLE 13.6
Example Tasks and Durations for a Typical Patient Seen in a Primary-Care Setting

Task Description	Duration (min)
Check-in Clerk	
Answering a patient's question	9.533
Ask patient about visit type (urgent/return)	0.933
Call patient for check-in	2.867
Check out Prescriptions	3.583
Enter/update visit and demographic data	7.733
Fill out paperwork	11.283
Free time	26.883
Insurance	5.817
Non patient time	25.817
Patient sign-in	1.467
Place chart on stack for nurse triage	3.233
Stuff charts	3.983
Take co-pay money	3.117
Total	106.250

(Continued)

TABLE 13.6 (*Continued*)
Example Tasks and Durations for a Typical Patient Seen in a Primary-Care Setting

Task Description	Duration (min)
Check-out Clerk	
Checkout prescriptions	9.33
Free time	22.00
Hand scripts to pt.	9.00
Stuff charts	15.00
Work on release	22.00
Total	77.33
Doctor	
Document release orders in computer	43.517
Free time	2.517
Non patient time	19.817
Review computer records	15.433
Review old paper chart	4.833
See patient	46.900
Wait for printer and hand in report	14.467
Total	147.483
Nurse	
Draw blood	7.783
Free time	0.317
See patient and document notes	2.950
Tasks on computer	21.033
Wait for doctor signature	3.387
Total	35.450
Nurse's Aid	
Call patients into the triage area	2.333
Draw blood	3.483
Free time	56.133
Get patients into rooms	0.533
Give vision test	3.750
Non patient time	16.133
Take vital signs (includes Accu check/finger prick)	7.633
Total	90.000

goal, the interpersonal communication needed to be efficient, so that each participant had the necessary information at the correct time to determine what needed to be done and how to do it properly, and so that delays and errors at subsequent stages of the process would be avoided. As discussed in the section, A Case Study for Mobile Device Implementation in Clinical Settings, this requirement created numerous difficulties interacting with the HIT located on a stationary computer, and suggested that use of a mobile device for entering and retrieving information would be helpful for addressing this issue.

Another issue is that type of visit may be varied from time to time. For example, the patient may come in for a general care visit to their primary care physician (PCP). These visits typically last 15–20 minutes, and a provider may see 12–15 patients in a half-day. During those 15–20 minutes the following needs to happen:

1. Provider looks at the chart to familiarize themselves with past notes, history, labs, visits, discharge summaries, ER and other visits, consultant notes, and medication lists (5 minutes).
2. Provider visits with the patient, takes a history, performs an exam, and makes decisions regarding diagnosis and treatment (10 minutes).
3. Provider documents the encounter (5 minutes).

For a moderate-complexity patient, such as above (Table 13.5), addressing nine medical problems in a visit is very challenging. Similarly, addressing any acute care concerns during that same visit is even more burdensome. As a consequence, the provider may elect to address a few of the nine problems and leave the other problems for a subsequent visit. They may deal with the problems that require fine tuning and adjustment to treatment and leave the other chronic problems alone. And if they need to perform any procedures (Pap smear, EKGs, X-rays, lab work) in the clinic, that takes extra time. If they are too busy they may end up sending the patient to the ER or urgent care center for any acute care issues, such as chest pain. Thus, triage of problems becomes very important for preserving the flow of care in a clinic. More time spent with one patient means less time with the next, and this can have deleterious consequences.

Unintended Consequences of HIT Implementation: Cultural Issues

Recent evidence suggests that unless careful attention is paid to human factors and cultural issues, implementation may lead to faulty systems that could cause additional harm that these systems are intended to prevent. The types of unintended consequences have been categorized previously into several classes (Ash, Berg, and Coiera 2004; Harrison, Koppel, and Bar-Lev 2007). Broadly speaking, these classes include (1) changes in existing workflow patterns that undermine vital interactions among care providers, (2) user-interface issues that lead to faulty or inefficient interaction with the systems for data entry and data access, and (3) changes in power structures that reduce provider autonomy and impose practice restrictions, and which can lead to confusion among less able personnel making higher-level decisions. We next describe our experiences with each of these.

First, there are interactions within an existing system that are often vital to the successful completion of an intended task. If the task network is perturbed by the HIT implementation then these interactions may disappear with the result that the intended task cannot be successfully completed. For example, in one large academic medical center, providers were used to looking in a basket for paper-based cardiac echo and EKG reports for their primary care patients. One day the basket disappeared, and the providers were poorly trained to access the same results online.

This resulted in critical echo and EKG results being ignored for a period of time until the requisite training was completed asynchronously.

In another medical setting, the nurses would alert providers to critically abnormal lab results returned by fax on paper and then would schedule the patients for urgent return visits for result adjudication. When the same critical results were re-routed for computer display, the provider was required to log in and note the critical results. This task shifting then caused delays in adjudication of abnormal results and the nurses had to be approached asynchronously for rescheduling these urgent appointments.

Second, user-interface problems abound with the current generation of HIT systems. In one esteemed medical center, a new inpatient provider order-entry system was implemented that had a major flaw. It permitted orders to be entered without first verifying the currently selected patient. In one case, a blood-thinner called Warfarin was improperly administered to a patient who was not on it. In another case, a blood transfusion was given to the wrong patient, resulting in an adverse reaction. Use of bar codes on medications and strict verification systems can reduce these types of problems.

In another case, a specialist consult request and lab and radiology test order-entry system was in effect an order-communication system where orders entered electronically were not "sent anywhere" but just stored. This was because the intended receivers of the orders had strict control over schedules and would not let a computer system change schedules. A human operator on the other end had to physically extract the orders and make telephone calls to set up appointments for tests and consults. This resulted in ordered tests and consults never occurring, with the patient coming back to their primary care provider several months later without the requisite tests or consult visits completed.

Third, well-meaning efforts to ensure that practitioners follow clinical guidelines have led to "alert fatigue" by users of HIT. From a clinician's perspective there are simply too many alerts, many of which may not be relevant. There is no easy answer to the problem of alert fatigue, but it is essential to find a balance between providing too many alerts and missing critical alerts. This often comes with experience and fine-tuning of the HIT system in the use environment. Some insight into how to address this issue can be found from a recent study that explored methods of improving the computerized clinical reminder (CCR) used throughout the Veterans Administration (VA) (Wu et al. 2010). The study found that a redesign of the system that included features such as a risk factor repository, role-based filter, and prioritization by due date and risk factors were useful for the physicians. These features allowed physicians to quickly screen out less-relevant reminders and improved their ability to prioritize reminders.

Fourth, power-structure preservation is important for maintaining workflows. In one example, an older physician shied away from entering orders in a computer and had an inpatient ward clerk enter these orders. The physician would write orders on paper and have the ward clerk "translate" them into the computer system. In many cases, these orders would not translate uniformly and slightly different orders were entered. This resulted in delays in care and missed opportunities for treatment. This is an example of a shift in power structure where a less competent person is making decisions about how the orders translated into the order types the order entry system could understand.

These examples highlight the need to perform a comprehensive human factors analysis prior to implementing HIT systems. This entails understanding the entire task pipeline or network, understanding roles and responsibilities, and then mapping out how the new HIT system would be overlaid on the task networks and role maps. Replacing a human operator by a computer equivalent often equates to a shift in the power structure and unless the newly assigned roles in this shift can be adjudicated, accepted, and understood, errors will undoubtedly occur. In the next section we look at some guiding principles for HIT system implementation that can help mitigate some of these problems.

Lessons Learned about Best Practices for Implementing HIT

There are some simple guiding principles that can be applied during the technology needs assessment and implementation processes that can help mitigate problems down the road. These principles are time tested in the medical informatics literature and can prove useful when dealing with the complex cultural, financial, and technical challenges encountered during HIT implementation.

First, it is helpful to view the implementation process as a cycle of needs assessment, implementation, and evaluation. Technology implementation should be driven by a need to solve real-world problems and not as an end in itself. The key types of problems for which HIT solutions exist today include the following:

1. Inadequate communication between providers, that is, lack of relevant clinical data at the point of care in a fragmented healthcare delivery system, leading to duplicate testing, increased cost of care, missed diagnoses, and poor patient satisfaction.
2. Inadequate knowledge of current clinical practice guidelines and research results leading to poor quality and geographic disparities in care.
3. Lack of knowledge about drug-drug, drug-lab, and drug-diagnosis interactions and inadequate monitoring of treatments, leading to adverse drug events.

Second, formative and summative evaluation is a critical ingredient for success, both for the technology selected and for the implementation process itself. The evaluation drives the quality and process improvement in a continuous manner. Evaluation strategies exist for many types of HIT implementations (Cusack et al. 2007–2009), ranging from simple electronic medical record systems to full regional health information exchange. Evaluation provides the answer to the question of need assessment and whether the technology solution is actually doing what it was intended for. Evaluation need not be rigorous (such as a randomized controlled trial of interventions) but in most cases a simple before-after observational study will suffice. Evaluation should encompass both quantitative and qualitative measures of performance. Satisfaction by users is as important as cost reduction or medical errors prevented.

Third, understanding the current workflows using a human factors approach and predicting how the HIT system will interact with, augment, or replace elements of the workflow is of critical importance. An HIT system should be well integrated into the existing workflow, or it will not be accepted (Poon et al. 2004). Furthermore, the degree to which the information provided by an HIT system is integrated into the workflow can influence the quality of provider decision making (Wu et al. 2009). Understanding workflows requires tracking of data elements, task lists, and roles and responsibilities over time, along with the relevant technology pieces that are interfaced with at each step. Task analysis is an important part of this. Inefficient processes that are currently done manually should not be automated as they will likely result in bottlenecks in the workflow that could prove difficult to circumvent. Rather, these processes should be replaced by more efficient automatic processes. An example of this is when handwritten medication prescriptions are replaced with electronic versions. In one environment, the new HIT system allowed providers to type prescriptions into the system, which would then print out on paper. The clinic staff still had to hand-carry these prescriptions to the pharmacy, and in many cases would lose the scripts. By replacing the printed prescriptions with an ePrescribing system that sent electronic messages to the pharmacy, this problem disappeared.

Fourth, the users should be involved early. This cannot be stressed enough. There are numerous cases of spectacular systems failures because users were "surprised" by how the systems were implemented—in one case a $20-million HIT system had to be turned off after 2 months of use because the users complained so much. The stakeholders include not only the healthcare providers but all of the other support staff that may interface with the system at one time or another. Bringing all stakeholders together at the outset helps offset cultural issues down the line.

Fifth, small and pilot test implementations should be started before "going live" in the care environment. A small set of dedicated users and promoters can go a long way towards enticing less enthusiastic users to try the system. In one case, a senior physician used to work with a nurse practitioner who was an enthusiastic user. The senior physician had very little computer experience, so the nurse practitioner was the on-the-ground trainer for the physician.

Sixth, the data flows need to be understood. You should be able to account for how every piece of data entered into, or retrieved from, a system gets from point A to point B. Data quality issues abound in electronic systems, and as the adage suggests, "garbage-in, garbage-out." Alert fatigue is a real issue, where too many "warnings," often based on incorrect or out-of-date information, are displayed by a clinical decision support system. This irritates and frustrates users, and so they ignore all warnings, including the critical ones. As an example, at one hospital, the system would alert providers to provide flu shots for many of the eligible patients, despite the fact that these patients already had their flu shots. What had happened was that these shots were being administered at another site and the information about those "events" was not uploaded into the database. As a result, a faulty alert was being generated that irritated users. In another case, medications and diagnoses from several years ago persisted in the system, so that incorrect alerts were being generated about drug-drug interactions for drugs that had long been discontinued.

Change Management

It was the genius Niccolo Machiavelli who once remarked, "There is nothing more difficult to take in hand, more perilous to conduct, or even more uncertain in its success than to take the lead in the introduction of a new order of things." This dogma is all the more relevant today as the complexity of things is much greater than the fifteenth century world of Machiavelli.

Change management refers to the process whereby an organization achieves its goals and vision by facilitating change, rather than just delineating the steps along the way (planning). This process requires early involvement of important stakeholders and careful consideration of many different issues at the organization, technical, and process levels (Table 13.7). Thus, creating change starts with a *vision* and then empowering individuals to help create that vision. These individuals act as agents of change and need to have a 360-degree appreciation of the process. These agents need to be total systems oriented, future directed, and realistic about what is possible.

TABLE 13.7
A General Framework for Change Management in Health Care

Organizational Level
- Establish management, clinical, and technical leadership groups and a process to monitor the people, process, and technology
- Early involvement of all stakeholders

Technical Level
- Understand information flows, establish the data standards and data models, and pilot test it all with real users in real settings

Process Level
- Clearly define the objectives, roles/responsibilities (especially who is in charge, and name people to head up specific change management objectives in order to create a sense of ownership)
- Establish clear communication channels between these parties
- Establish efficient processes for coordination
- Create a process for dealing with midcourse changes and requests

Training and Educational Needs
- Orientation and establishing readiness for change
- Computer use and information technology
- Benefits of HIT
- User-centered case studies
- Establish a clear educational plan for all parties involved

Financial Concerns
- Need to obtain financial support early on
- Continuously monitor financial resource use
- Have contingency plans and address sustainability issues

Legal Issues
- Establish clear standard operating procedures, formal agreements, and policies early on

Political
- Assess the "climate" for change to see where the "pockets of resistance" may be and address them early
- Identify *all* possible stakeholders (very granular)
- Establish a climate of trust with the stakeholders
- Involve all people to some degree early on so people do not feel they are just "along for the ride"

When implementing change, it is critical to understand which types of change are happening and for whom. Change in healthcare systems can be distinguished in at least four different levels:

- Operational change: Affects the way operations occur (automation).
- Strategic change: Shift in strategy, for example, from an inpatient to outpatient focus.
- Cultural change: Transformation in basic organizational philosophy (e.g., implementation of a continuous quality improvement program).
- Political change: Modifications in staffing or work roles for political reasons (e.g., job level change).

At each level, situational awareness is a key requirement to ensure that change is managed properly. The level or extent of change is also an important consideration (Table 13.8). There are microchanges (first order change) and megachanges (second order change). Microchange is change with differences in degree, for example,

TABLE 13.8
Levels of Change in HIT Implementation

First-Order Change[a]
- A variation in the way processes are done in a system
- Leaves the system itself relatively unchanged
- Example: Creating new reports, new ways to collect the same data

Second-Order Change[a]
- The system itself is changed
- Occurs as a result of strategic shift or threat against survival
- Involves a redefinition or recapitualization of basic business processes and procedures
- Example: Paper to electronic medical record, ATMs in banking versus human teller agents

Middle-Order Change[b]
- Individuals stand midway between two positive forces
  - Example: A choice between two good systems to purchase
- Individuals stand midway between two negative forces
  - Purchase power for a "good product" may be limited by economics, available technologies, and organizational issues
  - Example: Decision makers must choose an information system that they *know* will not completely meet their needs, that is, choose the lesser of two evils
- Individuals are exposed to positive and negative forces
  - This type of conflict is very common in health care
  - Example: Conflict between system users and the IT/financial people

[a] Watzlawick, Weakland, and Fisch 1974.

[b] Golembiewski, Billingsley, and Yeager (1976) augmented these definitions with middle order change that lies somewhere in between first- and second-order change. Deutsch (1965) described three fundamental types of conflicts in a system that is being perturbed that occur at this level.

improvements or upgrades to a system. Megachanges are changes with differences in kind, for example, changing from paper to EMR.

One person's microchange can be another person's megachange. To avoid problems, it is critical to understand roles and responsibilities. Successful implementation requires identification of what role you (as an implementer) play and how it differs from other roles, and to identify carefully which role is speaking when you communicate. Organizational change almost always involves some type of personal loss—be prepared for this and warn users in advance. Learning a new system requires time, energy, and motivation. Many resist this process unless they see disadvantages with the current system or significant advantages to the new one. People want to feel good about themselves, so allow that to happen. Organizational culture may inhibit this "feel good" experience, so people often invent strategies to make themselves feel good. This may involve opposition to management. Change initiatives often require larger losses for the "middle manager" because they perceive that the change allows top executives to exert more direct control by knowing in more detail what is going on—so managers need to be assured that this is not the case.

Solutions to Cultural Issues in HIT Implementation

The hardest part about any HIT implementation process is that it requires significant cultural change. Implementing the technology is the easy part. A general rule of thumb is that about 80% of the effort in successful implementation of HIT will have to be directed towards culture change, and less than 20% to implementing the technology itself (GAO 2003). To gain some perspective into what this takes, each year there is a Davies Award given to successful national implementers of HIT. When comparing successful recipients of this award with unsuccessful implementations, a set of common themes have emerged.

First, almost all successful implementers approached change management incrementally, by focusing on a specific barrier to be addressed by HIT. All winners had to re-engineer some workflow process. A simple rule of thumb was to not automate a manual process that occurs commonly but does not work. Another common theme in successful implementations was to conduct frequent, sustained, end-user orientations and feedback with demonstrated responsiveness to the feedback. Examples include weekly pizza meetings at the Regenstrief Health Center, the use of physician focus groups at Kaiser Permanente, and weekly feedback with "supplements" at Northwestern Medical Center. System developers were also the salespeople, troubleshooters, coaches, and colleagues. Importantly, plans were put in place for system evaluation and monitoring, and the HIT systems were viewed as *tools* to enable care process improvement and were not an end in themselves.

Involving all relevant stakeholders early on was another key to success. In summary, the keys to success in achieving cultural change necessary for successful implementation of HIT include (1) frequent open communication; (2) taking steps to ensure management sponsorship to avoid competing agendas; (3) early delineation of partner responsibilities; (4) making sure all stakeholders feel they are contributing; (5) tackling political issues early on, including getting buy-in from the state and local government; (6) establishing standard operating procedures; and (7) understanding that ROI is a long-term benefit—that there is intrinsic value in the data shared.

The importance of taking steps prior to deciding upon implementation of HIT to ensure the key stakeholders are ready for change is another lesson learned from successful implementations. Readiness for change refers to both a shared commitment within the organization to implement a change and a shared belief the organization has the capability to do so (Weiner 2009). To gain commitment to change it is critical that the stakeholders understand why the change is necessary and how the changes will affect them. To address this issue, steps should be taken well before implementation, to address user concerns about issues such as how much learning is necessary to use the new system, who will have what responsibilities, training and support, workflow interruptions, and data security. One of the most important requirements is to take proactive steps to minimize the disruption that will occur when the system is brought into use. Failures at this point can cause neutral or ambivalent stakeholders to oppose the HIT. The latter issue is critical to success, as the introduction of HIT will in most cases result in some disruptions, while users develop confidence and skill in using the system and following the new workflows (e.g., a need to "co-sign" verbal orders electronically), as well as trust in the "back-end" processes, for example, what happens to the "order" once it is entered? How does it get done? Who is responsible and when and how can you see a result?

Several strategies can be followed to minimize disruption during implementation and more rapidly gain the advertised benefits of HIT. Following these approaches can at the same time encourage user acceptance and lead to positive cultural change. Up-front, ongoing, training, and retraining are important. Our experience is that it is essential to avoid underestimating the amount of training required in order to avoid disruptions and encourage positive cultural change. In most cases, the organization will need full-time support staff to adequately address this need. Up to 20% of the IT budget will have to be spent on training alone. Innovative approaches can also be helpful to move resisters towards acceptance, such as the use of "surrogate" trainers (e.g., the Cleveland Clinic CPOE example). The latter strategy involved extensively training staff that worked the closest with the individuals who were resistant and then using these "surrogates" to monitor, train, and support the resistant providers. To implement this strategy, workflows can be observed during the early stages of implementation to identify both resisters and potential training surrogates. The culture change may take years to develop, so it should not be rushed; work more closely with the enthusiasts and early adopters (20%) and let them lead the others.

Another lesson learned is that it is essential to first extensively pilot test and troubleshoot the HIT system before going live. Going through this process can yield benefits such as (1) ensuring the "vocabulary" is as close to that of a user as possible so that a seamless transition occurs between the paper and electronic worlds; (2) identifying "catch-on" features that seem to help users understand how the system works, and advertising them well and to everyone; (3) flagging potential problems with the user interface; and (4) identifying workflow issues and suggesting how to better integrate HIT into the workflow. Steps also must be taken prior to going live to anticipate and prepare answers to user questions, understand the full closed-loop system, train the support people, have contingency plans ready in case of disaster, and have a method

TABLE 13.9
Network Security Tools and Personnel

Do you have security personnel?
  Password security rules and change schedules
  HIPAA training and certification
Do you use secure communications channels?
  SSL certificates
  https (128-bit encryption)
  VPNs (virtual private networks)—quite safe
  Peer-to-peer connections—safest
Do you have a firewall, virus protection, and intrusion detection capabilities and competent people to oversee them?
Educate the users well! It takes a lot of time, effort, and patience for the physicians to "accept" the security equation

in place for continuously benchmarking the system (use/acceptance, percent orders, etc.). Once everything is in place, the system should then go live in stages (by care units, staff types, institutions, etc.).

Lack of trust in data security is another commonly voiced barrier to adoption of HIT systems. Despite the fact that good technical solutions are available for computer data security issues, and paper-based systems are in fact often much less secure, many providers still worry about what will happen if the system goes down or is hacked, and will ask questions such as "Do you need paper backups? Redundant servers? How much will this cost?" This issue is confusing to some providers because, despite the fact that vendors advertise that they are "HIPAA compliant," a lot of the burden for HIPAA compliance is actually on the provider, and the vendor can do nothing about that. Vendors are responsible for making sure that their application is HIPAA compliant (i.e., uses login/passwords, has automatic signouts, uses secure messaging such as https, SSL, and so on, and is backed up). However, they have no control over the network architecture used by the provider. For small practices this could be a problem, as it may create a need to invest in security tools and personnel (Table 13.9), and a Web-based application service provider (ASP) model may be a more attractive option.

Case Study for Mobile Device Implementation in Clinical Settings

Clinical workflows can be depicted as complex, intercalating networks of directed graphs of task sets. They are highly interruptive and require frequent signing in and signing out of a computer system for simple tasks such as results look up or order entry. Furthermore, these workflows occur in highly mobile settings and systems that incorporate fixed, stationary HIT systems that can impede the workflows. For example, in one study we found that "walk-time" in a busy emergency room cost the system $70,000 per year per provider (Zafar and Lehto 2009). In another study described in the same paper, the rate of patient check-ins into a primary care clinic resulted in nurses aides sitting idle for 45% of their time due to bottlenecks generated elsewhere in the care pipeline.

Recent studies have also shown that unmet clinical information needs occur frequently during the routine care process. In one study, 154 unmet information need events occurred in 15.5 hours of observation for 35 clinicians. This translated to about 11 unmet information need events/hour (Currie et al. 2003). And 73% of these unmet needs related to questions about specific patients and the care domain including laboratory data, medical communication (notes, reports, etc.), and information in a paper chart or conversation.

Providing *mobile devices* for data access, data entry, and interpersonnel communication to providers could potentially alleviate many of these problems. Mobile technology has improved dramatically in recent years, and now includes features such as larger, more vibrant displays (tablet PCs); longer battery life (about 5 hours for a tablet PC); wireless connection (WiFi, Bluetooth); lighter weight and more portability; and good handwriting recognition capability.

Thus, mobile devices are a natural fit for the inherently mobile workflow of clinicians. However, the traditional keyboard-mouse interface commonly used for many EMRs does not translate well to mobile devices. Mobile devices use a pen interface, which is great for entering and looking up structured information but is not optimal for free-text data. As a starting point for our prototype, we proposed that the device should implement a *multimodal* approach to data entry, including context-specific prompts; efficient, intuitive information display; handwriting recognition; and structured pick-lists. These devices could also use push technologies for result notification that can reduce the need to constantly log in or log out to check whether results are available.

To examine the feasibility and potential benefit for using a mobile device of this type in HIT settings, we conducted a product development effort that resulted in a prototype interface running on a small handheld computer. The first step in the process was to conduct a formal requirements analysis to identify the features necessary for a mobile device and user interface that would be acceptable to clinicians. The results of this step are summarized in Table 13.10.

A focus group involving a panel of clinicians was then used to gather data about these requirements. At about the same time, we conducted an extensive literature search for relevant articles on previous efforts at creating mobile devices, such as PEN-Ivory among others. Based on the results of this our focus group and successful elements of existing mobile user interfaces, we developed a prototype user interface running on a tablet PC that addressed these important areas as well as our own vision for intuitive design characteristics. To test initial reactions to the prototype we held two focus groups (with informatics-trained and noninformatics-trained clinicians). The comments of the latter focus group are summarized in Table 13.11.

We were able to use the input from these providers to build a second prototype user interface that then garnered very positive reaction from both industry and academia at a meeting of the American Medical Informatics Association in 2005. This experience confirms our belief that many solutions are still out there waiting to be implemented that can smooth the transition to HIT in otherwise hostile environments containing users with computer anxiety and strong reservations about the learning curve.

TABLE 13.10
Desirable Features of a Mobile Device for Improving the Efficiency of Physician Data Entry to an HIT in Clinical Settings

Physical characteristics of the device: long battery life, lightweight, adequate screen size, "foldable" to fit into a white-coat pocket—"hands free workflows."

Limit the amount of "scrolling" that needs to be done to access the data entry screens. This scrolling introduces errors and the possibility for omissions.

Use a context-specific but fixed palette of modifiers for clinical terms.

Provide access to a comprehensive list of clinical terms but prioritize the list and show the more commonly used terms at the top.

Co-locate information access and data-entry elements based on the cognitive reasoning processes of clinicians. This permits faster data entry by reducing the time for context switches.

Fast response times: The mobile user is more focused on the people interaction than with computing as an end in itself, so the device is a tool rather than an environment and "wait times" are unacceptable and frustrating.

Macros, templates, gestures, selection lists, application specific keyboards (term keyboards), speech input, and other *multimodal* elements to help reduce keystrokes.

Customizable to the workflows and data element needs of the user.

Multiprocessing, that is, it should not "slow down" or "shut down" when busy with one task. Other tasks and functions should still be accessible.

Instant on/off: Significant boot-up times are unacceptable to the user who needs to log in and out frequently and quickly.

Use of convenient drug/clinical knowledge tools accessible from the same interface along with "recent" patient data (labs, test results, medication lists, etc.).

TABLE 13.11
Responses of Clinicians to a Prototype Mobile Computing Obtained in a Focus Group

Clinical Informatics Fellows (Part-Time Attending Physicians)	Noninformatics-Trained Full-Time Attending Physicians
Very interested in note-taking features, less enthusiastic about this being a "full EMR." Particularly excited about its potential to help document what was done and why.	This group in general was more enthusiastic about mobile devices being "full EMRs." Strongly suggested adding template-based note entry features.
Change the layout to fit the screen more appropriately, reduce border sizes on modal dialog boxes.	Also felt that changing the border sizes and dialog box locations would improve usability.
Very concerned about the ability to display so much information on a 640 × 480 screen size.	Not too concerned about readability on small form-factor screens.
Reduce the clutter on screen, display fewer types of information from certain screens.	Wanted to see specialized types of information (current patient locations, list of active patients to be seen, certain standing order sets for nursing staff, preventive care). These would help improve their workflows.

(Continued)

TABLE 13.11 (*Continued*)
Responses of Clinicians to a Prototype Mobile Computing Obtained in a Focus Group

Clinical Informatics Fellows (Part-Time Attending Physicians)	Noninformatics-Trained Full-Time Attending Physicians
Felt that only a subset of physicians would be willing to learn a new interface (as required for entering information using Graffiti or hierarchical template systems).	Felt that most physicians would find some aspects of a "comprehensive device" to be useful, if only for looking up test results, drug and clinical references, and so on.
Felt that a hierarchical selection list with modifiers is very time-consuming and hard to learn and would put off many users. Suggested voice recognition input as an alternative.	Felt that a template-based note-entry system, even if operated with a pen, would be time-saving. Strongly felt that template customization should be included.

CONCLUSION

The potential benefits of implementing HIT have led to a strong push for adoption in the United States by several groups. The economic benefits have been estimated to be in the order of hundreds of billions of dollars. The central question is, if HIT is so helpful in the context of clinical care, and can reduce the cost and improve the efficiency and safety of care, then why is it not being universally adopted? As summarized in this chapter, the answer is that the acceptance of HIT involves multiple players and a complex set of *cultural*, *political*, and *economic* issues. Part of the issue is that HIT is costly and technically difficult to implement and requires a shift in workflows. It requires IT expertise, something lacking in many healthcare settings, especially with solo or small group providers. It has a steep learning curve and busy providers are unmotivated to adopt and change habits, especially if that means reduced productivity and reimbursement during the implementation phase. Furthermore, the rewards are not immediately noticeable, and it takes a few years to financially recover from the investment. In addition, if improperly implemented, HIT actually has the potential to cause harm.

To help avoid these pitfalls, much can be learned from successful implementations of HIT. Involving all relevant stakeholders early on is a major key to success, as implementation will almost certainly involve re-engineering workflow processes, and this is best done collaboratively. Almost all successful implementers have approached change management incrementally by focusing on a specific barrier to care addressed by HIT. Another common theme in successful implementations has been to conduct frequent, sustained, end-user orientations, training, and responsive feedback. Plans were put in place for system evaluation and monitoring, and the HIT systems were viewed as tools to enable care process improvement.

We hope this chapter has provided a small glimpse of the complexity of health care, the cultural barriers to good-quality and high-efficiency care, and the role of health information technology. We are living in an ever-changing world with new systems and new standards emerging every year and more integration occurring alongside all of this as we realize that the way to combat complexity is to simplify the process and eliminate or automate unnecessary steps. In the end we all hope for the same outcomes, that of high-efficiency, high-quality health care that is accessible, cost-effective, and provides positive outcomes.

REFERENCES

Angst, C. M., and R. Agarwal. 2009. Adoption of electronic health records in the presence of privacy concerns: The elaboration likelihood model and individual persuasion. *MIS Q* 33:339–70.

Ash, J. S., M. Berg, and E. Coiera. 2004. Some unintended consequences of information technology in healthcare: The nature of patient care and information system-related errors. *J Am Med Inform Assoc* 11:104–12.

Berwick, D. M. 2003. Disseminating innovations in healthcare. *JAMA* 289:1969–75.

Blumenthal, D. 1999. The duration of ambulatory visits to physicians. *J Family Pract* 48(4):264–71.

Currie, L. M., M. Graham et al. 2003. Clinical information needs in context: An observational study of clinicians while using a clinical information system. In *Proceedings of AMIA 2003, AMIA Annu Symp Proc* 190–94.

Cusack, C. et al. 2007–2009. An evaluation toolkit for health IT implementation, an evaluation toolkit for health information exchange. AHRQ National Resource Center for Health IT website, http://healthit.ahrq.gov.

Davenport, T. H., and J. Glaser. 2002. Just-in-time delivery comes to knowledge management. *Harvard Business Review*, July 2002. Available at http://hbr.org/2002/07/just-in-time-delivery-comes-to-knowledge-management/ar/1.

Davis, F. D. 1986. A technology acceptance model for empirically testing new end-user information systems: Theory and results. PhD dissertation, Massachusetts Institute of Technology, Sloan School of Management.

Deutsch, M., and R. M. Krauss. 1965. *Theories in Social Psychology*. New York: Basic Books.

General Accounting Office (GAO). 2003. *Use of Information Technology for Selected Health Care Functions*. Washington, DC: Briefing for the Minority Staff of the Senate Committee on Health, Education, Labor, and Pensions.

Golembiewski, R. T., K. Billingsley, and S. Yeager. 1976. Measuring change and persistence in human affairs: Types of change generated by OD designs. *J Appl Behav Sci* 12:133–57.

Harrison, M., R. Koppel, and S. Bar-Lev. 2007. Unintended consequences of information technologies in health care—An interactive sociotechnical analysis. *J Am Med Inform Assoc* 14:542–9.

Holden, R. J. 2009. Beliefs about using health information technology: An investigation of hospital physicians' beliefs about and experiences with using electronic medical records. PhD dissertation, University of Wisconsin–Madison, Industrial Engineering and Psychology.

Johnston, D. E., J. Pan, D. W. Walker, D. Bates, and B. Middleton. 2003. *The Value of Computerized Provider Order Entry in Ambulatory Settings*. Wellesley, MA: Center for IT Leadership.

Kelly, W. N. 2001. Potential risks and prevention, Part 2: Drug-induced permanent disabilities. *Am J Health Syst Pharm* 58:1325–9.

Kohn, L. T., J. M. Corrigan, and M. S. Donaldson. 2000. To err is human: Building a safer health system. In *Committee on Quality of Health Care in America*. Washington DC: Institute of Medicine, National Academy Press.

Leap, L. L., D. W. Bates et al. 1995. Systems analysis of adverse drug events. *JAMA* 274:35–43.

L'Enfant, C. 2003. Clinical research to clinical practice—Lost in translation. *N Engl J Med* 349:868–74.

Lorenzi, N., and R. T. Riley. 2007. Managing change: An overview. *JAMIA* 7:116–24.

Louthan, M., S. Carrington, N. Bahamon, J. Bauer, A. Zafar, and M. Lehto. 2006. Workflow characterization in a busy urban primary care clinic. In *Proceedings of the AMIA Symposium*, 1015, Washington DC.

Meijden, M. J., H. J. Tange, J. Troost, and A. Hasman. 2003. Determinants of success of inpatient clinical information systems: A literature review. *J Am Med Inform Assoc* 10:235–43.

Moen, A., J. Gregory, and P. F. Brennan. 2007. Cross-cultural factors necessary to enable design of flexible consumer health informatics systems (CHIS). *Int J Med Inform* 76:168–73.

Overhage, J., S. Perkins, W. Tierney, and C. McDonald. 2001. Controlled trial of direct physician order entry: Effects on physicians' time utilization in ambulatory primary care internal medicine practices. *J Am Med Inform Assoc* 8:361–71.

Poon, A. D., and L. M. Fagan. 1994. PEN-Ivory: The design and evaluation of a pen-based computer system for structured data entry. *Proc Annu Symp Comput Appl Med Care* 447–51.

Poon, E. G., D. Blumenthal et al. 2004. Overcoming barriers to adopting and implementing computerized provider order entry systems in U.S. hospitals. *Health Aff* 23:184–90.

Shortliffe, E. H. 2005. Strategic action in health information technology: Why the obvious has taken so long. *Health Affairs* 24(5):1222–33.

Simon, R., Steven, R. Kaushal, P. D. Cleary, C. A. Jenter et al. 2006. Correlates of electronic health record adoption in office practices: A statewide survey. Presented in abstract form at the 29th Annual Meeting of the Society of General Internal Medicine, April 2006, and at the HMO Research Network Conference, May 2006. *Journal of the American Medical Informatics Association* 14(1), January-February 2007, 110–17.

Stafford, R. S., D. Saglam et al. 1999. Trends in adult visits to primary care physicians in the United States. *Arch Family Med* 8:26–32.

Timmons, S. 2001. How does professional culture influence success or failure of IT implementation in health services? In *Organizational Behavior in Health Care: Reflections on the Future*, ed. L. Ashburner, 218–31, Basingstoke: Palgrave.

Valdez, R. S., and P. F. Brennan. 2008. Medical informatics. In *eHealth Solutions for Healthcare Disparities*, ed. M. C. Gibbons, 93–108. New York: Springer-Verlag.

Walker, J. et al. 2005. The value of healthcare information exchange and interoperability. *Health Aff* 19 (January). http://content.healthaffairs.org/content/suppl/2005/02/07/hlthaff.w5.10.DC1.

Watzlawick, P., J. H. Weakland, and R. Fisch. 1974. *Change: Principles of Problem Formation and Problem Resolution*. New York: Norton.

Weiner, B. J. 2009. A theory of organizational readiness for change. *Implement Sci* 4:67.

Wu, S., M. R. Lehto, Y. Yih, M. Flanagan, A. Zillich, and B. Doebbeling. 2006. A logistic regression model for assessing clinicians' perceived usefulness of computerized clinical reminders. In *Proceeding of the 36th International Conference on Computers and Industrial Engineering*, 3028–38. Taipei, Taiwan.

Wu, S., M. R. Lehto, Y. Yih, J. J. Saleem, and B. Doebbeling. 2009. On improving provider decision making with enhanced computerized clinical reminders. In *Lecture Notes in Computer Science*, 569–77. Berlin: Springer.

Wu, S., M. R. Lehto, Y. Yih, J. J. Saleem, and B. Doebbeling. 2010. Impact of clinical reminder redesign on physician's priority decisions. *Int J Med Inform* In press.

Zafar, A., and M. Lehto. 2009. Workflow characterization in busy urban primary care and emergency room settings: Implications for clinical information system design. *Int J Ind Ergon* In review.

Section V

Specific Applications

14 Cultural Factors and Information Systems

An Application to Privacy Decisions in Online Environments

Fariborz Farahmand

CONTENTS

INTRODUCTION

What determines how online users from different cultures assess consequences arising from privacy incidents in online environments? Is a privacy decision just the release of personal information in exchange for rewards? Can normative models of decision making be used to explain privacy decisions in online environments? Finding answers to these questions is essential for managing risks in online environments. This chapter combines approaches from fields of information systems, decision sciences, and social sciences to investigate the determinants and apparent inconsistencies of privacy decision making and behavior in online environments.

First, I define privacy, and its relation to the context and to the security culture. I provide several real-world examples of decisions about privacy and its variation with the global and cultural differences. The concept of procedural fairness and its impact on perception of privacy risks are then explained. User perceptions of privacy risks are described, as are how these perceptions are influenced by affect heuristics and how user perceptions will become proxies of actual risk. Finally, I explain the shortcomings of normative decision models in describing cultural factors in privacy decisions and suggest behavioral economics and interdisciplinary approaches to understand the variations of privacy decisions by users from different cultures.

Privacy: A Process in a Cultural Context

Individuals' concepts of privacy are tied to concrete situations in everyday life (Laufer and Wolfe 1977). These situations can be described in terms of three dimensions: self-ego, environmental, and interpersonal. A key to understanding privacy is in relation to the individual from birth to death. Cultural and sociophysical environmental limits on possible privacy experiences will apply differently to individuals at various stages of the life cycle. Furthermore, even the properties of the life cycle are not static. Periods of time devoted to specific activities (e.g., childbearing and child-raising, employment) will vary as a function of changing technology, changing sociocultural patterns, and historical environment.

Westin's (1967) theory of privacy speaks of ways in which people protect themselves by temporarily limiting access to themselves by others:

> Privacy is the claim of individuals, groups, or institutions to determine for themselves when, how, and to what extent information about them is communicated to others. Viewed in terms of the relation of the individual to social participation, privacy is the voluntary and temporary withdrawal of a person from the general society through physical or psychological means, either in a state of solitude or small group intimacy or, when among large groups, in a condition of anonymity or reserve. (p. 7)

The fundamental observation of Altman (1975) is that privacy regulation is neither static nor rule-based. He conceptualizes privacy as the "selective control of access to the self or to one's group" (p. 18), regulated as dialectic and dynamic processes that include multiple mechanistic optimizing behaviors. Altman describes privacy's dialectic and dynamic nature by departing from the traditional notion of privacy as a withdrawal process where people are to be avoided. Instead, Altman conceptualizes privacy as a boundary regulation process where people optimize their accessibility along a spectrum of "openness" and "closedness" depending on context. Privacy is not monotonic, that is, more privacy is not necessarily better (Palen and Dourish 2003).

Both theories are examples of limited-access approaches, that is, both emphasize controlling or regulating access to self. Both address process and employ classifications of privacy. However, process is primary in Altman's (1975) theory, whereas classification is primary in Westin's (1967) theory.

Privacy is not context-free. By context we refer to the structural social settings characterized by canonical activities, roles, relationships, power structures, norms (rules), and internal values, and rooted in specific times and places (Nissenbaum 2010). The results of many empirical studies indicate that privacy should be considered as a process and not as an object. For example, Adams (2000) conducted a study of privacy concerns in an environment capturing audio/video. Results showed that users' perceptions of privacy were shaped by the perceived identity of the information receiver, the perceived usage of the information, the subjective sensitivity of the disclosed information, and the context in which the information was disclosed.

Relating Security Culture and Privacy

Security culture is defined as "the totality of patterns of behavior in an organization that contribute to the protection of information of all kinds" (Dhillon 1995, p. 90). Chia, Maynard, and Ruighaver (2002) identified eight information security topics that could be used to describe the security culture of an organization:

1. The basis of truth and rationality that security is important
2. The balance between short-term and long-term security goals in an organization
3. Rewards, punishment, and motivation structure in the organization installed to motivate employees' behavior towards information system security
4. Inclination towards risk by the management and employees of the organizations
5. Pervasiveness of information system security in the daily work practices of the employees
6. Level of involvement of employees in managing information system security in an organization
7. Empowerment of employees, so as to instill responsibility towards information system security related actions
8. Level of balance maintained to satisfy external and internal influences on information system security

Security culture varies with the type of profession in organizations. For example, the study of Ramachandran, Rao, and Gols (2008) on security cultures of four professions—those of information systems, accounting, marketing, and human resources—indicated that the accounting profession has a strong security culture and the marketing profession a weak security culture, with the information systems and human resources professions lying between the two. The cultural differences lead to different behavior towards privacy.

Security in information systems, from the technical perspective, refers to many aspects of protecting a system from unauthorized use (e.g., authentication of users, information encryption, firewall policies, and intrusion detection). But, the concept of privacy goes beyond security to examine how well the use of information that the system acquires about a user conforms to the explicit or implicit assumptions regarding that use associated with the personal information. For present purposes I adopt Westin's (1967) definition of privacy as "the ability of individuals to control the terms under which their personal information is acquired and used" (p. 7).

Karat et al. (2009) provide a detailed discussion on relating security and privacy. They explain that, from an end-user perspective, privacy can be considered to be preventing storage of personal information, or it can be viewed as ensuring appropriate use of personal information. They explain that security involves technology to ensure that information is appropriately protected. Security is a required building block for privacy to exist. Privacy involves mechanisms to support compliance with some basic principles and other explicitly stated policies.

Cultural Factors and Adoption of Information Technologies: Cases of Mobile Banking and Electronic Health Records

Review of the literature indicates that the adoption of information technologies is highly influenced by the cultural factors. For example, by presenting illustrative data from exploratory work with small enterprises in urban India, Donner and Tellez (2008) explain the cultural contexts surrounding the use of mobile banking. Even in the United States, cultural factors play an important role in adoption of mobile banking. According to the Tower Group (2007), more than 40 million U.S. consumers will be using mobile banking by 2012; however, due to cultural divides, consumers do not yet strongly associate their wireless service and mobile device capability with their financial institution and banking capability. Although a number of studies (Laukkanen and Pasanen 2008; MacGregor and Vrazalic 2006) show that in Western cultures males are more likely than females to adopt e-service, in Saudi Arabia females are more likely to adopt over males (Siddiqui 2008). Al-Ghaith, Sanzogni, and Sandhu (2010) relate this to the nature of the Saudi society, which can be described as a conservative society framed within the culture that females tend and prefer to achieve their needs from their homes by using the Internet.

Cultural factors also play a significant role in adoption of electronic health records. Meijden et al. (2003) identify culture and characteristics of the organization as determinant of success of inpatient clinical information systems. Timmons (2001) discusses the influence of professional cultures on the acceptance of new health technologies. Boddy et al. (2009) explain the importance of culture and the influence of context and process when implementing e-health. Smedley, Stith, and Nelson (2003) argue that the major determinant of differences, associated with healthcare disparities, is race and ethnicity. Moen, Gregory, and Brennan (2007) contend that being mindful of sociocultural environments, age groups, and health conditions will enhance the design and use of consumer health informatics systems. Based on data from a survey from 38 countries, Bellman et al. (2004) make a case that the differences in Internet privacy concerns reflect and are related to differences in cultural values. Angst and Agrawal (2009), using a Web-based survey with 366 subjects, claim that in adoption of electronic health records even when people have high concerns for privacy, their attitudes can be positively altered with appropriate message framing. Valdez and Brennan (2008) maintain that use of culturally informed patient-centered information technologies can reduce disparities between persons of varying races, ethnicities, and cultures and improve patients' health.

Procedural Fairness and Privacy Calculus

Business experts in information systems argue that organizations, by addressing risks to privacy and trust, can gain business advantage (for example, through customer recruitment or retention) by observing procedural fairness—that is, the perception by an individual that a particular activity in which he or she is a participant is conducted fairly (Culnan 1993; Smith, Milberg, and Burke 1996). In this sense, "fairness" means the Principles of Fair Information Practice suggested by a federal commission in 1973 that provides the basis for the U.S. federal privacy law enacted in 1974 (Ware 1974). The practices include:

1. Notice/awareness: Consumers should be informed of an organization's information practices before being asked to supply any personal information. The scope of the notice should include:
   - Who will collect the data?
   - How the data will be collected?
   - Is supplying the data mandatory or voluntary, and what are the consequences of refusal?
   - To what uses the data will be put?
   - Who might receive the data?
   - How will the data's confidentiality, integrity, and quality be protected?
2. Choice/consent: Presenting options about how the collected information may be used, including secondary uses.
3. Access/participation: How an individual can access the data, not only to view information but also to contest its accuracy and completeness.
4. Integrity/security: The data must be accurate and secure, with protection not only against loss but also against unauthorized access, modification, destruction, or use.
5. Enforcement/redress: The mechanisms available to enforce the principles and provide redress.

Procedural fairness is perceived as providing the consumer with a voice, and giving a consumer control over actual outcomes (Lind and Tyler 1988). From this perspective, customers are assumed to be willing to disclose personal information and have that information used subsequently to create consumer profiles for business use when fair procedures such as these are in place to protect individual privacy. Studies indicate that individuals are less likely to be dissatisfied even with unfavorable outcomes if they believe that the procedures used to derive those outcomes are fair (e.g., Greenberg 1987). Pavlou and Gefen (2005) explain psychological contract violation as a buyer's perception of having being treated wrongly regarding the terms of an exchange agreement with an individual seller.

Studies on trust suggest that individuals are willing to disclose personal information in exchange for some economic or social benefit subject to the "privacy calculus," an assessment that their personal information will subsequently be used fairly and they will not suffer negative consequences (Milne and Gordon 1993).

In general, individuals are less likely to perceive information collection procedures as privacy-invasive when (Smith, Milberg, and Burke 1996)

- Information is collected in the context of an existing relationship.
- They perceive that they have the ability to control future use of the information.
- The information collected or used is relevant to the transaction.
- They believe the information will be used to draw reliable and valid inferences about them.

Perceptions of Privacy Risks

Technology risks in the modern world are studied in two fundamental ways: (1) *risk as feelings*, which refer to our institutive reactions to technology hazards, and (2) *risk as analysis*, which uses logic and well-understood methods in different areas (e.g., engineering risk analysis) to measure and mitigate technology risks. Scientists and engineers normally prefer to use the latter approach.

With the advent of the nine Workshops on the Economics of Information Security (WEIS), held regularly since 2002, some progress towards cost-benefit and risk analysis of information security and privacy incidents has occurred. However, much work remains yet to be done with regard to

- Assessing the damages of past privacy incidents
- Evaluating the level of vulnerability to privacy incidents
- Preparing for privacy incidents by selecting appropriate control measures, toward the goal of allowing us to better protect ourselves

Related to this lack of formalisms, there are limited data to analyze the extent to which organizations around the world invest in information privacy. Breaches are known to be underreported, and therefore so are associated costs. This lack of data leaves a state of uncertainty with regard to risk levels. Risk is measurable in that risk measures are known probabilities of possible outcomes. Uncertainty, on the other hand, is by definition not knowing the probabilities of the possible outcomes and in some cases not even knowing what all the outcomes might be. In this regard, we could say that the current state in privacy risks is really a state of uncertainty. We do not know all the possible outcomes; nor do we know the likelihoods of those outcomes.

In addition to the lack of data on breaches, privacy risk research is unique in that user behavior is a factor in the level and nature of risk in a relatively unprecedented way when compared to other technologies. Because of the abundance and relative low cost of computers, their pervasive nature, as well as the malleability of computers and information systems, more users are more engaged in continually shaping the operation of information systems around the world than is the case with other technologies, such as nuclear technology, aviation, or motor vehicles.

The first step towards understanding how users from different cultures make decisions about information technologies is to understand how they perceive risk and benefit from those technologies. Previous studies show that perceived risk is quantifiable and predictable (Johnson and Tversky 1984; Slovic 1987). Psychometric techniques seem well suited for identifying similarities and differences among groups with regard to risk perceptions and attitudes. These studies have also shown that the concept of risk means different things to different people. When experts judge risk, their responses correlate highly with technical estimates of annual fatalities. Lay people can assess annual fatalities if they are asked to do so (and produce estimates somewhat similar to the technical estimates). However, their judgments of risk are related to more characteristics (e.g., catastrophic potential, threat to future generation). Fischoff et al. (1978) investigated perceptions of technology risks, and particularly ways to determine when a product is acceptably safe. Their model can be adopted and used to define risk perceived by online users:

1. Voluntariness: Does the user voluntarily get involved in an online activity?
2. Immediacy of effect: Is the risk of consequence from the user's actions (to him/her) immediate?
3. Knowledge about risk: Are the risks known (precisely) by the online user who is exposed to those risks?
4. Knowledge of science: Are the risks precisely known and quantified?
5. Control over risk: Can the online user, by personal skill or diligence, avoid the consequences to him/her while engaging in untoward activity?
6. Chronic or catastrophic: Does the risk affect the online user over time, or is it a risk that affects a larger number of users at once?
7. Newness: Are these risks new to the online user, or is there some prior experience/conditioning?
8. Common dread: Is this a risk that the information technologies have rationalized and can think about reasonably calmly?
9. Severity of consequences: When is the risk from the activity realized in the form of consequences to the user?

It has been shown that unknown risk and fear can be used to account for about 80% of the results generated by using all nine variables that were originally introduced by Fischoff and his colleagues (Johnson and Tversky 1984; Slovic 1987). My colleagues and I have revised the Fischhoff and Slovic model of risk perceptions—introducing ordinal scales to the identified characteristics of risk perceptions, and incorporating the dynamics of perception by including the important and neglected time element (Farahmand, Atallah, and Konsynski 2008).

Affect Heuristic

People from different cultures have different attitudes and *affective valuation* of information technologies. Slovic et al. (2007) defined affect as the specific quality of goodness or badness and explained that people use an affect heuristic to make judgments. That is, representations of objects and events in people's minds are tagged to varying degrees with affect. In the process of making a judgment or decision, people consult or refer to an "affect pool" containing all the positive and negative tags consciously or unconsciously associated with the representations. Using an overall, readily available affective impression can be far easier than weighing the pros and cons or retrieving from memory many relevant examples, especially when the required judgment or decision is complex or mental resources are limited. This characterization of a mental shortcut leads to labeling the use of affect a "heuristic." The affect heuristic also predicts that using time pressure to reduce the opportunity for analytic deliberation should enhance the inverse relationship between perceived benefits and risks.

These imply that people from different cultures judge risk of information technologies not only by what they think about a technology but also by how they feel about it. If their feelings toward digital media are favorable, they tend to judge the risks as low and the benefits as high; if their feelings toward the digital media are unfavorable, they tend to make the opposite judgment—high risk and low benefit. If affect guides perceptions of risk and benefit, then providing information about benefit of digital media should change people's perception of risk and vice versa.

I argue that *privacy decision making by online users from different cultures is colored by affect heuristics*. In the following sections I explain that heuristics and biases are not well defined in normative decision models that are based on theory of rational choice, therefore alternative methods are needed to understand cross cultural differences in decisions about privacy in online environments.

Normative Decision Models and Privacy Decisions

Classical decision theory (von Neumann and Morgenstern 1947; Savage 1954) frame the choice people make in terms of four basic elements:

1. A set of potential actions (A_i) to choose between
2. A set of events or world states (Ej)
3. A set of consequences obtained (C_{ij}) for each combination of action and event
4. A set of probabilities (P_{ij}) for each combination of action and event

According to classical decision theory, the expected value of an action is calculated by weighting its consequences over all events by the probability the event will occur (see Chapter 2). This theory can be applied to privacy decisions in online environments. For example, to address the privacy concerns, an online user may be deciding whether to install misuse detection software in his/her computer at home. Installing or not installing software responds to two actions A_1 and A_2. The expected consequences of either action depend upon whether misuse occurs. Misuse occurring or not occurring corresponds to two events E_1 and E_2. Installing misuse detection software may reduce the consequences (C_{11}) of misuse occurring. As the probability of misuse occurrence increases, use of software seems to be more attractive.

From probability theory, it can be shown that the return to the online user is maximized by selecting the alternative with the greatest expected value. The expected value of an action A_i is calculated by weighting its consequences C_{ik} over all events

k, by the probability P_{ik} the event will occur. The expected value of a given action Ai_i is therefore

$$EV[A_i] = \sum_k P_{ik} C_{ik}$$

More generally, the online user's preference for a given consequence C_{ik} might be defined by a value function $V(C_{ik})$, which transforms consequences into preference values. The preference values are then weighed using the same equation. The expected value of a given action A_i becomes

$$EV[A_i] = \sum_k P_{ik} V(C_{ik})$$

Another advance in utility theory was the extension to outcomes described by multiple conflicting attributes (Kenney and Raiffa 1976). For example, when choosing a privacy protection plan, one needs to consider not only the cost of the plan but also the breadth of the coverage, the quality of the technical support provided by the coverage, and other attributes. Thus, this decision involves evaluating consequences with respect to several conflicting objectives. Should I spend more money to achieve greater coverage or save money but take a risk with lower coverage? The most commonly used multiattribute utility model combines the values of the conflicting attributes according to a weighted additive rule (much like the utility theory for gambles), where the weights reflect the trade-offs among the attributes. The weighted additive rule is considered to be a compensatory rule that allows deficits on one attribute to be compensated by advantages on other attributes.

Using expected utility decision theory and drawing conclusions based on the theory of rational choice could be appropriate in transparent situations (e.g., where the relation of stochastic dominance is transparent) or in describing situations where decision makers have a well-articulated preference and they are familiar and experienced with the preference object. Even in such cases, however, situational factors may intrude (for details, see Bettman, Luce, and Payne 1998).

Tversky and Kahneman (1979) made a key contribution to the field when they showed that many of the previously mentioned discrepancies between human estimates of probability and Bayes' rule could be explained by the use of three heuristics:

1. *Representativeness*: In the representativeness heuristic, the probability that, for example privacy protection plan A is good is assessed by the degree to which he is representative of, or similar to, the stereotype of good privacy protection plans. This approach for estimating probability can lead to serious errors because similarity, or representativeness, is not influenced by several factors that should affect determination of probability.
2. *Availability*: There are situations in which an information online user conceptualizes the frequency of a class or the probability of an event by the ease with which past instances or occurrences can be brought to mind. For example, an online user may assess the risk of disclosure of information among

other users by hearing about such occurrences from one's acquaintances. Availability is a useful clue for assessing frequency or probability, because instances of large classes are usually recalled better and faster than instances of less frequent classes. However, availability is affected by factors other than frequency or probability, e.g., systematic nonreporting or underreporting of system penetrations within an industry. Consequently, the reliance on availability can lead to biases.

3. *Adjustment and anchoring*: In many situations, online users make estimates by starting from an initial value that is adjusted to yield the final answer. The initial value, or starting point, may be suggested by the formulation of the problem, or it may be the result of a partial computation. In either case, adjustments are typically insufficient. That is, different starting points yield different estimates, which are biased toward the initial values.

The notion of heuristics and biases has had a particularly formative influence on decision theory. A substantial body of work with applications in medical judgment and decision making, affirmative action, education, personality assessment, legal decision making, mediation, and policy making has emerged that focuses on applying research on heuristics and biases (Lehto and Buck 2008).

Although economic approaches, based on normative decision-making models, are appropriate models in some cases, but face several challenges—both conceptually and consequently—in describing privacy decision making in online environments. Studies of actual practice do not support the rational trade-offs that normative decision-making models suggest (Spiekermann, Grossklags, and Berendt 2001) and indicate that the traditional rational actor approaches fail to adequately account for every day behavior even within their own terms of reference (Rabin 1998). More importantly, the traditional economic models fail to acknowledge that privacy is a sociocultural practice and in decisions about privacy psychological and social factor seem to interfere with the mathematical principles of neoclassical economics.

Behavioral Economics

The traditional methods of engineering probabilistic risk analysis and expected utility decisions analysis assume that decision makers deal with risks by first calculating and then choosing among the alternative risk-return combinations that are available. These methods, despite all their differences, share a common core: *both rely on the assumption of complete rationality*. However, the results of studies by decision science researchers in the past four decades contrast with the outcomes of these traditional methods, which stem from the work of Daniel Bernoulli and Thomas Bayes in the seventeenth century.

For example, a team of anthropologists and one economist working in 12 countries on 4 continents, undertook a cross-cultural study of behavior in ultimatum, public goods, and dictator games in a range of small-scale societies exhibiting a wide variety of economic and cultural conditions (Henrich et al. 2005). They found that the canonical model—based on self-interest—failed in all of the societies studied. They also found substantially more behavioral variability across social groups than has been found in previous research. Results of a study by Levinson and Peng (2007) also underscore the importance of understanding the influence of cultural background on economic decision making. The results of these studies, along with many other studies, call for incorporating proximate-level decision-making models from behavioral economics, which have increasingly drawn insights on human motivation and reasoning from psychology and neuroscience (Camerer 2004; Sanfey et al. 2003).

Behavioral economics is modification of the traditional economic model that explains psychophysical properties of preference and judgment, which create limits on rational calculation, willpower, and greed (Camerer and Loewenstein 2004; Mullainathan and Thaler 2001). From a methodological perspective, behavioral economics is an approach to economics, which respects the comparative empirical advantages of neighboring social sciences and considers other disciplines (e.g., psychology and decision sciences) as trading partners. The empirical regularity and constructs carefully explored by those neighboring fields provide valuable to meet the requirements of mathematically elegant economic theories which are empirically unmotivated (Camerer and Malmendier 2007).

Future research on cultural factors and information systems will benefit from using decision theories that are able to reflect the role of behavioral biases in framing decision of stakeholders. Among those theories, researchers may use prospect theory developed by Tversky and Kahneman (1979), known to be among the most successful behavioral models of decision under risk (Trepel, Fox, and Poldrack 2005). In prospect theory, decision is modeled in two phases, a framing phase and an evaluation phase. It is in the framing phase that prospective outcomes—influenced by cultural factors—are evaluated as positive or negative deviations (gains or losses) in comparison with a neutral reference outcome. Attitudes toward risk are represented in an S-shaped value function that is concave above the reference point and convex below, indicating that subjective differences decrease as either losses or gains increase. I argue that shift of reference point—under positive affect—in the prospect theory value function may explain risk-taking behavior in electronic activities/technologies.

CONCLUSIONS AND FUTURE RESEARCH

Mental models of individuals are not just the outcomes of their individual cognition but also strongly related to the cultural contexts in which hazards of information systems. These contexts play a major role in the risk-taking behavior of individuals and framing their decisions. Future studies should devote more effort to the role of cultural context in privacy by taking into account how culture influences management of personal information. Different contexts within which users are embedded may lead to different interpretations of privacy and therefore different decisions.

Privacy may be viewed as part of a range of social practice rather than focusing on threats to some online activities (e.g., disclosure of credit card information) (Dourish and Anderson 2006). To have a meaningful understanding of the decision-making process in electronic activities, we need to consider the privacy context of a given transaction. This requires combination of expertise from different disciplines (e.g., computer and information systems, behavioral economics, psychology, sociology).

In this chapter, I provided the example of behavioral economics to understand the cross-cultural differences in decisions about privacy in online environments. I also suggest considering "computational social science"—a field that leverages the capacity to collect and analyze data at a scale that may reveal patterns of individual and group behaviors (Lazer et al. 2009)—for data gathering and analysis and modeling cultural attributes. Existing ways of conceiving human behavior were developed without access to terabytes of data describing real-time interactions and locations of populations and individuals. Computational social science enables us to understand cultural dynamics and cross-cultural differences by blending the behavioral and social sciences with technological fields such as computer science and operations research.

ACKNOWLEDGMENTS

Portions of this work were supported by CERIAS at Purdue University, and by the National Science Foundation under grant no. 0725152.

REFERENCES

Adams, A. 2000. Multimedia information changes the whole privacy ballgame. In *Proceedings of Computers, Freedom, and Privacy*, 25–32. Toronto, Canada: ACM Press.

Al-Ghaith, W. A., L. Sanzogni, and K. Sandhu. 2010. Factors influencing the adoption and usage of online services in Saudi Arabia. *Electron J Inf Syst Dev Ctries* 40:1–32.

Altman, I. 1975. *The Environment and Social Behavior*. Monterey, CA: Brooks/Cole.

Angst, C. M., and R. Agarwal. 2009. Adoption of electronic health records in the presence of privacy concerns: The elaboration likelihood model and individual persuasion. *MIS Q* 33:339–70.

Bellman, S., E. J. Johnson, S. J. Kobrin, and G. L. Lohse. 2004. International differences in information privacy concerns: A global survey of consumers. *Inf Soc* 20:313–24.

Bettman, J. R., M. F. Luce, and J. W. Payne. 1998. Constructive consumer choice processes. *J Consum Res* 25(3):187–217.

Boddy, D., G. King, J. S. Clark, D. Heaney, and F. Mair, F. 2009. The influence of context and process when implementing e-health. *BMC Med Inform Decis Mak* 9:1–9.

Camerer, C., and G. Loewenstein. 2004. Behavioral economics: Past, present, future. In *Advances in Behavioral Economics*, ed. C. Camerer, G. Loewenstein, and M. Rabin, 3–51. New York: Russell Sage Foundation Press and Princeton University Press.

Camerer, C. F., and U. Malmendier. 2007. Behavioral organizational economics. In H. Vartainnen and P. Diamond, *Frontiers of Behavioral Economics*. Princeton: Princeton University Press.

Chia, P. A., S. B. Maynard, and A. B. Ruighaver. 2002. Understanding organizational security culture. In *Pacific Asia Conference on Information Systems*. Taiwan: Association of Information Systems.

Culnan, M. J. 1993. How did they get my name? *MIS Q* 17:341–63.

Dhillon, G. 1995. *Interpreting the Management of Information Systems Security*. London: London School of Economics and Political Science.

Donner, J., and C. A. Tellez. 2008. Mobile banking and economic development: Linking adoption, impact, and use. *Asian J Commun* 18(4):318–32.

Dourish, P., and K. Anderson. 2006. Collective information practice: Exploring privacy and security as social and cultural phenomena. *Hum Comput Interact* 21:319–42.

Farahmand, F., M. Atallah, and B. Konsynski. 2008. Incentives and perceptions of information security risks. In *Proceedings of International Conference on Information Systems (ICIS)*. Paris. Taiwan: Association of Information Systems.
Fischoff, B., P. Slovic, S. Lichtenstein, S. Read, and B. Combs. 1978. How safe is safe enough? A psychometric study of attitudes towards technological risks and benefits? *Policy Sci* 9(2):127–52.
Greenberg, J. A. 1987. Taxonomy of organizational justice theories. *Acad Manage Rev* 12(1):9–22.
Henrich, J., R. Boyd, S. Bowles, C. Camerer, E. Fehr, H. Gintis, R. McElreath et al. 2005. Economic man in cross-cultural perspective: Behavioral experiments in 15 small-scale societies. *Behav Brain Sci* 28:795–855.
Johnson, E. J., and A. Tversky. 1984. Representations of perceptions of risk. *J Exp Psychol* 113:55–70.
Karat, J., C.-M. Karat, E. Bertino, N. Li, Q. Ni, C. Brodie, J. Lobo et al. 2009. Policy framework for security and privacy management. *IBM J Res Dev* 53(2):242–55.
Kenney, R. L., and H. Raiffa. 1976. *Decisions with Multiple Objectives: Preferences and Value Tradeoffs*. New York: John Wiley & Sons.
Laufer, R. S., and M. Wolfe. 1977. Privacy as a concept and a social issue: A multidimensional development theory. *J Soc Issues* 33(3):22–42.
Laukkanen, T., and M. Pasanen. 2008. Mobile banking innovators and early adopters: How they differ from other online users? *J Financial Serv Marketing* 13(2):86–94.
Lazer, D., A. Pentland, L. Adamic, S. Aral, A.-L. Barabási, D. Brewer, N. Christakis et al. 2009. Computational social science. *Science* 23:721–3.
Lehto, M. R., and J. R. Buck. 2008. *Introduction to Human Factors and Ergonomics for Engineers*. Boca Raton, FL: CRC Press.
Levinson, J. D., and K. Peng. 2007. Valuing cultural differences in behavioral economics. *CFAI J Behav Finance* 4:32–47.
Lind, E, A., and T. R. Tyler. 1988. *The Social Psychology of Procedural Justice*. New York: Plenum Press.
MacGregor, R. C., and L. Vrazalic. 2006. E-commerce adoption barriers in small businesses and the differential effects of gender. *J Electron Commer Organizations* 4(2):1–24.
Meijden, M. J., H. J. Tange, J. Troost, and A. Hasman. 2003. Determinants of success of inpatient clinical information systems: A literature review. *J Am Med Inf Assoc* 10(3):235–43.
Milne, G. R., and M. E. Gordon. 1993. Direct mail privacy-efficiency trade-offs within an implied social contract framework. *J Public Policy Mark* 12:206–15.
Moen, A., J. Gregory, and P. F. Brennan. 2007. Cross-cultural factors necessary to enable design of flexible consumer health informatics systems (CHIS). *Int J Med Inform* 76:168–73.
Mullainathan, S., and R. H. Thaler. 2001. Behavioral economics. In *International Encyclopedia of the Social & Behavioral Sciences*, eds. Neil J. Smelser and Paul B. Baltes, 1094–100. Oxford, UK: Pergamon Press.
Nissenbaum, H. 2010. *Privacy in Context, Technology, Policy, and the Integrity of Social Life*. Palo Alto, CA: Stanford University Press.
Palen, L., and P. Dourish. 2003. Unpacking "privacy" for a networked world. *Proceedings of the ACM Conference on Human Factors in Computing Systems CHI*. New York: Association for Computing Machinery.
Pavlou, P. A., and D. Gefen. 2005. Psychological contract violation in online marketplaces: Antecedents, consequences, and moderating role. *Inf Syst Res* 16(4):272–99.
Rabin, M. 1998. Psychology and economics. *J Econ Lit* 36:11–46. Pittsburgh, PA: Journal of Economic Literature.
Ramachandran, S., S. V. Rao, and T. Gols. 2008. Information security cultures of four professions: A comparative study. In *Proceedings of the 41st Hawaii International Conference on System Sciences*. http://www.hicss.hawaii.edu/

Sanfey, A. G., J. K. Rilling, J. A. Aronson, L. E. Nystrom, and J. D. Cohen. 2003. The neural basis of economic decision-making in the ultimatum game. *Science* 300:1755–8.

Savage, L. J. 1954. *The Foundations of Statistics*. New York: John Wiley & Sons.

Siddiqui, H. 2008. Investigation of intention to use e-commerce in the Arab countries: A comparison of self-efficacy, usefulness, culture, gender, and socioeconomic status in Saudi Arabia and the United Arab Emirates. PhD dissertation, Nova Southeastern University.

Slovic, P. 1987. Perceptions of risk. *Science* 236:280–5.

Slovic, P., M. L. Finucane, E. Peters, and D. G. MacGregor. 2007. The affect heuristic. *Eur J Oper Res* 177:1333–52.

Smedley, B. D., A. Y. Stith, and A. R. Nelson. 2003. *Unequal Treatment: Confronting Racial and Ethnic Disparities in Healthcare*. Washington, DC: The National Academy Press.

Smith, H., S. J. Milberg, and S. J. Burke. 1996. Information privacy: Measuring individual's concerns about organizational practices. *MIS Q* 20:167–95.

Spiekermann, S., J. Grossklags, and B. Berendt. 2001. E-privacy in 2nd generation e-commerce: Privacy preferences versus actual behavior. In *Third ACM Conference on Electronic Commerce*, 38–47. New York: Association for Computing Machinery.

Timmons, S. 2001. How does professional culture influence success or failure of IT implementation in health services? In *Organizational Behavior and Organizational Studies in Health Care*, ed. L. Ashburner. Basingstoke, UK: Palgrave.

Tower Group. 2007. U.S. mobile banking forecast: 2007–2012. http://www.ziffdavisenterpriseevents.com/event_files/US_Mobile_Banking_Forecast__2007_-_2012.pdf

Trepel, C., C. R. Fox, and R. A. Poldrack. 2005. Prospect theory on the brain? Toward a cognitive neuroscience of decision under risk. *Cogn Brain Res* 23:34–50.

Tversky, A., and D. Kahneman. 1979. Prospect theory: An analysis of decisions under risk. *Econometrica* 47(2):263–91.

Valdez, R. S., and P. F. Brennan. 2008. Medical informatics. In *eHealth Solutions for Healthcare Disparities*, ed. M. C. Gibbons, 93–108. New York: Springer.

von Neumann, J., and O. Morgenstern. 1947. *Theory of Games and Economic Behavior*. Princeton, NJ: Princeton University Press.

Ware, W. 1974. Records, computers, and the rights of citizens: Report of the secretary's advisory committee on automated personal data systems. DHEW Publication (OS) 73–94. U.S. Dept. of Health, Education and Welfare.

Westin, A. 1967. *Privacy and Freedom*. New York: Athenaeum.

15 Factors Influencing the Decisions and Actions of Pilots and Air Traffic Controllers in Three Plausible NextGen Environments

Kim-Phuong L. Vu, Thomas Z. Strybel, Vernol Battiste, and Walter Johnson

CONTENTS

INTRODUCTION

The word *culture* is often used to refer to shared ways of thinking and behaving among members of any group (Kurosu and Hashizume 2009). Cultures are also made up of subgroups, each with its own unique roles and obligations. Differences can thus arise between subgroups, as each one will have its own shared values, roles, and responsibilities. Cultures can change through time as the roles and responsibilities of the subgroups are altered. In this chapter, we explore differences between two professional groups of human operators in the air traffic management (ATM) system: pilots and controllers. We examine how their interactions change when the current modes of operation are altered such that the pilots and controllers have to share responsibility for separation assurance with each other or with automation.

Air transportation is one of the safest modes of transportation, resulting in only two fatal accidents in 2008 for U.S. air carriers (operating under 14 CFR 121) (National Transportation Safety Board—Aviation Branch 2010) compared to 34,017 motor vehicle fatal crashes (National Highway Traffic Safety Administration 2010) for that same year. Although the topic of safety is a concern to all operators in the ATM system, context and occupational culture are factors that influence decision-making and risk perception of operators. Mearns, Flin, and O'Connor (2001) studied pilots engaging in crew resource management training and noted "Sources of conflict arise where different professional or occupational cultures clash in the view of how the workload should be managed, who should make decisions or how the situation should be resolved" (p. 378). This quote captures the importance of examining the roles and responsibilities of operators within a system. Understanding how differences in professional cultures influence operator decision making and action is even more important when new technologies or procedures are being introduced because these technologies change the roles and responsibilities of the operators employing them.

The impressive safety record of the National Airspace System is achieved in the current-day ATM system through a centralized system that is ground-based and human intensive. The primary operators in this system are those of air transport pilots (ATPs) and air traffic controllers (ATCs). As shown in Table 15.1, the roles and responsibilities, activities, primary displays, and training requirements for the two types of operators are very different. ATCs are responsible for the safe and expeditious movement of traffic. ATCs are trained on flight rules, air traffic management procedures, weather, communications, and operational procedures that allow them to plan the movement of aircraft and transmit instructions to ATPs for carrying out their clearances. Because of the lateral separation precision needed to do their job, ATCs rely on a two-dimensional radar display that is best suited to show aircraft locations in their airspace. ATPs, on the other hand, are responsible for transporting passengers and cargo in a safe and expeditious manner. ATPs process instructions received from the ATC and act on them. The primary flight deck displays are designed for vertical profile planning, fuel management, and meeting flight scheduling requirements (Kerns 1999). ATP training involves basic airmanship, aircraft systems operation, and navigation. As this brief description indicates, the primary operators in the National Airspace System use different procedures to accomplish their tasks, and the type of information and training they are given are not designed to achieve common tasks. Note also that the primary means of communication and coordination is through voice. A formalized language ("standard phraseology") was developed for this purpose, but in busy traffic situations, voice communications may not follow these standard operating procedures.

The Next Generation Air Transportation System (NextGen) is a nationwide transformation of the ATM system in the United States that will modernize the system and increase its capacity (Joint Planning and Development Office 2010). To achieve these goals, NextGen will use advanced technologies such as (1) controller-pilot datalink communications, (2) cockpit displays of traffic, weather, and terrain, (3) conflict

TABLE 15.1
Differences between Current-Day Air Traffic Controllers and Air Transport Pilots in Roles and Responsibilities, Primary Tasks and Primary Displays, and Training Requirements

Air Traffic Controller	Air Transport Pilot
Roles and Responsibilities	**Roles and Responsibilities**
• Responsible for safe and expeditious movement of air traffic • Pivotal role is planning movement of air traffic and transmitting instructions to carry out plan	Managing flight deck systems to • Transport people and cargo in a safe, efficient manner • Role is processing controller advisory information, accepting and acting on controller instructions • Managing relative to environmental conditions
Primary Display Characteristics	**Primary Display Characteristics**
• Two-dimensional display with directional traffic symbols with data tags (speed and altitude) for three-dimensional radar separation procedures and representations of vector solutions to separation and ascending/descending spacing profiles	• Flight management and display systems support lateral and vertical profile planning in all phases of flight, fuel management, system assessment, and flight schedule management
Major Training Topics	**Major Training Topics**
• Basic flight, weather • Flight rules, traffic control procedures and management, and sector-specific information	• Ground school basic flight (Visual Flight Rules, Instrument Flight Rules, Parts 91 & 121) • Basic airmanship, aircraft systems operation, navigation, weather

Source: Adapted from Kerns, K. 1999. Human factors in air traffic control/flight deck integration: Implications of data-link simulation research. In D. J. Garland, J. A. Wise, and V. D. Hopkin (eds.), *Handbook of Aviation Human Factors*, 519–546. Mahwah, NJ: Earlbaum Associates; Pohlman, D. L. and J. D. Fletcher. 1999. Aviation personnel selection and training. In D. J. Garland, J. A. Wise, and V. D. Hopkin (eds). *Handbook of Aviation Human Factors*, 277–308. Mahwah, NJ: Erlbaum Associates.

alerting and resolution tools, and (4) semiautonomous automated agents. It will also involve adopting new procedures and operating concepts such as trajectory-oriented operations and performance-based navigation procedures that will optimize the traffic flow (see, e.g., NASA 2007). NextGen is expected to benefit everyone. The Federal Aviation Administration (FAA 2010) estimates that by 2018, the NextGen program will reduce delays by 21%, save 1.4 billion cumulative gallons of fuel, and eliminate 14 million tons of CO_2 emissions. However, for NextGen to fulfill its promise, its concepts of operation and technologies must be evaluated for their effect on human operator performance, as these operators must interact safely and effectively within the system.

One important cultural change expected from NextGen is distributed decision making. Critical decisions that are currently the responsibility of the ATC will be made at a more local level by the operator with the most dependable information. Coordination between all operators in the system will be facilitated by real-time and shared information regarding traffic, weather, and aircraft intent. Because of the revolutionary changes being brought about by NextGen, how operator performance will change with the adoption of specific concepts and technologies needs to be determined. Current-day ATC performance is measured in terms of safety and efficiency (e.g., Rantanen 2004). Safety metrics include number of incidents and violations (e.g., losses of separation) and accidents (e.g., fatal versus nonfatal); efficiency metrics include average delays, and distance/time traveled per aircraft (e.g., Pierce et al. 2008). ATP performance is often measured by aircraft state (e.g., airspeed, descent rate, and glide slope) (Gawron 2000). However, both ATC and ATP performance is affected by the interaction of system and cognitive factors, and the cognitive constructs must be measured to evaluate how they are affected by the potential NextGen concepts of operation. Two of these critical operator cognitive factors are situation awareness and workload.

Situation awareness can be defined as an operator's "perception of the elements in the environment…the comprehension of their meaning, and the projection of their status in the near future" (Endsley 1995, p. 36). Situation awareness is particularly important to operators who work in a complex system where the environment is constantly changing and where there is a great deal of information to keep track of in order to anticipate future events (Durso and Gronlund 1999). Poor situation awareness can lead to decisions that have fatal consequences. For example, Rodgers, Mogford, and Strauch (2000) examined ATC errors and found that the origins of many of the errors were in failures of situation awareness (see also Jones and Endsley 1996). Although most of these studies are of pilot error in the United States, failures of situation awareness have also been tied to operator error in Eastern cultures, such as India (e.g., Kumar and Malik 2003).

Mental workload refers the cognitive demands placed on an operator relative to the operator's processing capacity (Hart and Staveland 1988). Although situation awareness is considered to be a separate construct from mental workload, it is related to workload (Pierce, Strybel, and Vu 2008): changes in workload can affect situation awareness and vice-versa. Jentsch et al. (1999), for example, searched NASA's Aviation Safety Reporting System (ASRS) database for aviation incidents in which loss of situation awareness was known to be a contributing factor. They identified 221 such cases and found that in 142 of them, the captain was flying the aircraft and in 79 the first officer was flying the aircraft. Jentsch et al. (1999) also found that the captain exhibited lower situation awareness and made more errors when she or he was flying the aircraft during critical periods compared to when the first officer was flying the aircraft. They attributed the captain's lower level of situation awareness to the additional workload produced by flying the plane while simultaneously engaging in critical, decision-making activities. In other words, the captain's capacity to engage in cognitively demanding, decision-making tasks was reduced by the increase in workload associated with flying the plane.

Because NextGen will employ automation to help operators perform tasks that they may not have the capacity to do currently (see Prevot et al. 2009), the effect of automation on operator situation awareness, and workload must be evaluated. The costs and benefits of automation have been explored in aviation and other task domains (see Parasuraman and Wickens 2008). Automation can reduce operator workload and allow the operator to take on additional roles and responsibilities; however, complete automation can also decrease the operators' decision-making capabilities by reducing their situation awareness. It is known that humans do not perform well on vigilance tasks (Mackworth 1969), making them poor monitors of automation, yet, automation often puts the human operator in a supervisory or fail-safe (i.e., back-up) role, in which the operator is essentially performing a vigilance task. Moreover, early work on NextGen automation concepts has shown that attempts to reduce workload may have the opposite effect. For example, in a study of pilot acceptance of automated conflict resolutions in 3x traffic density, Battiste et al. (2008) found that pilots wanted to contact a human ATC to discuss and clarify 30% of the auto-resolutions sent to the flight deck. This number of inquiries would greatly exceed the controllers' capability, especially if put in a back-up mode. These results suggest that automating conflict resolutions for reducing ATC workload could create more workload if pilots are not able to interact with a human operator.

Although, separation responsibility is currently allocated to the controller, in NextGen, pilots can also play a critical role in ATM decision making. Dao et al. (2009) examined the effect of automation level on pilot situation awareness in a laboratory task. Pilots were presented short, three-minute scenarios in which their aircraft was in conflict with another aircraft and were asked to perform a conflict resolution task with or without the support of automation. On trials in which automation was available, pilots were able to modify proposed automated clearances in one block of trials, but only execute the automated clearances in another block. After resolving the conflict, pilots were probed for situation awareness. With the scenario frozen, but all displays still visible and active, pilots were asked questions relating to the scenario (i.e., about past events, current information displayed, or about future states if the scenario was to be resumed). Dao et al. found that situation awareness was lowest in the fully automated condition in which pilots evaluated and executed automated clearances for conflict resolution, regardless of the acceptability of the resolution. Situation awareness was higher when pilots created resolutions to the conflicts on their own or when they interacted with automation to modify initial clearances suggested by automation. These findings suggest that automation tools used to aid decision making in conflict resolution, that is, that leave the pilots in the loop, may be more effective in maintaining pilot situation awareness than the implementation of automated clearances.

In the remainder of the chapter, we provide an overview of a distributed air-ground simulation study that examined changes in performance, workload, and situation awareness as a function of three strategies for traffic separation. These strategies differed in the operator most responsible for separation, either human pilots, human air traffic controllers, or automation, in a high-density traffic environment. It has been shown that current-day air traffic controllers are not likely to be able to manage this level of traffic density (3x) without alleviation (Prevot et al. 2009). Thus, the three concepts were chosen to partially reallocate separation responsibility to pilots or to automation. In concept 1, responsibility was shared between pilots and controllers, with pilots assigned the majority of conflicts. In concepts 2 and 3, responsibility was shared between controllers and automation, with more conflicts allocated to controllers in concept 2 and more to automation in concept 3. It should be noted that these concepts of operation are not necessarily being endorsed by any agency as promising NextGen operational concepts. Rather, they were selected because they alter the decision-making characteristics of the pilots and controllers, and should therefore result in changes in performance, workload, and situation awareness.

All scenarios for the three concepts required the pilots we were testing to accomplish two tasks. First, the pilots had to develop a trajectory modification that would take them around en route weather systems that lay between them and their arrival airport. Second, pilots were required to engage in a spacing task, where they had to follow a designated lead aircraft by 105 seconds at the merge point and to maintain that interval to the runway. Because spacing instructions were given before pilots modified their flight plans to go around weather, these modifications could make it impossible to space behind their originally assigned lead aircraft. Therefore, pilots sometimes needed to make subsequent requests to be resequenced (assigned a new lead aircraft) by the controllers, who were responsible for resequencing the aircraft. Because separation responsibility for experimental aircraft was allocated to pilots in concept 1 and to automation in concepts 2 and 3, we were able to examine how controllers responded to these requests as well as assess the controller's awareness of the aircraft making these requests.

SIMULATION ENVIRONMENT AND OPERATOR TOOLS

The simulation was conducted over the Internet, distributed across four research labs: the Flight Deck Display Laboratory (FDDRL) at NASA Ames Research Center, the Center for Human Factors in Advanced Aeronautics Technologies (CHAAT) at California State University Long Beach (CSULB), the Systems Engineering Research Laboratory (SERL) at California State University–Northridge, and the Human Integrated Systems Engineering Laboratory (HISEL) at Purdue University. Participant controllers were located at CSULB, and participant pilots were located at NASA Ames. Confederate pilots were located at all of the sites and confederate controllers were located at CSULB.

The simulation was run using the Multiple Aircraft Control System (MACS) and Cockpit Situational Display (CSD) software. MACS was developed by the Airspace Operations Laboratory at NASA Ames Research Center (see Prevot 2002), and CSD was developed by the Flight Deck Display Research Laboratory at NASA Ames (see Granada et al. 2005). Both MACS and the CSD can be configured to run current-day or future distributed air-ground ATM operations. Both included algorithms that supported conflict detection and manual resolutions via pilot/controller modification of flight plans. Both the CSD and MACS also incorporated an auto-resolver tool, based upon the work of Erzberger (2006), that would automatically generate a suggested resolution upon request from the pilot or controller. Finally, the conflict detection and auto-resolution capability was the core of the ground-based auto-resolver agent, which autonomously uplinked resolutions to some of the flight decks in concepts 2 and 3.

The simulation environment mimicked Kansas City Air Route Traffic Control Center, Sector 90, and Indianapolis Air Route Traffic Control, Sector 91, but consisted of a larger area of airspace (i.e., combining a high and super-high sector to form a super sector) to load the air traffic controller with three times current day traffic. Surrounding airspace was also included in the scenario (called ghost sectors because they were not being managed by a participant controller but by students and staff working in the CHAAT lab). Aircraft populating the simulation were designated as TFR (trajectory flight rules) or IFR (instrument flight rules). TFR aircraft were never directly managed by the human controller and, depending on the concept of operation being employed, may also have had conflict alerting and resolution tools (concepts 1 and 2) or were capable of interacting with an auto-resolver agent to negotiate flight plan changes independent of a human controller's involvement (concepts 2 and 3). IFR aircraft were always managed by the human controller, had neither onboard conflict alerting nor the capability to interact with the auto-resolver agent. All experimental participants flew simulated desktop stations of TFR aircraft.

The controller's scope mimicked an Air Route Traffic Control Center (ARTCC) radar scope, with the active (controlled) sector highlighted. The controller radar display consisted of traffic and static weather. IFR aircraft were displayed at full brightness while TFR aircraft were low-lighted unless they were in conflict with an IFR aircraft. The weather display was static and depicted 2D Nexrad weather. Controllers were given the following advanced tools: conflict alerting (that alerts controllers about an upcoming loss of separation up to 8 minutes in advance; see Figure 15.1), trial planner (that provides controllers with a method of visually altering aircraft trajectory; when coupled with the conflict probe, allows controllers to

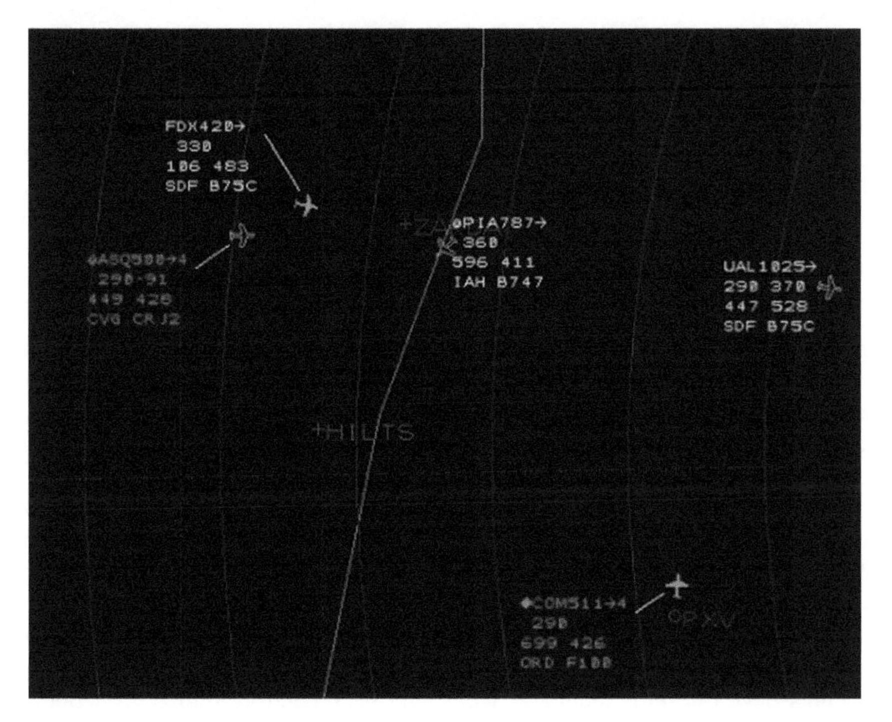

FIGURE 15.1 Illustration of conflict detection tool on the controller's scope. Conflicted aircraft flash red (illustrated as dimmer data tags above) up to 8 minutes prior to loss of separation.

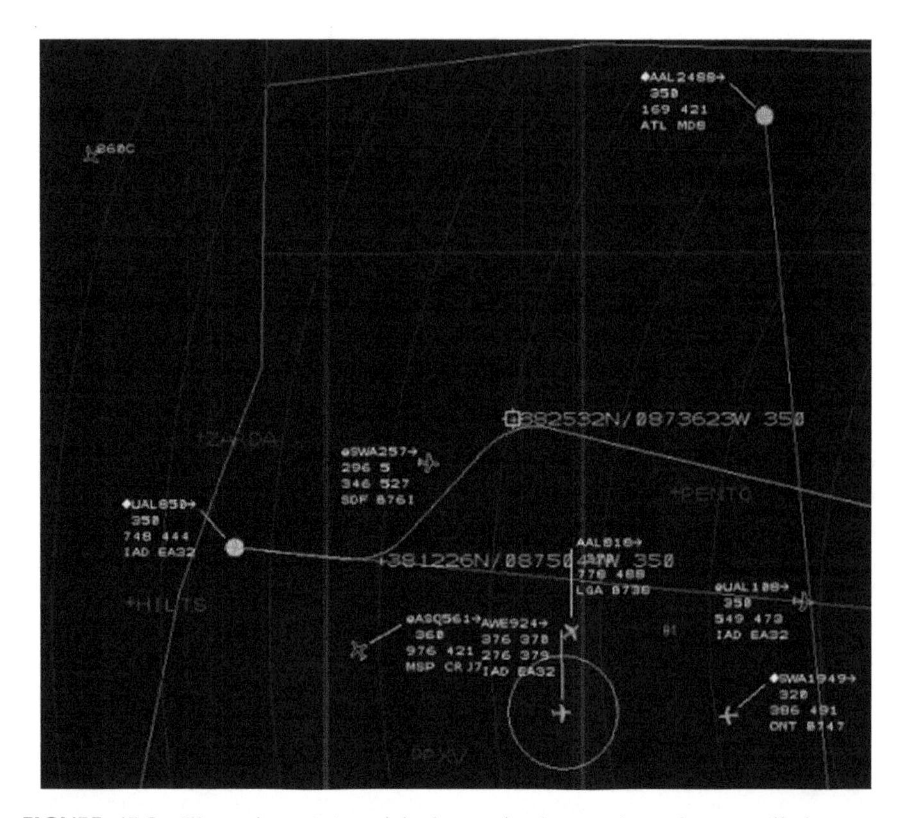

FIGURE 15.2 Illustration of the trial planner implemented on the controller's scope. Controllers can click on a portal (the arrow next to the call sign in the data tag) to bring up the current flight plan trajectory (illustrated for UAL850). The controller can modify the flight plan by clicking on the path to insert a waypoint and drag the waypoint across the scope to make a new path. Although not illustrated, the controller can click on the altitude in the data tag to select alternative altitudes for conflict resolution as well.

create conflict-free trajectory changes within 8 minutes; see Figure 15.2). Controllers were also equipped with an auto-resolver tool that, at the request of the controller, brought up a suggested resolution to the conflict being probed. In concepts 2 and 3 (to be described in the Concepts of Operation section), a proportion of the conflicts was delegated to an auto-resolver agent at the start of the scenario. Once it was delegated responsibility, the auto-resolver agent acted autonomously to resolve conflicts between designated aircraft (see Figure 15.3). Because of the high volume of traffic, several tools for alleviating controller workload were also included: auto-handoff, autofrequency change, no radio check-in for TFR aircraft (IFR aircraft performed radio checked in), and TFR radio contact only when discontinuing spacing or requesting conflict resolution (in an emergency).

Pilots flew a desktop simulator with flight deck controls similar to a Boeing 757 using the MACS interface. Pilots also interacted with the CSD (see Figure 15.4), which is a 3D volumetric display of traffic and weather information. The

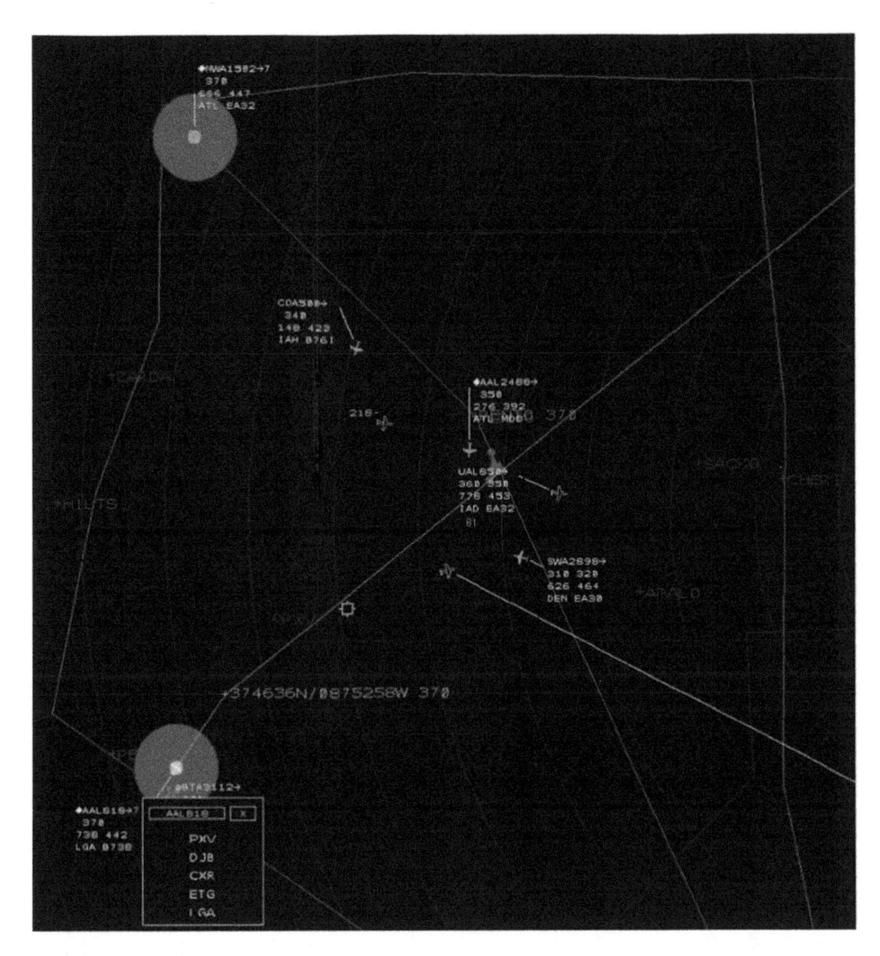

FIGURE 15.3 Illustration of the auto-resolver agent implemented for the ground. The auto-resolver agent detects and sends up a resolution to the flight deck without intervention of a human operator.

CSD provided pilots with the location of ownship and surrounding aircraft. It also allowed the pilots the capability to view the 4D trajectories of ownship and all traffic (Granada et al. 2005). The CSD also provided a route assessment tool (RAT; Figure 15.5), that allowed pilots to manually create flight plan changes around weather. In addition, in concepts 1 and 2 (to be described in the Concepts of Operation section), pilots were given access to advanced conflict detection and resolution tools on the CSD. These included (1) a conflict probe that detected and alerted pilots of an upcoming loss of separation, and which, when coupled with the RAT, allowed pilot-generated resolutions or deconflictions for traffic; and (2) an auto-resolver tool (see Figure 15.6) that at the request of the pilot finds a resolution to the current conflict. When possible, more than one alternative was presented to the pilot (i.e., a lateral versus vertical maneuver).

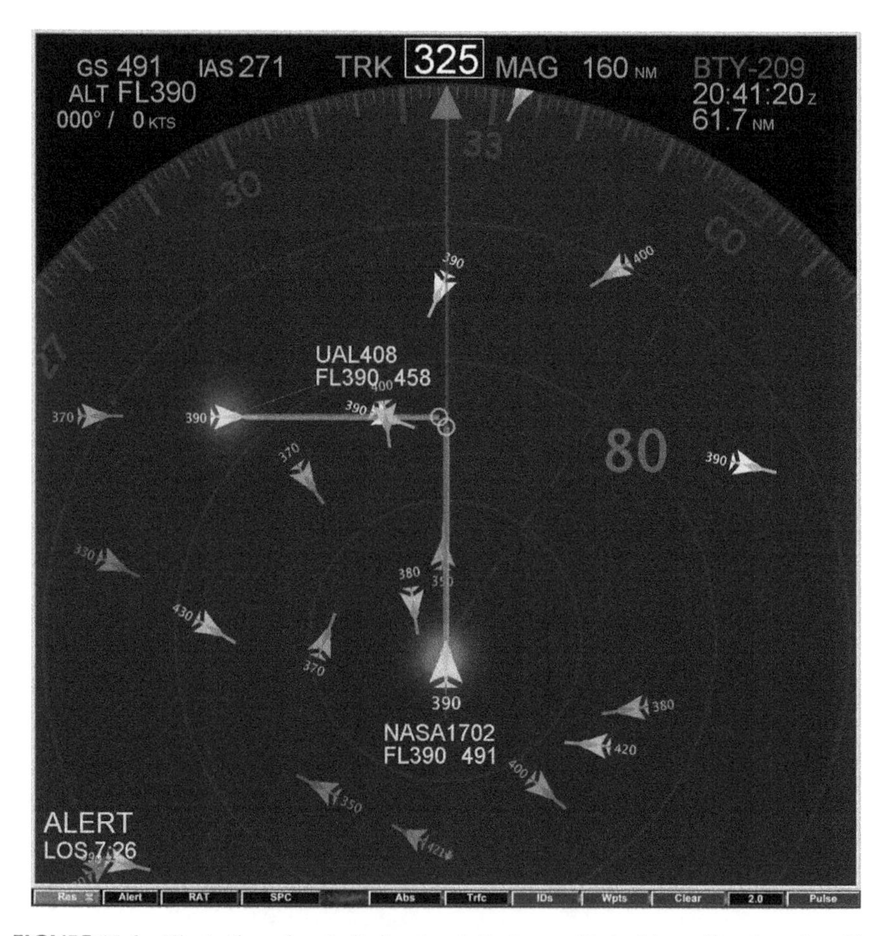

FIGURE 15.4 Illustration of cockpit situational display enabled with conflict detection. The system detects and alerts an upcoming loss of separation within 8 minutes. Conflicts are highlighted in amber (shown as lit data tags for NASA1702 and UAL408 above) and the data tags of conflicting aircraft are expanded.

To assess situation awareness and workload, a separate touch screen tablet computer was used to administer online queries (probe questions to pilots and controllers). Administration of the probe questions followed the situation present awareness method (SPAM) (Durso and Dattel 2004). A ready prompt appeared approximately every 3 minutes. The ready query was accepted by operators when they felt that the workload was reasonable enough for them to be able to read the question (and answer it). The time to accept a ready query can be used as a measure of workload because if the operator is busy at the time of the probe, it should take him or her longer to accept the probe. Following the acceptance of the ready query, a probe question was immediately presented that asked the operator about his/her workload (workload probe) or situation awareness (situation awareness probe). The

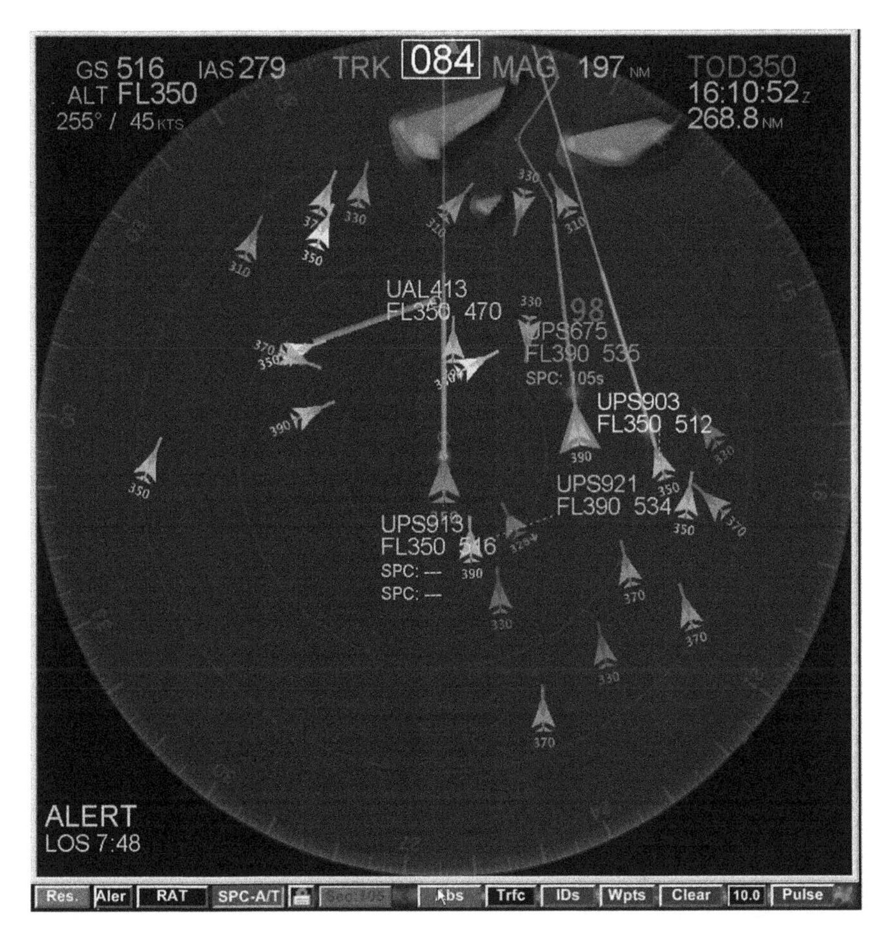

FIGURE 15.5 Illustration of the route assessment tool (RAT), a feature of the cockpit situational display that allows pilots the capability to manually create flight plan changes. When the RAT is used with the auto-resolver tool, the pilot can engage in conflict resolution and weather avoidance.

workload probe asked the operator to "rate your workload" on a scale of 1 (low) to 5 (high). The situation awareness probes varied in content, from information about the conflicts (e.g., will UPS549 be in conflict with another aircraft within the next 5 minutes?), sector status/traffic or effect of weather avoidance (e.g., in which direction did your lead aircraft deviate to avoid weather?), and commands/communications (e.g. was the last command you issued a frequency change?). If the time to accept a ready probe does in fact measure workload, we expect those times to correlate with the workload ratings given for the workload probes. The time to answer the situation awareness probes (as well as accuracy) was our measurement of operator situation awareness, with faster reaction times indicating greater situation awareness.

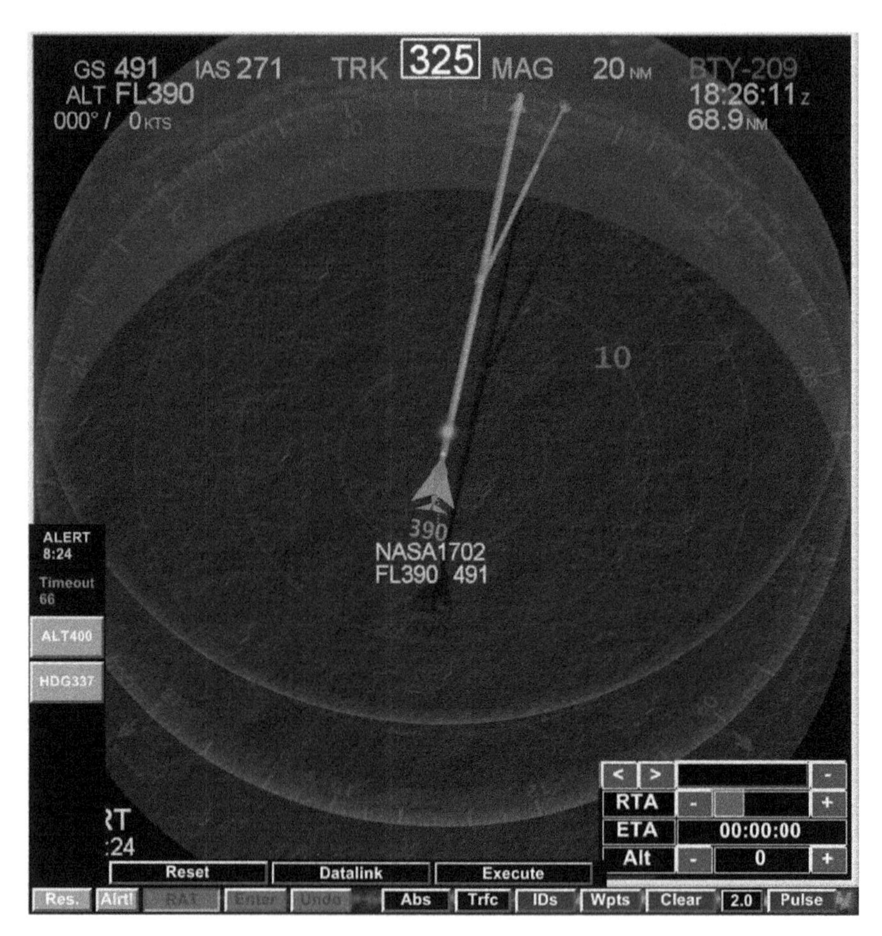

FIGURE 15.6 Illustration of the auto-resolver tool implemented in a cockpit situational display. The tool uses algorithms to find a resolution for conflicting traffic at the request of the pilot. The proposed solutions are graphically displayed to the operator.

CONCEPTS OF OPERATION

In concept 1, pilots were given the primary responsibility for separation assurance. TFR aircraft had on-board conflict alerting, the RAT, and auto-resolver tool. IFR aircraft did not have on-board conflict alerting and could not interact with the auto-resolver tool; the RAT was available for modifying ownship's route for weather avoidance for both IFR and TFR aircraft. TFR pilots were responsible for managing separation between TFR-TFR and TFR-IFR aircrafts (75% of the conflicts). For TFR-TFR conflicts, *rules of the road* were built into the system, where the burdened aircraft (i.e., the aircraft that has responsibility for moving) appeared in amber and the nonburdened aircraft appeared in blue on the CSD tool bar. For TFR-IFR conflicts, pilots of TFR aircraft were instructed to maneuver their aircraft; the IFR pilot was not responsible for resolving conflicts. The human ATC was equipped with conflict alerting, the trial planner, and auto-resolver tool. ATC was responsible for IFR-IFR conflicts and for conflicts with any experimental aircraft on their CDA, giving the experimental arrival aircrafts priority. The auto-resolver agent has no responsibility.

In concept 2, controllers were given the primary responsibility for separation assurance. As in concept 1, TFR aircraft had on-board conflict alerting, the RAT, and the auto-resolver tool. The IFR aircraft did not have on-board conflict alerting and were not able to interact with the auto-resolver tool. IFR pilots were able to use the RAT to modify ownship's route for weather avoidance. TFR pilots were not responsible for managing separation, but because they have on-board alerting, TFR pilots could interact with auto-resolver agent and propose conflict resolutions for consideration by the controller/auto-resolver agent. At the beginning of the trial, the auto-resolver agent was allocated responsibility for managing TFR-TFR conflicts (25% of conflicts). The human ATC was equipped with conflict alerting, trial planner, and the auto-resolver tool. ATC was responsible for TFR-IFR and IFR-IFR conflicts (75% of conflicts). For TFR-IFR conflicts, the controller was instructed to move the IFR aircraft. As in concept 1, ATC was responsible for conflicts with any experimental aircraft on a continuous descent approach (CDA).

In concept 3, the automation was given the primary responsibility for separation assurance. Both TFR and IFR aircraft were not equipped with on-board conflict alerting. However, the RAT tool was available to aid rerouting for weather avoidance. TFR and IFR pilots were not responsible for managing separation. At the beginning of the trial, the auto-resolver agent was allocated responsibility for managing TFR-TFR and TFR-IFR conflicts (75% of conflicts). For TFR-IFR conflicts, the auto-resolver agent was programmed to move the TFR aircraft. The human ATC was equipped with conflict alerting, the trial planner, and auto-resolver tool. ATC was responsible for IFR-IFR conflicts (25% of conflicts). In addition, ATC was responsible for conflicts with any experimental aircraft on their CDA, giving the experimental aircraft arrival priority.

OPERATOR DEMOGRAPHICS AND TASKS

The simulation was run over a two-week period. During each week, eight pilots and two controllers participated as test participants (referred to as experimental pilots/controllers). The remaining pilot and controller duties were carried out by our confederates (mainly students and staff working in the various labs; referred to as pseudo-pilots and ghost controllers). Due to some technical difficulties, though, we were not able to run all the test trials in week 1. Thus, the performance data described are only from the experimental pilots and controllers from the second week. However, because all participants from both weeks were able to experience the variety of tasks and concepts, data described from the post-experimental questionnaires and debriefing consist of comments and ratings from all participants.

For experimental pilots in week 2, the professional position most recently held was captain by five of the pilots and first officer by the remaining three. None of the participants had any prior experience with merging and spacing operations, and only three had flown a CDA. Four of the pilots had between 3000 and 5000 flight hours, and three had over 5000 flight hours. All pilots had glass cockpit experience, with three having over 3000 hours of glass cockpit experience. The experimental controllers were both retired, radar certified, ARTCC controllers. One controller had 34 years of experience, and one had 25 years of civilian center control experience.

Pilots flew and controllers managed traffic in twelve, 90-minute scenarios, each concept of operation appearing in four of the scenarios. Again, we emphasize that these concepts of operation, which allocate separation responsibility across pilots, controllers, and automation, are not necessarily being endorsed by any agency for implementation as NextGen operational concepts. They were selected, however, because they should produce measurable changes in operator performance, workload, and situation awareness. In all scenarios, experimental pilots flew arrival aircraft into Louisville International-Standiford Field (SDF) Airport under one of the three different concepts of operation. At the start of the simulation, they were assigned a lead aircraft and given spacing instructions. Prior to entering sector 90, the first active sector controlled by an experimental controller, pilots encountered convective weather and had to reroute around weather by using the RAT tool. Both controllers and pilots were asked to answer the situation awareness and workload probes.

EFFECT OF OPERATIONAL CONCEPTS ON OPERATOR DECISION MAKING AND ACTION

This section summarizes the findings from the simulation in terms of how the three concepts of operation influenced pilot and controller decision making and action. More details regarding operator performance, workload, and situation awareness under the three concepts of operation can be found in papers by Dao et al. (in press), Ligda et al. (2010), Vu et al. (2010), and Strybel et al. (2010).

PERFORMANCE

Although there were many different performance metrics that could have been analyzed as an indicator of safety (e.g., violations of separation and collisions) and efficiency (e.g., distance traveled and fuel consumption), in this chapter we focus on loss of separation (i.e., aircraft within 5 nautical miles and 1000 feet of altitude) as the safety metric of interest in evaluating pilot and controller performance. The experimental (TFR) pilots lost separation more often when they were responsible for separation of ownship from surrounding traffic in concept 1 than in concept 2 when the human controller was responsible, and in concept 3 when the auto-resolver agent was responsible.

Cultural differences can be seen in how pilots resolved conflicts compared to controllers. Although all operators tended to rely on lateral maneuvers to resolve conflicts (about 60% of conflicts), the ground operators (human controllers and auto-resolver agent) were more likely to use altitude (i.e., vertical) resolutions compared to pilots. One reason for why pilots would prefer to make lateral resolutions is that pilots would typically be flying at their optimal cruise altitude, making them more resistant to deviate from that altitude when given a choice. Controllers, on the other hand, would be more willing to use altitude resolutions when two aircraft are in conflict. Moreover, controllers were the only operators to utilize speed and combinations of lateral and vertical tactics to resolve conflicts. The auto-resolver tool implemented in the simulation did not have the algorithms to perform such resolutions.

Pilot comments in a post-simulation questionnaire and debriefing session yielded qualitative information about factors that influenced their decision making for separation assurance under the different concepts of operation. When given the role for separation, experimental pilots indicated that they needed information regarding the nature of surrounding traffic. This information was provided by the CSD, which aided their decision-making capabilities. Pilots also indicated that they wanted intent information from surrounding aircraft (i.e., their plans to alter a trajectory before actually executing the change). This intent feature was not implemented in the present simulation. Although the sharing of information in real time is one of the features that will be characteristic of the NextGen environments, it is acknowledged that there may be delays in communication due to the transmissions needed to exchange information.

In general, pilots thought all three concepts were workable in theory. Pilots were not apprehensive about taking responsibility for separation. However, they did want ATC involvement as either a backup or a consultant for resolutions, and as a monitor to oversee the traffic. This finding is not unexpected given the pilots' limited experience with resolving traffic conflicts. Currently, pilots are not given the responsibility for resolving conflicts, so this additional responsibility changes how they would respond to conflicts when they arise. Many pilots expressed concerns that the human controller lacked awareness of their aircraft when requesting help (this concern was confirmed when controllers acknowledged that they did not have much awareness of what the experimental aircraft were doing, as they were not given responsibility for TFR aircraft at the onset of the simulation). Pilots also indicated that knowing the goals of the surrounding aircraft was an important factor in their decision making. Although the data tags of the aircraft had information about destination (which allowed pilots to project the general direction where the aircraft was heading and general goals of the aircraft, such as aircraft near their destinations are likely to descend when in conflict), many pilots indicated that additional information about other pilots' goals would have been helpful as well. For example, one pilot indicated that she or he was a cargo pilot and was willing to go through mild weather because passenger comfort is not a concern, while cargo delivery time was a central concern. This type of information could help pilots, especially those who are engaging in self-spacing responsibilities to anticipate maneuvers and preferences of lead and surrounding aircraft.

The human controllers lost separation of aircraft in their sector more often in concept 2, that is, when they were responsible for 75% of the conflicts, than in concepts 1 and 3, when they were responsible for only 25% of the conflicts. Although the number of LOS may appear to be high (i.e., ideally LOS should be 0), we want to emphasize that the traffic volume was extremely high (3x), so these numbers are not unexpected for the types of scenarios tested in this simulation. For example, Prevot et al. (2009) found that once traffic levels exceed 1.5x, controller performance decreases. The controllers exhibited different strategies for resolving conflicts as a function of the concept of operation. In concepts 1 and 3, where they were only responsible for resolving 25% of conflicts, controllers used lateral and vertical tactics at a similar rate. However, in concept 2, where they were responsible for resolving the majority of conflicts, controllers relied on vertical separation strategies more often than lateral ones. At the same time, it should be noted that these findings regarding controllers should be taken with caution, as only two controllers were tested and each one was managing a different sector.

Similar to the pilots, the participant controllers indicated that all three concepts were workable in theory. However, in the post-simulation questionnaire and debriefing, they did identify some factors that influenced their actions and decision making under the different concepts of operations. For example, in concept 1, controllers commented on the lack of experience that experimental pilots had with resolving conflicts. Because pilots are not given this responsibility currently, this culture shift in allocation of separation assurance responsibility caused more workload for the controllers when pilots called in for help. Controllers also expressed concerns about the transitioning of responsibilities. In all concepts, the controllers were not generally responsible for experimental aircraft until they started their descent and were on their CDA. Thus, when TFR aircraft became their responsibility in "emergency" situations (i.e., when conflicts arose), the controllers reported that it was difficult for them to gain an understanding of the situation in order to help the pilots. Indeed, controllers indicated that they did not have much awareness of the TFR aircraft because they were not responsible for them. Thus, having the controller as a backup conflict resolver did not work well in the scenarios tested.

Workload and Situation Awareness

Pilot and controller workload and situation awareness were also measured in the simulation. For pilots, workload measures from four separate metrics (latency to ready prompts, workload ratings on workload probes, post-trial questionnaire and post-simulation questionnaire) all indicated that workload was relatively low and did not differ significantly between concepts (see Ligda et al. 2010 for more details). In contrast, workload measures from the same four metrics indicated that workload was moderate to high, and that it changed as a function of sector properties and concept of operation (see Vu et al. 2010 for more details). The controller for sector 90 showed higher levels of workload compared to the controller for sector 91. This difference is related to sector properties of the simulation. Sector 90 was a larger sector and was affected by the weather reroutings made by pilots; the sector 90 controller was also responsible for resequencing of experimental pilots if they had to discontinue spacing. The top of descent point was also located in sector 90, which is where the pilots begin their CDA. The merge point for the experimental aircraft arriving into SDF was located in sector 91. The differences in workload for the two groups of operators (pilots versus controllers) and within a group of operators (e.g., controllers managing different sectors) illustrate the importance of testing the effects of new concepts and technologies across a range of operators and air traffic situations.

Our situation awareness metrics (latency to situation awareness probes and accuracy to the probe questions) indicate that pilots were more aware of conflicts, surroundings, and communications/commands issued in concept 1, when they were actively engaged in separation responsibility (see Dao et al. [in press] for details). However, as noted above, in concept 1 they actually lost separation more often than in concepts 2 and 3. This indicates that situation awareness, although required for good decision making (Rodgers, Mogford, and Strauch 2000), does not ensure good performance. Controller situation awareness, as measured by our metrics, showed a more complex picture. Because we only have data from two controllers, each managing a different sector, we will not characterize controller situation awareness here. Rather, we note that the two controllers showed different levels of awareness based on the information requirements of the sector and concepts of operation (see Vu et al. 2010 for a descriptive summary of their behavior).

SUMMARY AND FUTURE RESEARCH DIRECTIONS

We started this chapter by noting how cultural differences between professional groups can affect operator decision making and actions. The Situation Awareness in Trajectory Oriented Operations with Weather simulation reported in the chapter examined how the performance of pilots and controllers were affected by three concepts of operations that uniquely allocated separation responsibility between pilots, controllers, and automation. Implementing a strategy where pilots or automation are primarily responsible for maintaining separation, would involve an important cultural shift from the present situation, where controllers are primarily responsible for separation assurance. Both groups of human operators indicated that all concepts tested were workable in theory; however, the data showed that the three concepts produced different effects on workload, and situation awareness. Pilots showed little change in workload across the three concepts, but higher awareness when actively engaged in concept 1. Controllers showed sector-specific changes in awareness and workload depending on concept of operation.

Because the current ATM system allocates separation responsibility to the human controllers, they appear to be a natural backup to pilots and automation when the pilots are not able to find acceptable solutions to conflicts. Pilots used the auto-resolver tool to resolve conflicts when they were responsible in concept 1. However, they had a tendency to revert to controller intervention when their solutions failed (or when the auto-resolver agent did not come up with a resolution in a time frame comfortable for pilots). This finding of reliance on the human controller is consistent with the findings of Battiste et al. (2008), where pilots wanted to coordinate with the controller on 30% of automated resolutions provided to them. However, when pilots called the controllers for help, the controllers were not always aware of the situation and thus unable to help the pilots. Controllers showed little awareness of aircraft for which they were not responsible from the outset, which included the experimental pilots in all concepts tested. Thus, the notion of using the human controller as a backup to automation may not work in real-life settings unless controllers can become more aware of pertinent information regarding these aircraft when emergency situations arise. Controllers did use the auto-resolver tool to bring up potential conflict resolutions, but they used the technology merely as a tool and not as a backup to any decision-making fails.

Although NextGen applies to the ATM system in the United States, similar efforts are being made in other regions of the world (e.g., EUROCONTROL 2011, The Single European Sky program). Thus, the influence of geographical and ethnic differences in decision making should be explored in future studies. Li,

Harris, and Chen (2007) conducted a meta-analysis of human factors mishaps in three regions of the world: Taiwan, India, and the United States. They specifically examined 18 categories of error outlined by the Human Factors Analysis and Classification System and found that national culture was associated with the management style of the crew. Crews in the United States, whose national culture is more individualistic and less authoritarian compared to Indian and Taiwanese culture, resulted in fewer aviation accidents. Li et al. (2007) attributed this finding to the crew members being more comfortable discussing problems/solutions with the captain rather than following orders. Because Li et al. noted that "global aviation is strongly influenced by the United States and Western Europe" (p. 420), it is important to understand the factors that affect human performance among pilot and controller cultures in the United States and in other regions around the world.

Future studies should also examine the leadership role of the individuals to determine the effect that different roles and responsibilities have on decision making, performance, and situation awareness. For example, Thunholm (2009) found that military leaders reported higher levels of naturalistic decision making (e.g., more spontaneous and less rational) compared to their team members and that this style resulted from the fact that the leader should display decisive and action-oriented decision-making qualities. Finally, research should be conducted to examine the effect of new technologies and procedures not only on current day operators, but on future ones as well. The ATC strike of 1981, where the Professional Air Traffic Controllers Organization (PATCO) union declared a strike was considered illegal because it violated a law {(5 U.S.C. (Supp. III 1956) 118p)} that banned strikes by government unions. At that time, U.S. president Ronald Reagan ordered those remaining on strike to return to work within 48 hours or forfeit their jobs. Not all controllers returned to work and then-President Reagan fired the striking controllers. The FAA had to replace these controllers over the following decade, and these replacement controllers are now at or close to retirement age (currently, controllers must be hired prior to their thirty-first birthday and retire by their fifty-sixth birthday).

Another important factor to examine has to do with the subculture of student versus expert controllers. We have conducted a simulation study comparing these two groups, in which we found that highly trained students on a sector did not differ much from actual ATCs on many of the sector performance variables measured in the simulation (Vu et al. 2009). However, there were several differences evident between students and controllers. In particular, students were more compliant with experimental procedures and were more willing to use new technology and procedures (prior knowledge did not prevent them from taking on the new roles). Students also reported that their SA and workload was more negatively affected by scenario difficulty than did controllers. Finally, students relied more on automation and alerting tools. These differences need to be taken in account when evaluating new technologies using current operators as they may not be viewed in the same way by the group that will inherit the system.

NextGen will change how pilots and controllers perform their tasks by incorporating advanced technologies, employing new procedures, and having operators engage in collaborative decision making. Thus it is important to evaluate cultural factors influencing the decisions and actions of NextGen operators with respect to their profession (e.g., pilots versus controllers), their roles and responsibilities, geographic region and culture, and their age group (e.g., cohort effects).

ACKNOWLEDGMENTS

The simulation described in the chapter was supported in part by NASA cooperative agreement NNA06CN30A, *Metrics for Situation Awareness, Workload, and Performance in Separation Assurance Systems* (Walter Johnson, technical monitor). Preparation of this chapter was supported by NASA cooperative agreement NNX09AU66A, *Group 5 University Research Center: Center for the Human Factors in Advanced Aeronautics Technologies* (Brenda Collins, technical monitor).

REFERENCES

Battiste, V., W. Johnson, S. Brandt, A. Q. Dao, and N. H. Johnson. 2008. Assessment of flight crew acceptance of automated resolutions suggestions and manual resolution tools. In *Proceedings of the 26th International Congress of the Aeronautical Sciences (ICAS).* Anchorage, AS: Optimage Ltd.

Dao, A.-Q. V., S. L. Brandt, P. Bacon, J. Kraut, J. Nguyen, K. Minakata, H. Raza, D. Rozovski, and W. W. Johnson. In press. Conflict resolution automation and pilot situation awareness. *HCII 2011, Lecture Notes in Computer Science.*

Dao, A.-Q. V., S. L. Brandt, V. Battiste, K.-P. L. Vu, T. Strybel, and W. Johnson. 2009. The impact of automation assisted aircraft separation on situation awareness. In M. J. Smith and G. Salvendy (eds.), *Human Interface, Part II, HCII 2009, Lecture Notes in Computer Science* 5618:738–47.

Durso, F. T., and A. Dattell. 2004. SPAM: The real-time assessment of SA. In *A Cognitive Approach to Situation Awareness: Theory and Application,* ed. S. Banbury and S. Tremblay, 137–54. Aldershot, UK: Ashgate.

Durso, F. T., and S. D. Gronlund. 1999. Situation awareness. In *Handbook of Applied Cognition,* ed. F. T. Durso, R. S. Nickerson, and R. W. Schvaneveldt, 283–314. New York: John Wiley and Sons.

Endsley, M. R. 1995. Toward a theory of situation awareness in dynamic systems. *Hum Factors* 37(1):32–64.

Erzberger, H. 2006. Automated conflict resolution for air traffic control. *Proceedings of the 25th International Congress of the Aeronautical Sciences (ICAS)*, Germany: Optimage Ltd.

Eurocontrol. 2011. Single Europen Sky. http://www.eurocontrol.int/dossiers/single-european-sky (accessed July 2, 2011).

FAA. 2010. NextGen Implementation Plan, March, 2010 [PDF File]. http://www.faa.gov (accessed June 1, 2010).

Gawron, V. J. 2000. Human peformance measures handbook. Mahwah, New Jersey: LEA Publishers.

Granada, S., A. Q. Dao, D. Wong, W. W. Johnson, and V. Battiste. 2005. Development and integration of a human-centered volumetric cockpit display for distributed air-ground operations. In *Proceedings of the 12th International Symposium on Aviation Psychology*. Oklahoma City, OK: Wright State University.

Hart, S. G., and L. E. Staveland. 1988. Development of NASA TLX (Task Load Index): Results of empirical and theoretical research. In *Human Mental Workload,* ed. P. A. Hancock and N. Meshkahi, 139–83. Amsterdam, Netherlands: North-Holland.

Jentsch, F., J. Barnett, C. A. Bowers, and E. Salas. 1999. Who is flying this plane anyway? What mishaps tell us about crew member role assignment and air crew situation awareness. *Hum Factors* 41:1–14.

Joint Planning and Development Office. 2007. *Concept of operations for the next generation air transportation system version 1.2.* http://www.jpdo.gov/library/NextGenConOpsv12.pdf (retrieved October, 06, 2010).

Jones, D. G., and M. R. Endsley. 1996. Sources of situation awareness errors in aviation. *Aviat Space Environ Med* 67(6):507–12.

Kerns, K. 1999. Human factors in air traffic control/flight deck integration: Implications of data-link simulation research. In D. J. Garland, J. A. Wise, and V. D. Hopkin (eds.), *Handbook of Aviation Human Factors*, 519–46. Mahwah, NJ: Earlbaum.

Kumar, U., and H. Malik. 2003. Analysis of fatal human error aircraft accidents in IAF. *Indian J Aerosp Med* 47:30–6.

Kurosu, M., and A. Hashizume. 2009. Culture and communication behavior: A research based on the artifact development analysis. In M. Kurosu (ed.), *Human Centered Design*, HCII 2009, LNCS 5619, 468–75. Berlin, Germany: Springer-Verlag.

Li, W.-C., D. Harris, and A. Chen. 2007. Eastern minds in western cockpits: Meta-analysis of human factors in mishaps from three nations. *Aviat Space Environ Med* 78:420–5.

Ligda, S. V., A.-Q. V. Dao, T. Z. Strybel, K.-P. L. Vu, V. Battiste, and W. W. Johnson. 2010. Impact of conflict avoidance responsibility allocation on pilot workload in a distributed air traffic management system. In *Proceedings of the Human Factors and Ergonomics Society's 54th Annual Meeting*. San Francisco, CA: Human Factors and Ergonomics Society.

Mackworth, J. F. 1969. *Vigilance and Habituation*. Middlesex, England: Penguin.

Mearns, K., R. Flin, and P. O'Connor. 2001. Sharing 'worlds of risk'; improving communications with crew resource management. *J Risk Res* 4:377–92.

NASA. 2007. Next Generation Air Transportation System White Paper. http://www.aeronautics.nasa.gov/nextgen_whitepaper.htm (accessed June 13, 2010).

National Highway Traffic Safety Administration. 2010. http://www-fars.nhtsa.dot.gov/Main/index.aspx (accessed August 25, 2010).

National Transportation Safety Board—Aviation Branch. 2010. http://www.ntsb.gov/aviation/Stats.htm (accessed August 25, 2010).

Parasuraman, R., and C. D. Wickens. 2008. Humans: Still vital after all these years of automation. *Hum Factors* 3:511–20.

Pierce, R., T. Strybel, and K.-P. L. Vu. 2008. Comparing situation awareness measurement techniques in a low fidelty air traffic control simuluation. In *Proceedings of the 26th International Congress of the Aeronautical Sciences (ICAS)*. Anchorage, AS: Optimage Ltd.

Pierce, R., K.-P. L. Vu, J. Nguyen, and T. Strybel. 2008. The relationship between SPAM, workload, and task performance on a simulated ATC task. In *Proceedings of the Human Factors and Ergonomics Society 52nd Annual Meeting,* 34–8. New York: Human Factors and Ergonomics Society.

Pohlman, D. L., and J. D. Fletcher. 1999. Aviation personnel selection and training. In D. J. Garland, J. A. Wise, and V. D. Hopkin (eds.), *Handbook of Aviation Human Factors*. Mahwah, NJ: Lawrence Erlbaum Associates.

Prevot, T. 2002. Exploring the many perspectives of distributed air traffic management: The Multi Aircraft Control System: MACS. In *International Conference on Human-Computer Interaction in Aeronautics, HCI-Aero 2002*, 23–5. Cambridge, MA: MIT.

Prevot, T., J. Homola, J. Mercer, M. Mainini, and C. Cabrall. 2009. Initial evaluation of air/ground operations with ground-based automated separation assurance. In *Proceedings of the 8th USA/Europe Air Traffic Management Research and Development Seminar.* Napa, CA: Kluwer Publishers.

Rantanen, E. 2004. Development and Validation of Objective Performance and Workload Measures in Air Traffic Control. Technical Report AHFD-04-19/FAA-04-7, Federal Aviation Administration.

Rodgers, M. D., R. H. Mogford, and B. Strauch. 2000. Post hoc assessment of situation awareness in air traffic control incidents and major aircraft accidents. In *Situation Awareness Analysis and Measurement,* ed. M. Endsley and D. Garland, 73–112. Mahwah, NJ: Lawrence Erlbaum Associates.

Strybel, T. Z., K. Minakata, J. Kraut, P. Bacon, V. Battiste, and W. Johnson. 2010. Diagnosticity of an online query technique for measuring pilot situation awareness in NextGen. In *29th Digital Avionics Systems Conference.* Salt Lake City, Utah: IEEE.

Thunholm, P. 2009. Military leaders and followers—do they have different decision styles. *Scand J Psychol* 50:317–24.

Vu, K.-P. L., Minakata, K., Nguyen, J, Kraut, J., & Raza, H. 2009. Situation awareness and performance of students versus experienced air traffic controllers. *Human Interface, Part II, HCII 2009, Lecture Notes in Computer Science*, 5618, 865–74.

Vu, K.-P. L., T. Z. Strybel, J. Kraut, P. Bacon, J. Nguyen, K. Minakata, A. Rottermann, V. Battiste, and W. Johnson. 2010. Pilot versus controller workload and situation awareness under three traffic management concepts. In *29th Digital Avionics Systems Conference.* Salt Lake City, Utah: IEEE.

Section VI

Conclusion

16 Cross-Cultural Decision Making and Action

Issues, Challenges, and Prospects

Robert W. Proctor, Kim-Phuong L. Vu, and Gavriel Salvendy

CONTENTS

> The Washington Post today is full of stories from Egypt ... Both the Egyptians and Americans are apparently craving for that light on the hill—for freedom, and for sustaining the fruits of democratic freedom in today's tough and uncertain times. While I was thinking about the issues of culture and of Obama walking a fine line ostensibly to disrupt the making of that perfect storm which could engulf the whole region and the world, one should pay heed to culture barriers and to Mubarak's reported remark that Obama does not understand the "Egyptian Culture" and how it really plays out there ... And yes, culture barriers do exist, be it a simple question of marriage or international politics.
>
> **R. Piruhit**
> *Petition Writer Blog, Feb. 6, 2011*

INTRODUCTION

This opening statement, which appeared as demonstrations were being held in Egypt demanding that President Mubarak step down, illustrates the importance of understanding the role of cultural factors in decision making and the ensuing actions that are taken. These factors influence everything from the acts of individuals in their personal lives to international policies enacted by countries and businesses, with the choices having potentially long-lasting and far-reaching consequences. As aptly stated by the editors of a special journal section devoted to culture and psychological science,

> Culture, as a key source of sustained experience, intricately affects such phenomena as neural structure and function; memory; decision making; the self; personality; and developmental, group, organizational, and national processes. In turn, such processes reinforce and sustain the very cultural contexts in which they are embedded. (Gelfand and Diener 2010, p. 390)

In the global economy of the modern world, the success of virtually all aspects of businesses, industries, and other organizations depends on consideration of cultural factors and diversity. People's knowledge, practices, and beliefs differ across countries and between various racial, ethnic, religious, and occupational groups within countries. Systems designed for widespread use by, or interactions among, individuals from different cultures must accommodate the varied backgrounds that may affect the users' decisions and actions. Although there are barriers to accommodating cultural differences, there are often significant benefits that arise from doing so (Crisp and Turner 2011). Much is known about decision processes, culture and cognition, design of products and interfaces for human interaction with machines, and organizational processes. Yet, this knowledge is dispersed across several disciplines and research areas, making it difficult to use the knowledge for particular applications.

The goal of this book, and the symposium that led to it, is to cultivate innovations in understanding how cultural differences influence decision making and action by bringing together experts with diverse knowledge and experience. The primary theme motivating this project is that it is essential to adopt a multidisciplinary, multimethod approach (e.g., Eid and Diener 2005) to maximize progress toward solving problems and reaping the benefits associated with cultural influences on decision making and action. This theme is illustrated in several chapters from the book that emphasize the need to incorporate concepts and methods from social psychology (see Chapter 1), industrial management (see Chapter 11), and human factors (see Chapters 12 and 15), among other disciplines, in solving problems associated with international programs for economic development, healthcare safety, and air traffic management systems. The chapters in this book, as well as the invited talks at the symposium, provide coverage of cultural factors in decision making and action from the perspectives of the different authors, with their varied backgrounds and experiences. In this chapter, we seek to integrate the main ideas of the contributors and highlight specific areas in which research and development are needed.

At the conclusion of the symposium, many of the participants took part in a roundtable session that had the goals of organizing and discussing many of the expressed views, identifying emerging issues, and developing a research agenda to advance knowledge in the area. The outgrowth of that session was a white paper, titled "Understanding and Improving Cross-cultural Decision Making in Design and Use of Digital Media: A Research Agenda," which focused on understanding and improving cross-cultural decision making in the design and use of digital media

(Proctor et al. 2011). In this chapter, we take a broader perspective that extends beyond digital media to the cultural factors that operate in decision making and action more generally. However, we restrict our coverage to the issues discussed by participants in the roundtable session, which was structured around a questionnaire sent in advance to the invited speakers. The questions from that survey provide a convenient framework around which to organize the issues. Therefore, we will describe the questions and summarize some of the answers provided to them during the course of the roundtable session, in the symposium talks, or in the authors' contributions to this book. More detailed treatments of the issues are included in many of the book's chapters that we reference and in the white paper.

IMPACT OF CULTURE ON DECISION MAKING

The first question focused on determining the impact of culture on decision making in general and in the areas of management, health care, and technology. From the prior chapters in this volume, it is apparent that the contributors are in accord that culture exerts a strong influence on all aspects of decision making and action. Beliefs and values held within various cultures influence, among other things, risk perception and preferences (Chapter 1), affective reactions to events (Chapter 14), interaction styles among people (Chapter 8), and interactions between people and technology (Chapters 9 and 15). As noted by Tong and Chiu (Chapter 4), culture provides guidance on decision making and action when complex coordination problems are involved and when an individual needs to draw on collective wisdom to address her or his personal needs for physical, cognitive, social, and existential security.

In general, which of several problem-solving strategies will be adopted in specific situations varies across cultures, resulting in individuals and groups adopting different mentalities that ultimately lead to choices between distinct courses of action (Strohschneider 2008). One need only consider the decisions and actions of people from different countries following natural disasters to appreciate the extent to which culture plays a role. More broadly, culture is an important part of an individual's self-concept, influencing almost everything that she or he does (Spencer-Rodgers et al. 2009). Cultural differences can also make cross-cultural communication and collaboration difficult (Chapters 1 and 12), due to distinct assumptions and objectives of people from different cultures. Knowledge of the ways in which individuals from various cultures think and approach problems is essential for situations in which people from different cultures must work together.

MANAGEMENT

Failure to take into account adequately the culture in which a product, program, or system will be implemented may lead to unanticipated problems and failures (Chapters 1 and 5). For example, Rouse asks the question of why targeted populations often are unable to take advantage of programs and ventures intended to stimulate economic development in their region. He provides answers to this question by drawing upon his personal experiences regarding how cultural and situational factors affect people's reception of entrepreneurial opportunities and training programs, and the extent to which knowledge from research in the social sciences can predict or explain these outcomes. Rouse indicates a variety of cultural and situational factors that may influence the likelihood of success of a program and advocates use of systematic modeling that takes into account these factors. He emphasizes that near-term situational changes need to be balanced with longer-term cultural changes. A major point is that models can clarify the desired behaviors and goals that are to be achieved, reveal the possible barriers to success, and identify the courses of action that are most likely to overcome those barriers and increase the probability of the desired sets of behaviors.

Another approach for developing successful international programs is to rely on best practices and lessons learned from prior ventures. Upton (Chapter 6) provides two case studies of how new programs were successfully implemented in urban and rural areas of India. The first example involves farmers in India who were transitioned from an ancient farming model to a more modern one by introducing systems that reflected existing communal structures. The second example describes how culturally based decisions were made for urban and rural communities when developing the largest school meals program in the world. The primary lesson learned from these examples that Upton provides is that culturally sensitive decision making can allow people to benefit from advantages of modern technology within a context of their cultural traditions.

In his talk at the symposium, John Edwardson (2010), chairman and CEO of CDW, Inc., described the best practices that have helped make his company a leading global provider of technological products and services for business, government, and education. Edwardson emphasized his use of the scientific method for developing and evaluating programs. Another best practice advocated by Edwardson is to promote diversity within the company culture at every level and to implement diversity initiatives in a manner that promotes self-actualization (Maslow 1954) of the employees (referred to as coworkers at CDW). He noted that cultural diversity can be measured and that diversity increases the likelihood that cultural differences will be respected within the organization and incorporated appropriately when marketing or designing products for persons from various cultural groups. In doing so, people will fully appreciate the contribution of taking a diverse approach to decision making and actions within the organization.

In Chapter 7, de Bedout provides another example of how a company can successfully expand into the global market to cultivate talent around the world that can result in development of innovative products. As global technology director of the Electrical Technologies and Systems division of GE Global Research, de Bedout underscores that there is increasing interest in developing product lines locally to serve specific needs of the local markets. As an example he describes development of GE Healthcare's portable ultrasound product for the rural market in China. He points out that although the initial rationale was to target the rural China market, the portable ultrasound device is now widely used in developed regions for emergency response teams. There are many benefits to global research and development teams, including access to the most talented individuals within a region and having technology development teams with awareness of global market trends. However, de Bedout also indicates that there are many challenges.

Of the five dimensions that Hofstede (2001) identified for distinguishing cultures, de Bedout indicates that two, power distance (hierarchical versus distributed power relations) and individualism (strong versus weak relationships between individuals), have seemed to be most relevant to differences in development of GE's international research and development centers. He describes the best practices in which organizations should engage for developing global teams in terms of three successive phases: (1) building credibility, (2) building ownership, and (3) providing thought leadership. Of GE's global sites in Munich, Shanghai, and Bangalore, de Bedout points out that the progression through the phases was fastest in Munich, in part because the lower power hierarchy in Germany allowed employees to approach their manager more directly when encountering problems. Knowing that the Chinese and Indian cultures have a stronger power hierarchy, extra steps needed to be taken to encourage their employees to be more open in terms of communication so that problems and issues could be dealt with earlier rather than later. Knowledge of how power hierarchy influence employees' decisions and actions can be applied to the development of new research and development (R&D) teams in various countries around the world.

Health Care

Health care is an exceedingly complex system involving many stakeholders, concerns about cost, and other factors (e.g., Perry and Wears 2011). Thus, many of the issues described in the preceding section apply to healthcare systems as well. Villa and Bellomo (Chapter 11) describe the levels of health care that are provided in the territorial healthcare network of Italy and demonstrate how these levels determine the services and expectations of patients receiving the services. They highlight the need to systematically evaluate the costs and services associated with service centers in the network in order to justify decisions about investments in the network. These authors advocate application of industrial management methods and illustrate how those methods can be used for service planning and control in the healthcare sector. They note that the methods provide a good assessment tool for evaluating the cost effectiveness of healthcare systems, as long as the needs of the end users are given appropriate consideration.

As the example provided by Villa and Bellomo (Chapter 11) indicates, in the area of health care, decisions are made at many levels. These include not only the organizational levels that they emphasize, but also the individual doctors and healthcare workers (van Velden, Severens, and Novak 2005), along with the patients and their families. Cultural factors should be examined and considered at each of these levels. The prevalence of medical errors relating to poor decision making of healthcare professionals has led to considerable research being conducted with the goal of reducing the errors (e.g., Alonso et al. 2006). Less attention has been paid to cultural factors that influence the healthcare professionals' decisions and actions, including societal and professional differences, as well as ethnic and national differences. Also, patients and their families regularly make decisions about whether to seek treatment, what treatment or intervention to consider, and whether to comply with a physician's recommendations, and this process needs to be understood as well (e.g., Fraenkel 2011).

In Chapter 10, Flynn, Khan, Klassen, and Schneiderhan emphasize the decisions made by individual healthcare workers and patients take place within a social context. As such, they draw upon what is known about decision making from the social science literature and apply that knowledge to healthcare decisions. Flynn et al. take the position that the theoretical foundations provided by social science research need to be given greater consideration in the field of healthcare decision making, and that researchers need to integrate the empirical database regarding healthcare decisions with those theories. They underscore the need to develop a better understanding of the explicit deliberative processes in which people engage when making healthcare decisions. Because choices are also influenced by implicit processes (e.g., Marquardt and Hoeger 2009), many of which may be culturally based, it is important as well to understand the influence of those processes (Chapter 4).

With the ongoing interest in reducing medical errors, recent years have seen increasing attempts to perform human factors evaluations in medical settings (Carayon 2007). In Chapter 12, Carayon and Xie point out that there is a significant communication problem because members of the human factors culture hold beliefs and values that are different from those held by members of the healthcare culture. They describe four values of the human factors approach, which include thinking in terms of systems, viewing people as at the center of the system, focusing on continuous improvement of human aspects of the system, and balancing multiple objectives. They attribute four different values to the healthcare culture: integrating scientific evidence with clinical expertise; responsibility of the individual healthcare worker for patient care; autonomy of the individual worker to exercise decision making and control; and striving toward excellence. Carayon and Xie see a need to train people in both human factors and health care to understand the alternative approaches, including their rationales and benefits. Bridging of research and practice in human factors is a general concern that extends beyond the healthcare field and that requires exposure to multidisciplinary perspectives (Proctor and Vu 2011).

In Chapter 13, Zafar and Lehto describe the most significant barrier for the introduction of new technology into the healthcare system as cultural. They present a model that attributes the acceptance of new technology to three factors: (1) the shared values and beliefs of the stakeholders' cultures; (2) consideration of cultural constraints involving the design of technology in such a way that it appeals to the stakeholders and corresponds with the way in which they actually use it; (3) designing the technology to be usable. Zafar and Lehto propose solutions that can improve the effectiveness of healthcare technology and its acceptance, providing a case study concerning outfitting providers with mobile devices for data access, data entry, and interpersonnel communication.

Technology

Technology plays an important role in everyday life and interacts with the cultural factors influencing decisions made at all levels. In this section, we discuss how cultural differences influence a variety of interactions that involve both people and technology. These include cultural differences in preferred modes of interaction with other people in collaboration support systems (Chapter 8) and with social robots intended to assist and support people (Chapter 9). We also discuss how technology influences risk perception for online transactions (Chapter 14) and the communication protocols between pilots and air traffic controllers in potential future flight environments (Chapter 15).

In Chapter 8, Nof considers how collaboration, which by its nature involves cultural factors, influences decision making. He points out that the networked environment has opened the way to a variety of collaborative strategies that were not possible before (see, e.g., Coffey, Hoffman, and Novak 2011). These strategies include, among others, effective task presentation and delegation, information filtering, and decision and collaboration support. He enunciates the benefits and costs of collaborative e-work and describes ways to measure the influence of cultural factors on decision making based on collaborative control theory. Cultural factors play a role in expectations concerning availability, dependability, integrability, viability, autonomy, and agility. Nof's recommendation to apply collaborative control theory to analysis of collaborative e-work is analogous to Villa and Bellomo's (Chapter 11) recommendation to apply the methods of industrial management to the evaluation of healthcare systems.

In Chapter 9, Rau and Li focus on cultural differences in self construal, or views of self, and communication style that influence individual and group decision making. They indicate that in interactions with social robots, trust is a critical requirement. Rau and Li's main proposition is that culturally consistent trust-warning signs need to be implemented in the robots in order for people to place much trust in them. They illustrate in two experiments the different reactions that users in China have from users in Germany and the United States in terms of their preference for an indirect versus direct communication style with the robot.

Privacy is a significant concern for online transactions. In Chapter 14, Farahmand makes the argument that cultural differences are of consequence with regard to privacy decisions made in online environments. He notes that in order to understand how users make decisions about how they will use various information technologies, it is necessary to understand how the users perceive risks in relation to the perceived benefits of those technologies. Because both perceived risk and benefit are subjective, they are influenced by a variety of factors, including culture (e.g., Lane and Meeker 2011). Security is necessary to ensure privacy, and the term "security culture" is used to refer to the attitudes that users within an organization have toward security (Kuusisto and Kuusisto 2009). Farahmand points out that different stakeholders within an organization (e.g., designers versus marketers) may have distinct security cultures, which means that the attitudes toward security measures may vary across different stakeholders.

In Chapter 15, Vu et al. describe the decision-making cultures of pilots and air traffic controllers in terms of their roles and responsibilities. They note that there is likely to be a shift made to the existing structure of controller-pilot communication and responsibility as a function of the next-generation air transportation system (NextGen). In their chapter, they illustrate how changing the roles and responsibilities of pilots and controllers by allocating changing the task and separating aircraft across pilots, controllers, and automation will affect operator awareness and workload. In a simulation, Vu et al. found that pilots welcomed the added responsibility of maintaining self-separation from other aircraft. However, the pilots want the controller to be a back-up/consultant to their decision making. Similarly, when automation fails, it is assumed that controllers will be able to resolve the conflicts. Yet, under high traffic loads and high levels of automation, controllers showed no awareness of air traffic for which they were not initially responsible.

STATUS OF THEORETICAL FOUNDATION FOR CULTURAL DECISION MAKING

The second question related to the state of knowledge in identifying the roles of cultural factors in decision making and what factors are lacking in the theoretical foundation of decision making to enable better incorporation of cultural factors. Traditionally, models of decision making have been static, in that they have not captured the time course of decision making or the wavering between alternatives over time that often occurs when the decision process is extended. Decision-making models have begun to incorporate dynamic processing aspects (e.g., Chapters 2 and 3), but further development is needed. In Chapter 2, Matthew and Busemeyer point out that current versions of expected utility models can accommodate some differences in decision making between East Asians and Westerners through disparities in subjective decision weights and utilities, but they cannot explain other differences such as those associated with holistic versus analytic information-processing styles. The authors stress the importance of applying models of decision making that depict the underlying cognitive processes. They propose more specifically that multiattribute decision field theory (MDFT), a dynamic approach to theorizing about the psychological processes involved in decision making, provides the best model for formally depicting cultural influences on decision making. In this model, information accumulates through a sequential sampling process toward thresholds for the choice alternatives, with a decision made when a threshold is reached. The probability of attending to an attribute or relation among attributes at each step in the process varies.

In Chapter 2, Matthew and Busemeyer indicate that several of the parameters in MDFT could be influenced by culture. They also demonstrate that MDFT makes unique predictions with regard to the effects of context on persons from East Asian and Western cultures, who adopt relatively holistic and analytic decision-making strategies, respectively. Matthew and Busemeyer establish that the model predicts differences in magnitude of several specific context effects and implicates different mechanisms as contributing to the effects.

Another dynamic decision-making model is instance-based theory (IBL) favored by Gonzalez and Martin (Chapter 3). The IBL model stores instances of actions and their payoffs in memory. Choices among alternatives in the future depend on the value of the payoff associated with the outcome of an action and the probability of that outcome being retrieved from memory, which is an increasing function of the number of prior instances. Gonzalez and Martin also advocate an experimental approach to understanding decision-making processes. The value of such an approach is that the researcher gains control over many variables, which can be systematically manipulated and measured. However, as indicated by the authors, there is difficulty in studying cultural influences on decision making in laboratory contexts. To overcome this difficulty, Gonzalez and Martin use generic decision-making tasks and a context-specific computer game based on historical events. Through a study examining the role of group identification in decision making within the microworld game, PeaceMaker, set in the Israel-Palestine conflict, they show how participants' biases vary as a function of the cultural roles they take on in the game.

Time is important not only from a theoretical perspective, but also from an applied perspective. Many decisions and products in industry are time-dependent (Edwardson 2010). Time is critical as well for consumer-related services such as providing help to customers (Chapter 5), for making medical decisions (Chapter 13), and in other business-related contexts. Consequently, decisions in many time-constrained circumstances will benefit from technology that speeds up the process for appropriate decisions/communications to be made. However, as Balasubramanian emphasizes, cultural differences exist in time perception and expectations of what will be accomplished in a given amount of time, and these differences need to be taken into consideration.

Decision dynamics associated with time need to be better understood in a variety of other contexts. For example, understanding the ways in which cultures and situations interact over time to affect attitudes such as those involved in risk is needed. With respect to team decision making, knowledge of how the team structure and communication style influence the decision-making dynamics is a critical component. Within the context of collaborative e-work, it is even more essential to understand the dynamics of the decision-making process due to the lack of face-to-face communication among the collaborators. Because team members often work in parallel, it is crucial that they receive information in a timely manner to prevent duplication of work and miscommunication of goals or deadlines, and that all team members have adequate time to respond to queries and comments on deliverables before a deadline (van Tilburg and Briggs 2005).

In cross-cultural decision-making research, a current focus is on differences between populations with different shared knowledge traditions. In Chapter 4, Tong and Chiu describe results showing that, despite large differences between the American and Chinese cultures in drawing spontaneous trait inferences, the psychological processes that contribute to these inferences are the same in the two cultures. Their general point is that the underlying cognitive processes of different cultural groups are similar, although the relative contribution of different facets differs across cultures. An implication of Tong and Chiu's view is that there will be both individual and situational variations within cultural groups. A potentially valuable direction to take is to expand the current research agenda to consider how people in a particular culture react to the traditions of another culture. For example, Sam and Berry (2010) report that contact between people from different cultures results in cultural and psychological changes to members of both cultures. Understanding in detail the dynamics of such changes is important to forecasting the long-term consequences of the intercultural contact. Another possible direction is to expand research that has focused on national cultural differences (e.g., Dinev et al. 2009) to include differences and interactions among distinct functional groups within the national cultures (e.g., Jeong, Salvendy, and Proctor, in press).

In Chapter 5, Balasubramanian describes various beliefs and values that play a significant role in decision making and are critical to the success or failure of implemented systems, using examples of help-desk system design and healthcare delivery process design for customers in India and the United States. In the help-desk example, he makes the point that Western cultures tend to value quick service such that the wait time and time spent resolving the problem are minimized. In India, in contrast, help-desk callers do not mind a long wait as long as the quality of the service they receive when it is their turn is high. For Indian users, quality of service extends beyond the specific question they had in mind when contacting the help desk and tends to include additional services that reflect the operator's knowledge of the case and the beneficial services that the operator can offer to the caller. Thus, the cultural values held by the caller will determine their judgments of the quality of service they receive. Conversely, the understanding that the operator has of the culture from which the caller is coming will determine what type of services should be provided and how they should be provided.

Although progress is being made in this area of cross-cultural decision making and several success stories have emerged from taking a cultural perspective on decision making, not enough is currently known about decision making as a practical process in which individuals in many organizations from different disciplines across the world engage. In particular, it is important to understand how cultural values may affect the practical decisions that leaders make regarding changes in systems and processes that may have wide-ranging effects inside and outside of the organization.

CONTEMPORARY ISSUES IN CROSS-CULTURAL DECISION MAKING

The final question area related to contemporary issues in cross-cultural decision making, including identification of barriers and potential solutions for overcoming them. This topic was the focus of the previously published white paper (Proctor et al. 2011), and readers are referred to it for a more detailed treatment of these issues within the context of design and use of digital media. Here, we provide a general overview of challenges and prospects for improving the quality of decisions and actions from a cultural perspective.

One barrier for addressing contemporary issues concerning the role of culture in decision making is that much of the research has not included culture as a variable of concern (see Weber and Hsee 2000, for a detailed assessment of this matter as of that date). As such, many models of decision making do not include a cultural component. Of the research that has considered cultural differences, for the most part the studies have compared only two distinct cultures, typically one from the East and one from the West. Exactly how culture should be incorporated into models of decision making is an issue that continues to need more attention.

Weber and Morris (2010) recently summarized a number of different influences that culture has on the information processing involved in decision making. These influences include the features of the situation to which attention is directed, which particular judgment or action schemas (i.e., knowledge structures) are activated, and what decision strategies are employed. Weber and Morris also indicated several societal structures as being important, including the normative patterns actions and responses, prevailing forms of interpersonal interaction, the types of social networks, the cultural frameworks adopted and expressed by organizations and institutions. The ways in which the models incorporate these cultural factors need to be validated, which requires valid and reliable measures for capturing the behaviors and processes of interest.

Examples of cultural factors determining which action schemas are activated and thus influence decisions can be seen in population stereotypes that are known to exist in the operation of controls for producing changes to a display or system (Proctor and Vu 2010). As implied by the term "population stereotypes," these actions are learned from experience and therefore will vary across groups of individuals who have different experiences. A well-known example is operation of light switches, which are flipped upward to turn on a light in the United States but downward to do so in the United Kingdom. The training associated with a professional culture may also influence actions. Hoffmann (1997) had engineering students and psychology students select whether to rotate a knob in the clockwise or counterclockwise direction to increase the value of a linear display indicator. The stereotypical action for engineering students was to rotate the knob in the direction corresponding to the direction of movement of the side of knob nearest to the display (known as Warrick's principle), whereas the stereotypical action for psychology students was to rotate the knob clockwise regardless of its placement relative to the display. Hoffman suggested that this difference in stereotypical actions for the two groups was due to the engineers basing their decisions on mechanical principles with which they were familiar but with which the psychology students were not. Thus, a person's expectancies about what actions to take in a situation, and what the outcomes will be, depends on many factors, including the person's national and ethnic background, as well as their professional knowledge and experience.

With respect to contemporary issues, it is essential that a more principled understanding of the relationships between cultures, situations, and attitudes be developed (e.g., Oetzel, Garcia, and Ting-Toomey 2008). It is clear that broad cultural distinctions do not adequately capture the characteristics of decision making in a variety of contexts (Chapter 4). Also, as Flynn et al. (Chapter 10) emphasized with respect to healthcare issues, compilations of empirical results are not sufficient; theoretical understanding is necessary. Issues related to how people react to cultures that are different from their own in different social, regional, and technical environments are also important. How to manage the cultural conflicts that emerge when different cultures interact is in need of more research.

Unification of research on decisions from different domains, including applied experience, cultural studies, experimental investigations, and organizational analyses, is needed. Such unification will allow a better understanding of the dynamic, bidirectional influence between an individual and his or her environment, and the way that feedback about past actions informs future decisions. It is also essential to develop more concrete cognitive and social models of how people learn from their experiences. These models can suggest ways for people to improve decision making, such as through public interventions and improvement of organizational communication processes. Rather than assume an optimal mode of decision making, these models should conform to the natural way that people incorporate personal cultural experiences into their cognitive judgments and social interactions.

Another barrier to understanding cultural issues in organizational decision making is that oftentimes the decision-making process is not documented for the public or for researchers to study. As such, the best practices that have been revealed by case studies such as those provided by Upton (Chapter 6), Balasubramanian (Chapter 5), and de Bedout (Chapter 7), are not known and put into practice. Although we realize that keeping such information proprietary can give a company an advantage over its competitors, in the long run it stunts the development of the entire enterprise. The lack of data in the decision-making process of organizations as a function of successful failed projects also makes it difficult to verify whether particular processes are indeed effective.

Even if cultural factors have been taken into account adequately in the design of products and systems, cultural issues must be addressed in the implementation and marketing of products and services. With regard to implementation, care must be taken to ensure that the technology being provided does not impose a threat to cultural traditions of the area. The introduction of any technology will lead to different groups of users/adopters, and this differentiation of users can cause conflicts within a group. As Rouse (Chapter 1) notes, the conflicts caused by differences in viewpoints can lead to in-group/out-group distinctions, and formation of new groups. The danger of these group formations is that decisions may be driven by the characteristics that most saliently distinguish members of the groups, rather than more rational evaluation of the relevant dimensions.

With regard to the marketing group, its culture needs to be distinguished from that of research and development (Proctor et al. 2011). These two cultures need to be able to communicate effectively to promote the products and services. However, it is up to the development team to ensure that quality of the products and services is not being lost in the marketing itself and that marketing is not driving the research. That is, although emphasis should be placed on features that make the product or service effective, user preferences also need to be taken into account. However, those preferences need to be evaluated in the overall research and development scheme. It is the responsibility of the research team to provide the marketing team with information regarding the unique features of the product or service, and it is up to the marketing team to make sure that those features get emphasized. Moreover, marketers should take caution not to offend potential users by promoting using irrelevant information in the promotion of the product.

One general step to overcoming cultural barriers is to construct a broader social/cognitive framework to understand the social and psychological contexts of cultural processes, for example, how culture evolves to meet societal and psychological needs. Another step is to make the cultural characteristics more explicit. In many contexts, such as that of healthcare redesign and transformation, there is little discussion of the role of culture. With regard to health care, many human factors specialists assume that human factors is a tool that can be applied directly to redesign of the healthcare systems and processes, and therefore, improve quality of care and patient safety. However, this view is too simplistic because the tools come with a set of values that may conflict with the values of the healthcare culture (Chapter 12).

For example, checklists have been shown to be of value in reducing human error in many areas, including aviation. Knowing this benefit of using checklists, human factors professionals may want to implement this type of procedure in medical settings. Intuitively, this is a good idea because medical procedures are well documented and could easily be represented in a checklist format. However, complications arise in medical settings that may nullify some of the effectiveness of checklists. These include emergencies that arise on the operation table when an emergency occurs that could not have been anticipated in the checklist. Also, because the procedures involved in making diagnoses and other decisions are dependent on many factors, medical professionals may not like the stringent constraints placed on them by checklists. The general point is that the implementation is unlikely to be successful unless the human factors specialists consider the unique problems and perspectives of the medical field.

CONCLUSION

This chapter has summarized much that is known about cultural factors in decision making and the barriers to advancement in the field. One of the biggest barriers is that of communication between parties. Although much effort needs to be placed on improving communication and knowledge of the different components of decision making within a multicultural context, there is much to be gained by doing so. We have provided examples of success stories in this chapter, as well as throughout the book, as with potential solutions for overcoming the communication barrier and fostering an environment for successful innovations and implementations.

Although efforts are required to promote cultural diversity within an organization, the benefits to learning about other cultures, including the various viewpoints and approaches to solving problems associated with them, are substantial. This chapter has also highlighted different approaches to studying and modeling decision making. As with the emphasis on taking a diverse, multicultural approach to decision making conveyed in the book, we strongly advocate such an approach to decision-making research as having the greatest promise for producing significant increases in our knowledge of cultural factors in decision making.

REFERENCES

Alonso, A., D. P. Baker, A. Holtzman, R. Day, H. King, L. Toomey, and E. Salas. 2006. Reducing medical error in the military health system: How can team training help? *Hum Res Manag Rev* 16:396–415.

Carayon, P., ed. 2007. *Handbook of Human Factors and Ergonomics in Health Care and Patient Safety*. Mahwah, NJ: Lawrence Erlbaum Associates.

Coffey, J. W., R. R. Hoffman, and J. D. Novak. 2011. Applications of concept maps to Web design and Web work. In *Handbook of Human Factors in Web Design*, ed. K.-P. L. Vu and R. W. Proctor, 2nd ed., 211–30. Boca Raton, FL: CRC Press.

Crisp, R. J., and R. N. Turner. 2011. Cognitive adaptation to the experience of social and cultural diversity. *Psychol Bull* 137:242–66.

Dinev, T., J. Goo, Q. Hu, and K. Nam. 2009. User behaviour towards protective information technologies: The role of national cultural differences. *Inf Syst J* 19:391–412.

Edwardson, J. A. 2010. How decision making in companies has changed with technology, corporate culture, and diversity. Paper presented at the First Gavriel Salvendy International Symposium on Frontiers in Industrial Engineering, W. Lafayette, IN.

Eid, M., and E. Diener. ed. 2005. *Handbook of Multimethod Measurement in Psychology.* Washington, DC: American Psychological Association.

Fraenkel, L. 2011. Uncertainty and patients' preferred role in decision making. *Patient Educ Couns* 82:130–2.

Gelfand, M. J., and E. Diener. 2010. Culture and psychological science: Introduction to the special section. *Perspect Psychol Sci* 5:390.

Hoffmann, E. R. 1997. Strength of component principles determining direction of turn stereotypes—Linear displays with rotary controls. *Ergonomics* 40:199–222.

Hofstede, G. 2001. *Culture's Consequences: Comparing Values, Behaviors, Institutions and Organizations Across Nations.* 2nd ed. Thousand Oaks, CA: Sage.

Jeong, K.-A., G. Salvendy, and R. W. Proctor. In press. Smart-home interface design: Layout organization adapted to Americans' and Koreans' cognitive styles. In *Human Factors in Manufacturing and Service Industries.*

Kuusisto, R., and T. Kuusisto. 2009. Information security culture as a social system: Some notes of information availability and sharing. In *Social and Human Elements of Information Security: Emerging Trends and Countermeasures*, ed. M. Gupta, and R. Sharman, 77–97. Hershey, PA: Information Science Reference/IGI Global.

Lane, J., and J. W. Meeker. 2011. Combining theoretical models of perceived risk and fear of gang crime among Whites and Latinos. *Vict Offender* 6:64–92.

Marquardt, N., and R. Hoeger. 2009. The effect of implicit moral attitudes on managerial decision-making: An implicit social cognition approach. *J Bus Ethics* 85:157–71.

Maslow, A. H. 1954. *Motivation and Personality.* New York: Harper.

Oetzel, J., A. J. Garcia, and S. Ting-Toomey. 2008. An analysis of the relationships among face concerns and facework behaviors in perceived conflict situations: A four-culture investigation. *Int J Confl Manag* 19:382–403.

Perry, S. J., and R. L. Wears. 2011. Large-scale coordination of work: Coping with complex chaos within healthcare. In *Informed by Knowledge: Expert Performance in Complex Situations*, ed. K. L. Mosier, and U. M. Fischer, 55–67. New York: Psychology Press.

Proctor, R. W., S. Y. Nof, Y. Yih, P. Balasubramanian, J. R. Busemeyer, P. Carayon, C.-Y. Chiu et al. 2011. Understanding and improving cross-cultural decision making in design and use of digital media: A research agenda. *Int J Human Comput Interact* 27:151–90.

Proctor, R. W., and K.-P. L. Vu. 2010. Universal and culture-specific effects of display-control compatibility. *Am J Psychol* 123:425–35.

Proctor, R. W., and K.-P. L. Vu. 2011. Convergence of basic and applied research in human factors and ergonomics. *Theor Issues Ergon Sci.*

Purohit, R. 2011. Culture barriers and decision making. Petition Writer. http://www.petition-writer.com/2011/02/culture-barriers-and-decision-making.html (accessed February 22, 2011).

Sam, D. L., and J. W. Berry. 2010. Acculturation: When individuals and groups of different cultural backgrounds meet. *Perspect Psychol Sci* 5:472–81.

Spencer-Rodgers, J., H. C. Boucher, S. C. Mori, L. Wang, and K. Pen. 2009. The dialectical self-concept: Contradiction, change, and holism in East Asian cultures. *Pers Soc Psychol Bull* 35:29–44.

Strohschneider, S. 2008. Strategies in complex decision making: Economics, problem solving, and culture. In *Impact of Culture on Human Interaction: Clash or Challenge?* ed. H. Helfrich, A. V. Dakhin, and I. V. Arzhenovskiy, 319–30. Ashland, OH: Hogrefe & Huber Publishers.

van Tilburg, M., and T. Briggs. 2005. Web-based collaboration. In *Handbook of Human Factors in Web Design*, ed. R. W. Proctor and K.-P. L. Vu, 551–569. Mahwah, NJ: Lawrence Erlbaum Associates.

van Velden, M. E., J. L. Severens, and A. Novak. 2005. Economic evaluations of healthcare programmes and decision making: The influence of economic evaluations on different healthcare decision-making levels. *Pharmacoeconomics* 23(11):1075–82.

Weber, E. U., and C. K. Hsee. 2000. Culture and individual judgment and decision making. *Appl Psychol Int Rev* 49:32–61.

Weber, E. U., and M. W. Morris. 2010. Culture and judgment and decision making: The constructivist turn. *Pers Psychol Sci* 5:410–19.

Index

D

T

U

V

W

For Product Safety Concerns and Information please contact our EU representative GPSR@taylorandfrancis.com
Taylor & Francis Verlag GmbH, Kaufingerstraße 24, 80331 München, Germany

www.ingramcontent.com/pod-product-compliance
Ingram Content Group UK Ltd.
Pitfield, Milton Keynes, MK11 3LW, UK
UKHW021834240425
457818UK00006B/190

* 9 7 8 1 4 3 9 8 4 6 4 6 9 *